lectures on statistical mechanics

lectures on statistical mechanics

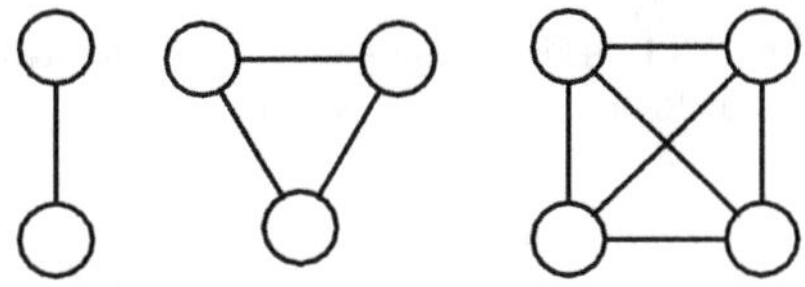

berthold-georg englert
National University of Singapore, Singapore

World Scientific
NEW JERSEY • LONDON • SINGAPORE • BEIJING • SHANGHAI • HONG KONG • TAIPEI • CHENNAI • TOKYO

Published by

World Scientific Publishing Co. Pte. Ltd.
5 Toh Tuck Link, Singapore 596224
USA office: 27 Warren Street, Suite 401-402, Hackensack, NJ 07601
UK office: 57 Shelton Street, Covent Garden, London WC2H 9HE

Library of Congress Control Number: 2020940850

British Library Cataloguing-in-Publication Data
A catalogue record for this book is available from the British Library.

LECTURES ON STATISTICAL MECHANICS

ISBN 978-981-122-457-7 (hardcover)
ISBN 978-981-122-554-3 (paperback)
ISBN 978-981-122-458-4 (ebook for institutions)
ISBN 978-981-122-459-1 (ebook for individuals)

For any available supplementary material, please visit
https://www.worldscientific.com/worldscibooks/10.1142/11945#t=suppl

To my teachers, colleagues, and students

Preface

Statistical Mechanics is one of the strong pillars on which the house of physics stands. It is the branch of theoretical physics that deals with the bulk properties of systems composed of very many constituents. Statistical and probabilistic reasoning are the central tools that link the microscopic dynamics to the concepts of thermodynamics that provide the adequate description of the bulk properties.

These *Lectures on Statistical Mechanics* grew out of a set of lecture notes for a fourth-year undergraduate course that I taught at the National University of Singapore in recent years. The presentation is rather detailed and does not skip intermediate steps that — as experience shows — are not so obvious for the learning student.

Prior to this course, students would have gone through a second-year course on thermodynamics with a glimpse at statistical mechanics. Therefore, the review of basic thermodynamics at the beginning of these lecture notes omits some standard material, such as heat engines or the various equivalent versions of the Second Law, and emphasizes topics that relate more closely to the subsequent chapters on statistical mechanics proper.

While I was not following any particular textbook when preparing these lectures, I was certainly influenced by the splendid books on my shelf, in particular *Equilibrium and Non-Equilibrium Statistical Thermodynamics* by Michel Le Bellac, Fabrice Mortessagne, and G. George Batrouni, *Theorie der Wärme* by Richard Becker, *Thermodynamics and an Introduction to Thermostatistics* by Herbert B. Callen, *Introduction to Modern Statistical Mechanics* by David Chandler, *Thermodynamics and Statistical Physics* by Robert J. Finkelstein, *Statistical Mechanics* by Kerson Huang, *Statistical Mechanics* by Ryogo Kubo, *Statistical Physics* by Lev D. Landau and Evgeny M. Lifshitz, *Statistical Mechanics* by Raj K. Pathria and

Paul D. Beale, *Elements of Classical Thermodynamics for Advanced Students of Physics* by A. Brian Pippard, and *Thermodynamics and Statistical Mechanics* by Arnold Sommerfeld.

A set of lecture notes is not a monograph on the subject and is not meant to be one. Rather, its purpose is to give a solid introduction and prepare the student for further studies on her own. Accordingly, there is no ambition of, and no attempt at, treating each and every aspect of statistical mechanics in these notes — they just represent what I could and would deal with in one semester. The material of this book is my personal selection for that one-semester fourth-year course, presented in full during twenty-two two-hour lectures. Other lecturers will surely omit some of the material of my choice in favor of topics that I did not choose to include.

The feedback I received from students in class, from the Ph.D. students and postdoctoral fellows who conducted tutorial sessions, and from colleagues in Singapore and elsewhere was invaluable and led to many improvements of the text. While I am much obliged to all of them, I can only name a few: Georges BATROUNI, Ulrike BORNHEIMER, HAN Rui, Jun Hao HUE, LEN Yink Loong, Christian MINIATURA, Hui Khoon NG, Mikołaj PARANIAK, Jiaan QI, SEAH Yi-Lin, SHANG Jiangwei, SIM Jun Yan and Martin-Isbjörn TRAPPE, represent this large crowd.

I am sincerely grateful for the professional help by the staff of World Scientific Publishing Co., which was crucial for the completion; in particular, I acknowledge Peggy YEO's competent support.

Ian ANG Xing Yang and Darren TEO Kar Seng typed the original set of notes and thus produced the electronic version that I could then work on; LIM Zheng Liang and CHAI Jing Hao helped me much in getting the text into shape. I thank them cordially.

This book would not exist without the outstanding teachers, colleagues, and students who taught me so much. I dedicate these lectures to them.

I wish to thank my dear wife Ola for her continuing understanding and patience by which she is giving me the peace of mind that is the source of all achievements.

Singapore, June 2020 *BG Englert*

Contents

Glossary

Here is a list of the symbols and acronyms used in the text; the numbers in square brackets indicate the pages of first occurrence or of other significance.

Miscellanea

BE, FD, MB	Bose–Einstein, Fermi–Dirac, Maxwell–Boltzmann [103]
LHS, RHS	left-hand side, right-hand side
TF	Thomas-Fermi [133]
1D, 2D, 3D	one-, two-, three-dimensional
$\cdot$	indicates summation, as in $\mu \cdot \mathrm{d}n = \sum_j \mu_j \mathrm{d}n_j$ [7]
$\{\ ,\ \}$	Poisson bracket [225]
$\vert\ \rangle$, $\langle\ \vert$, $\vert\ \rangle\langle\ \vert$	ket, bra, ket-bra (quantum mechanics)
$\mathbf{1}$, $\mathbf{1}_D$	unit dyadic [108], unit $D \times D$ matrix [189]
$(\ )_X$	indicates that variables X are kept constant [7]
$X^{(k)}$; $\lceil X \rfloor$; $X^{(\mathrm{IF})}$	value of variable X in the kth phase [34]; in the coexistence region [58]; at the interface [66]
#(TDOF)	number of thermodynamical degrees of freedom [40]
$x!$	factorial of x [70]
$f'(x)$, $f''(x)$	first, second derivative of $f(x)$ with respect to its argument
$f^2(x)$, $f^{-1}(x)$	square, inverse of the function f: $f^2(x) = f\big(f(x)\big)$, $f^{-1}\big(f(x)\big) = x$
$f(x)^2$, $f(x)^{-1}$	square, reciprocal of the function value: $f(x)^2 = \big(f(x)\big)^2$, $f(x)^{-1} = 1/f(x)$
$\langle X \rangle$, $\langle X \rangle_t$	expected/average/mean value of variable X [70], at time t [215, 226]
$\langle X \rangle^{(0)}$, $\langle X \rangle_0$	zeroth contribution to $\langle X \rangle$ [171, 177]
$\overline{\cdots}$, $\langle \cdots \rangle$	long-time average, ensemble average [215]
$[x]_+$	positive values of x selected: $[x]_+ = x\eta(x)$ [135]
$a = \vert \boldsymbol{a} \vert$	length of vector $\boldsymbol{a}$
$\boldsymbol{\nabla}$, $\boldsymbol{\nabla}^2$	gradient vector differential operator, Laplacian

Latin alphabet

a, b	parameters of the van der Waals gas [50, 203]
a_0	Bohr radius, $a_0 = 0.529177\,\text{Å}$ [136]
$a_l(\beta)$	lth virial coefficient [198]
Å	angstrom, length unit, $1\,\text{Å} = 10^{-10}\,\text{m}$
A, B, C, ...	equilibrium states of the thermodynamical system [5]
b	scaled length unit of the TF model [137]
$b_l(\beta, V)$, $b_l(\beta)$	sum of l-clusters [194,195]
B	Baker's constant, $B = 1.58807$ [138]; mobility [215]
$\boldsymbol{B}$, $\boldsymbol{B}^{(\text{eff})}$	magnetic field [147], effective value [167]
c	speed of light [106], of sound [110]
c_V, c_P	specific heats for constant volume, for constant pressure [21]
$\cos, \sin, \ldots$	trigonometric functions; $\sin\alpha = \sin(\alpha)$, $\sin\dfrac{\beta}{2} = \sin(\frac{1}{2}\beta)$
$\cosh, \sinh, \ldots$	hyperbolic functions; $\cosh\vartheta = \cosh(\vartheta)$
C, C_V, C_P	heat capacity, for constant volume, for constant pressure [16]
C	coulomb, SI unit of electric charge
°C	degree celsius, unit of temperature
$\mathrm{d}X$	differential of quantity X
$(\mathrm{d}\boldsymbol{r})$, $(\mathrm{d}\boldsymbol{p})$	volume element in $\boldsymbol{r}$ space, in $\boldsymbol{p}$ space [74]
$(\mathrm{d}\boldsymbol{r}_j)(\mathrm{d}\boldsymbol{p}_j)$	phase-space volume element for the jth particle [77]
D	diffusion coefficient [217]
e	elementary charge, $e = 1.60218 \times 10^{-19}\,\text{C}$ [133]
e_0	energy parameter [173]
e	Euler's number, $\mathrm{e} = 2.71828\ldots$
eV	electron-volt, energy unit, $1\,\text{eV} = 1.60218 \times 10^{-19}\,\text{J}$
exp	exponential function, $\exp(x) = \mathrm{e}^x$; $\exp = \log^{-1}$
E	energy [73]
E_0	reference energy [11]
E_0, $E_0^{(\text{eff})}$	excitation energy (Ising model) [83,140], effective value [167]
E_k, $E_{k_j}^{(j)}$	energy of the kth microstate [80], of the jth paricle in its k_jth state [97]
E_k	kth eigenvalue of the Hamilton operator [174]
$E_k^{(0)}$, $E_k^{(1)}$	large, small contribution to E_k [171]
E_F	Fermi energy [121]

$E_{\text{kin}}[\rho]$; $E_{\text{int}}[\rho]$	density functional of the kinetic energy; of the interaction energy [133]
$E_{\text{TF}}[\rho]$	density functional of the TF energy [134]
$\boldsymbol{e}$	unit vector [147]
$\boldsymbol{E}(\boldsymbol{r})$	electric field [4]
f_j	force component in the work element [7]
$f(r)$, f_{jk}	reduced MB factor [192], abbreviates $f(r_{jk})$ [193]
$f(x,p,t)$	phase-space density (1D) [223]
$f(\cdot,p,t)$	marginalized over x [234]
$f_\alpha(z)$	fermion function [121]
$\text{func}(x,y)$	generic, unspecified function of variables x and y [9]
F; $\widetilde{F}$	Helmholtz free energy [18]; constrained free energy [165]
$F^{(0)}$	dominant contribution to F [172]
$F(x)$	universal TF function [136]
$\boldsymbol{F}$, $\boldsymbol{F}_{\text{random}}$	force vector, of random force [215]
$g(\omega)$	density of modes [109]
$g_n(y)$, $g_n(\phi)$	generating function [211,206]
$g_\alpha(z)$	boson function [114]
G	free enthalpy [19]
$G_t(a,b)$	generating function for expected values [229]
$\mathcal{G}(x,p,t;x',p')$	Green's function [233]
$h=2\pi\hbar$	Planck's constant, $\hbar = 1.05457\times10^{-34}\,\text{J s}$ [73]
H; H_0, H_1	Hamilton operator [174]; large, small contribution [175]
$H(\boldsymbol{r},\boldsymbol{p})$	Hamilton function [74]
$\text{H}_k(\ )$	kth Hermite polynomial [230]
$\boldsymbol{H}(\boldsymbol{r})$	magnetic induction field [5]
i	imaginary unit, $\text{i}^2=-1$
J	number of constituents [32]
J, $J_{jj'}$; J'	coupling constant (Ising model) [140], between the jth and the j'th site [167]; between next-next neighbors [183]
J	joule, SI unit of energy, $1\,\text{J} = 1\,\text{kg}\,\text{m}^2\,\text{s}^{-2}$
k_{B}	Boltzmann's constant, $k_{\text{B}} = R/N_{\text{A}} = 1.38065\times10^{-23}\,\text{J}\,\text{K}^{-1}$ [75]
k_{F}	Fermi wave number [124]
kg	kilogram, basic SI unit of mass
K	number of phases [34]
$K(t)$, $\widetilde{K}(\omega)$	autocorrelation function [218], Fourier transformed [219]
K; K_{c}	interaction strength [149]; critical value [163]

K', K'', K'''	renormalized interaction strength values [149]
$\widetilde{K}$	effective renormalized interaction strength [157]
K_T, K_S	isothermal [28], isentropic compressibility [30]
K	kelvin, SI unit of absolute temperature
$\mathrm{K}_n(\,)$	modified Bessel function [247]
$\boldsymbol{k}$	wave vector [106]
l, ℓ, L	length of an object
log	natural logarithm, $\log x = \log(x)$
m	mass [76]; electron mass, $m = 9.10938 \times 10^{-31}$ kg [127]
$\overline{m}_j$	effective mass of the jth normal mode [109]
$m_{jj'}$	element of the mass matrix [109]
$m_1, m_2, m_3, \ldots$	count of 1-clusters, 2-clusters, 3-clusters, ... [194]
m	list of cluster counts, $\mathrm{m} = (m_1, m_2, m_3, \ldots)$ [194]
m	meter, basic SI unit of length
mol	mole, SI unit of substance; see the note on units below
M	mass of pollen grain (Brownian motion) [215]
M, $M_{\pm\pm}$	partition-function matrix, its elements [143]
$\boldsymbol{M}(\boldsymbol{r})$	magnetization field [5]
(nn), (nnn)	next neighbor [152], next-next neighbor [157]
n; n_j, n_{tot}	number of moles, $n = N/N_\mathrm{A}$ [9]; of the jth component, total number of moles [32]
n_0, n_1, n_2, ...	occupation numbers [100]
N; N_j; N_c	number of particles, $N = nN_\mathrm{A}$; of the jth kind [70]; of volume cells [90]
N_0, $\langle N_0 \rangle$	ground-state occupancy (BE condensate) [115]
N_A	Avogadro's numbers, $N_\mathrm{A} = R/k_\mathrm{B} = 6.02214 \times 10^{23}$ [75]
$N_{(\mathrm{nn})}$	number of next-neighbor sites [167]
N	newton, SI unit of force, $1\,\mathrm{N} = 1\,\mathrm{kg\,m\,s^{-2}}$
$O(N)$	"of order N"
$p(X)$	probability of finding property X [70]
p_k	probability of the kth microstate [80]
$p_n(m)$, $p_m(t)$	probability that n jumps display by m units [205], that m events occur during time t [212]
$\widetilde{p}_m(\gamma)$	Laplace transform of $p_m(t)$ [213]
P; P_{ext}; P_c	pressure [4]; external pressure [17]; critical pressure [45]
$\overline{P}$	coexistence pressure [43]
Pa	pascal, SI unit of pressure, $1\,\mathrm{Pa} = 1\,\mathrm{N\,m^{-2}}$
$\boldsymbol{p}$, $\boldsymbol{p}_j$	momentum vector, of the jth particle [74]

$\boldsymbol{P}(\boldsymbol{r})$	polarization field [4]
q	latent heat [45]
q_j	jth displacement coordinate [108]
$q_1, q_2, q_3, \ldots$	pair variables (Ising model) [142]
$q(K)$	intensive logarithm of the canonical partition function [150]
Q	heat [4]; canonical partition function [80]; dissipation parameter in the Fokker–Planck equation [226]
$Q(t)$, $\mathsf{Q}(t)$	variances (Brownian motion) [229], [232]
$\widetilde{Q}$	constrained canonical partition function [165]
$Q^{(0)}$	dominant contribution to Q [171]
r; r_0	Poisson process rate [213]; distance parameter [192]
r_{jk}	distance between the jth and the kth particle [191]
R	universal gas constant, $R = N_{\mathrm{A}} k_{\mathrm{B}} = 8.31447\,\mathrm{J\,K^{-1}}$ [9]
Ry	rydberg, atomic-scale energy unit, $1\,\mathrm{Ry} = 13.6057\,\mathrm{eV}$
$\boldsymbol{r}$, $\boldsymbol{r}_j$	position vector, of the jth particle [74]
(sq)	square [157]
s	molar entropy [33]
$s_1, s_2, s_3, \ldots$	configuration variables (Ising model) [83, 140]
S; S_{A}; S_{B}; S_0	entropy; of state A [5]; of the bath [23]; at $T = 0$ [132]
$\mathrm{tr}\{M\}$, $\mathrm{tr}\{O\}$	trace of matrix M [144], of operator O [175]
t	time [106]
T; T_{B}	temperature [8]; of the bath [23]
T_{c}	critical temperature [45]
T_{D}; T_{F}	Debye temperature [111]; Fermi temperature [126]
T	momentum current density dyadic [108]
u	energy density [107]
U	internal energy [4]
v	molar volume [33]; specific volume [116]
v_{c}	critical molar volume [51]; critical specific volume [117]
V; V_0	volume [4]; reference volume [11], potential energy parameter [192]
$V(q)$, $V(r_{jk})$	potential energy [108, 191]
$\boldsymbol{v}$	velocity vector [215]
$w(p, p')$	momentum-change probability density [223]
W	work [4]; energy parameter [187]
X	set of external parameters [5]
X_k, y_k	kth extensive variable and its intensive partner [85]

Y	constraints [13]
z; $\overline{z}$	fugacity [112]; singular value [202]
z_{d}	location of the Gibbs dividing surface [65]
Z	generic partition function [86], partition function of the grand canonical ensemble [88]; atomic number [134]

Greek alphabet and Greek-Latin combinations

α	thermal expansion coefficient [26]; critical exponent [160]
β	reciprocal temperature, $\beta = (k_{\mathrm{B}}T)^{-1}$ [80]
β_{c}	critical β value [116]
$\underline{\beta}$	magnetization exponent [164]
γ	adiabatic index [9]
δ_{ab}; $\delta(x)$, $\delta(\boldsymbol{r})$	delta symbol [98]; delta function of x, of $\boldsymbol{r}$ [136]
δX	infinitesimal change of quantity X; spread of variable X [71]
ΔX	finite change of quantity X; across phase transition [44]
ϵ, $\boldsymbol{\epsilon}$	very small positive quantity, very short vector
$\overline{\epsilon}$	intermediate ϵ value [176]
$\boldsymbol{\epsilon}$	polarization vector [106]
ε_k, ε_0	kth single-particle energy [97], ground-state energy [99]
$\zeta(\ )$	zeta function [114]
$\eta(\ )$	step function [110]
H	enthalpy [19]
λ	scaling parameter [10]; thermal wavelength [113]
λ_{D}	Debye wavelength [110]
Λ	generic thermodynamical potential [86]; Landau free energy [89]
μ, $\overline{\mu}$	chemical potential [11,85], coexistence value [43]
$\boldsymbol{\mu}$	magnetic dipole moment vector [147]
ν_j, ν	molar fraction of the jth component, collection of ν_js [32]
ξ_j	jth normal-mode coordinate [109]
Ξ	partition function of the Maxwell's demon ensemble [246]
π	Archimedes's constant, $\pi = 3.14159\ldots$
ρ; ρ_{d}; ρ_{c}; ρ_{TF}	molar density [65], particle density [91]; discontinuous density [66]; critical particle density [117]; TF density [138]

ρ; ρ_β, $\rho_{0\beta}$	statistical operator [180]; thermal state [180, 182]
σ; σ_x, σ_z	surface area [64]; Pauli matrices [188]
τ	surface tension [64], time constant of dissipation [215]
$\chi(x)$	characteristic function for statement x [74]
ω, ω_j	circular frequency [106], of the jth mode [105]
ω_{D}	Debye cut-off frequency [110]
Ω	count of microstates [75]

A note on units

For us, Avogadro's number is the count of entities (atoms, molecules, ...) that make up one mole, that is $N_{\mathrm{A}} = 1\,\mathrm{mol} \cong 6 \times 10^{23}$. We do *not* write $N_{\mathrm{A}} \cong 6 \times 10^{23}\,\mathrm{mol}^{-1}$, although this is the accepted SI value of N_{A}, as that would just mean that $N_{\mathrm{A}} = 1$, exactly as $12\,\mathrm{dozen}^{-1} = 1$.

The universal gas constant is $R \cong 8\,\mathrm{J\,K^{-1}}$, it is *not* $R \cong 8\,\mathrm{J\,K^{-1}\,mol^{-1}}$. The per-mole version, the accepted SI value, would just state that R is equal to Boltzmann's constant k_{B}. For us, they are not the same but related to each other by Avogadro's constant, $R = N_{\mathrm{A}} k_{\mathrm{B}}$.

When stating the amount of substance of the jth kind by the value of n_j, we have $n_j = 1.5$ if there are 1.5 moles of the substance, *not* $n_j = 1.5\,\mathrm{mol}$. Equivalently, we can say that there are $N_j = 1.5\,N_{\mathrm{A}} \cong 9 \times 10^{23}$ entities, that is: $N_j = 1.5\,\mathrm{mol}$.

Chapter 1

Review of Selected Topics in Thermodynamics

Statistical Mechanics is concerned with physical systems that contain so many degrees of freedom that a detailed description is both impossible and useless. As Loschmidt* established, there are about 10^{20} molecules in a few cubic centimeters (or milliliters) of air at the usual room conditions, and each molecule needs six coordinates to specify its position and its orientation in space, plus as many velocities. Not only is it hopeless to attempt a description that takes full account of all these $\sim 10^{21}$ coordinates, it is also pointless because we would not have any use of such a lot of information if it were made available.

Rather, we adopt the strategy of Thermodynamics and describe the properties of the air sample by macroscopic quantities, such as volume, pressure, and temperature in addition to stating the amount of material involved. There are always very few of these macroscopic parameters, usually less than ten, but even in complicated situations where we would need more than that, their number is truly tiny compared with the number of variables needed for a detailed microscopic description.

It is clear, however, that the few macroscopic parameters cannot always capture the physical situation accurately. For example, if we pour boiling water into a glass and stir it vigorously, we'll have turbulent motion of the water and the steam bubbles moving around in the liquid. Any small number of parameters is not enough for a useful description of this situation. One knows from experience that the properties of the water in the glass become simpler after a while. The turbulent motion stops and the steam bubbles disappear as the water cools down and eventually reaches the room temperature of its surroundings.

*Johann Josef LOSCHMIDT (1821–1895)

1.1 Four laws of thermodynamics

1.1.1 *The Minus-First Law*

We have here a familiar illustration of a basic law of thermodynamics that we'll refer to as the *Minus-First Law* because the Zeroth, First, Second, and Third Law have agreed-upon traditional meanings and are later in a logical development:

Minus-First Law of Thermodynamics

A thermodynamical system relaxes to an equilibrium state when it is left alone long enough. (1.1.1)

Let us explain the various terms in this statement. A *thermodynamical system* is like the glass of hot water alluded to above, that is: a well-defined amount of material that is weakly interacting with its surroundings, or not at all.

The generic picture is this:

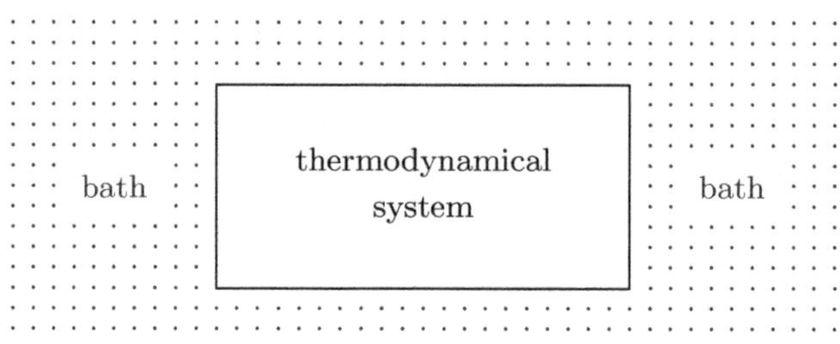

(1.1.2)

There is the system and its *bath*, which is the environment (usually with externally controlled temperature and possibly other adjustable parameters of relevance), and the coupling between the two is sufficiently weak to justify regarding them as separate objects. Should it happen that the coupling is strong, then the distinction between system and bath is meaningless.

Usually, we associate with a bath that it serves as a *reservoir*. By this we mean that the coupling between system and bath has noticeable and important effects on the system whereas the back action of the system on the bath is negligibly small. Think of putting a bottle with hot milk into a creek of cold water. After a while the milk will be cooled down to the water temperature but you won't register a noticeable increase in the water temperature.

The *equilibrium states* of a system are those states that appear stable and do not evolve into other states if the external parameters (such as the volume or the bath temperature) are maintained. When the system is in an equilibrium state, its properties are independent of the system history and describable by a few parameters as explained in the introductory remarks.

To illustrate this independence of the history, consider that — in addition to the boiling hot water in turbulent motion in one glass — you have a second glass which contains the same amount of water in an ice-water mixture. You put both glasses on a table and wait. Eventually there will not be any noticeable difference between the two glasses, and your friend, who did not witness the earlier stages, cannot tell which glass contained initially the boiling water and which the ice-water mixture.

These equilibrium states are central to thermodynamics and statistical physics. We describe them by a small set of parameters (volume, pressure, energy, temperature, applied electric and magnetic field, amounts of materials composing the system, and possibly others) and the so-called *equations of state* relate some of these parameters to each other, so that not all of them can be freely adjusted independently.

1.1.2 *The Zeroth Law*

If there are two thermodynamical systems that are coupled to each other, so that they can exchange heat, they will reach a common equilibrium state when left alone long enough. We then say that the two systems are in *thermal equilibrium.* The *Zeroth Law of Thermodynamics* states that this notion is transitive:

Zeroth Law of Thermodynamics

If one thermodynamical system is in thermal equilibrium with a second thermodynamical system, and the second system is in thermal equilibrium with a third thermodynamical system, then the first and the third systems are also in thermal equilibrium. (1.1.3)

If we focus our attention on the first and the third systems, we can regard the second system as providing the link between the first and the third system, through which they interact. In the course of reaching thermal equilibrium between the first and the third system, both will also get into equilibrium with the second system. In addition to heat, thermodynamical systems can also exchange particles or other attributes, and then there is a corresponding notion of transitive equilibrium for that.

As we'll see below, the Second Law that we'll come to shortly has implications for systems in equilibrium. Before we can discuss these matters, we need to prepare the ground.

1.1.3 *The First Law*

The *First Law of Thermodynamics* is about energy conservation:

First Law of Thermodynamics
The energy of a thermodynamical system can only change by either mechanical work on the system or by the exchange of heat with a bath or a heat source. (1.1.4)

We denote the *internal energy* of the system by U and express the First Law as

$$\mathrm{d}U = \mathrm{d}Q + \mathrm{d}W\,, \tag{1.1.5}$$

which says that a transfer of heat ($\mathrm{d}Q$) can change the internal energy or mechanical work ($\mathrm{d}W$) can achieve that.

It is important to appreciate that the internal energy U has a definite value in an equilibrium state whereas there is no well-defined heat content or work content of the system — there is no statement "$U = Q + W$" or similar. You can think of U as the analog of a bank account (in SGD, say) to which you can transfer money in various currencies (EUR and USD, say) but, once transferred, there is just a balance in SGD. Likewise, one may measure heat in calories and work in horsepower-minutes, while keeping the balance sheet of the internal energy in joules.

Both $\mathrm{d}Q$ and $\mathrm{d}W$ can, of course, be positive or negative, depending on whether the heat flow is from the bath to the system or in the reverse direction, and whether work is done onto the system or by the system. Further, there are different kinds of work, perhaps most importantly the work associated with a volume change resulting from a wall ("piston") yielding to the pressure P of the gas or fluid that makes up the thermodynamical system, for which

$$\mathrm{d}W = -P\mathrm{d}V\,. \tag{1.1.6}$$

In the presence of an external electric field $\boldsymbol{E}(\boldsymbol{r})$, there is the work element

$$\mathrm{d}W = -\int (\mathrm{d}\boldsymbol{r})\, \boldsymbol{E}(\boldsymbol{r}) \cdot \mathrm{d}\boldsymbol{P}(\boldsymbol{r})\,, \tag{1.1.7}$$

where $\boldsymbol{P}(\boldsymbol{r})$ is the polarization field of the material at position $\boldsymbol{r}$. Similarly, we have

$$dW = -\int (d\boldsymbol{r})\, \boldsymbol{H}(\boldsymbol{r}) \cdot d\boldsymbol{M}(\boldsymbol{r}) \tag{1.1.8}$$

for the work associated with an external magnetic induction field $\boldsymbol{H}(\boldsymbol{r})$ that couples to the magnetization field $\boldsymbol{M}(\boldsymbol{r})$.

1.1.4 *The Second Law*

Then, there is the *Second Law of Thermodynamics*, which has several equivalent formulations. The version we choose here is particularly useful in the context of statistical mechanics and the formalism we need to develop. We state the Second Law as follows:

Second Law of Thermodynamics
There is a function $S(U, X)$ of the internal energy U and the external variables X with these properties:
(i) $S(U, X)$ increases monotonically with U;
(ii) if the equilibrium state B is adiabatically accessible from the equilibrium state A, then $S_B \geq S_A$. (1.1.9)

Here, too, we need to explain the meaning of some terms. The *external variables* are those that we control from outside, including volume, amount of substance, electric charge and magnetic moment, and the like. State B is *adiabatically accessible* from state A, if it is true that the system, when prepared in state A, reaches state B without exchanging heat with the bath. We shall, indeed, take for granted that it is possible to isolate the system from the bath by an *adiabatic wall* — a wall that does not conduct heat:

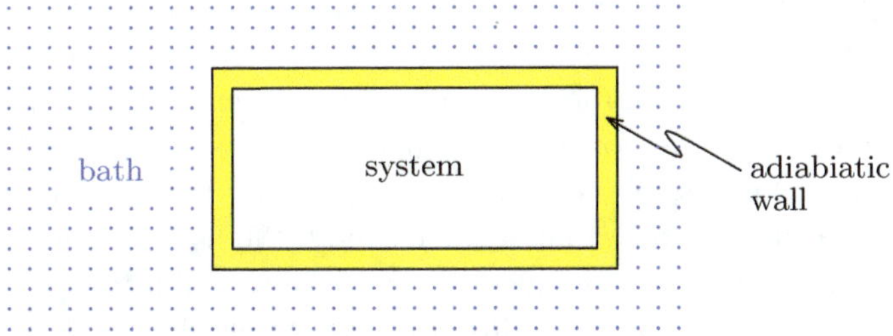

(1.1.10)

For example, all the gas inside a container could be pushed into a corner by an internal membrane:

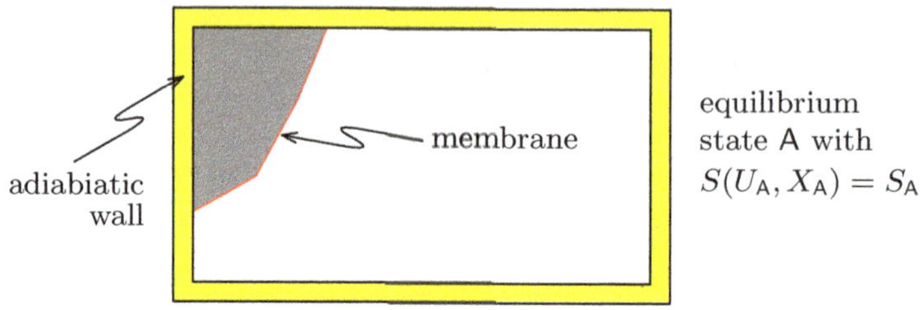

(1.1.11)

and when the membrane is removed, the gas spreads over the whole volume and eventually reaches another equilibrium state:

(1.1.12)

In this example, B is adiabatically accessible from A, but the reverse is not the case. You cannot prepare state B and then expect that the system evolves into state A. To drive it there, you need to exchange heat with a bath.

There are, of course, other situations in which B is adiabatically accessible from A, and also A is adiabatically accessible from B. In this case, we say that the process A $\to$ B is reversible, A $\leftrightarrow$ B, you can go back and forth.

We call $S(U, X)$ the *entropy* of the system, so that the Second Law states that the entropy cannot decrease when the system evolves from A to B in adiabatic isolation, that is: without heat exchange. Actual heat transfer can increase ($\mathrm{d}Q > 0$) or decrease ($\mathrm{d}Q < 0$) the entropy.

It is important to realize that $S(U, X)$ is a well-defined quantity for equilibrium states and, perhaps, states very close to equilibrium. Systems that are very far from equilibrium (the boiling water with moving steam bubbles is an example) do not have well-defined values of the usual parameters; it may not be possible to assign a temperature to the system and it may not have a meaningful value of the entropy. We have the desired simple description of a thermodynamical system in terms of a few parameters only when the system is in an equilibrium state or nearby.

1.1.5 *Temperature*

To find a first consequence of the Second Law, we consider adjacent equilibrium states that differ slightly in their values for U and X, so that

$$\mathrm{d}S = \left(\frac{\partial S}{\partial U}\right)_X \mathrm{d}U + \left(\frac{\partial S}{\partial X}\right)_U \cdot \mathrm{d}X\,, \tag{1.1.13}$$

where the $\cdot$ dot notation is short-hand for a summation over the various contributions for X_1, X_2, X_3, …,

$$\left(\frac{\partial S}{\partial X}\right)_U \cdot \mathrm{d}X = \sum_j \left(\frac{\partial S}{\partial X_j}\right)_U \mathrm{d}X_j\,, \tag{1.1.14}$$

and the $(\ )_X$ and $(\ )_U$ parentheses remind us of the quantities that are kept constant. This is a standard notational convention in thermodynamics; it is not really needed in the present context but very useful in many others, so let's get used to it. Now, the First Law tells us that

$$\mathrm{d}U = \mathrm{d}Q + \mathrm{d}W = \mathrm{d}Q + f \cdot \mathrm{d}X\,, \tag{1.1.15}$$

where we note that the work element $\mathrm{d}W$ is a sum of terms like $-P\mathrm{d}V$, $-\boldsymbol{E} \cdot \mathrm{d}\boldsymbol{P}$, $-\boldsymbol{H} \cdot \mathrm{d}\boldsymbol{M}$ that we have seen above; more generally, there is a "force" f_j associated with the "displacement" $\mathrm{d}X_j$; that is

$$\mathrm{d}W = \sum_j f_j\,\mathrm{d}X_j = f \cdot \mathrm{d}X\,. \tag{1.1.16}$$

Then,

$$\mathrm{d}S = \left(\frac{\partial S}{\partial U}\right)_X \mathrm{d}Q + \left[\left(\frac{\partial S}{\partial X}\right)_U + \left(\frac{\partial S}{\partial U}\right)_X f\right] \cdot \mathrm{d}X\,, \tag{1.1.17}$$

after combining the statements (1.1.13) and (1.1.15) of the Second and First Laws, respectively.

In particular, the adjacent equilibrium states can be reversibly reached from each other in adiabatic isolation, so that $\mathrm{d}S = 0$ and $\mathrm{d}Q = 0$, and

$$\left(\frac{\partial S}{\partial X}\right)_U = -\left(\frac{\partial S}{\partial U}\right)_X f \tag{1.1.18}$$

follows, where each ingredient — namely $\dfrac{\partial S}{\partial X}$, $\dfrac{\partial S}{\partial U}$, and f — is a function of the state, fully determined by the values of U and X.

At this point, we introduce the temperature T in accordance with

$$\left(\frac{\partial S}{\partial U}\right)_X = \frac{1}{T}. \tag{1.1.19}$$

This is consistent with other definitions of temperature and yields familiar relations, such as (1.1.21) below or (1.2.1) in the next section. Indeed, upon returning to (1.1.13), we have

$$\begin{aligned} \mathrm{d}S &= \underbrace{\left(\frac{\partial S}{\partial U}\right)_X}_{=\frac{1}{T}} \mathrm{d}U + \underbrace{\left(\frac{\partial S}{\partial X}\right)_U}_{=-\frac{1}{T}f} \cdot \mathrm{d}X \\ &= \frac{1}{T}(\mathrm{d}U - f \cdot \mathrm{d}X) \end{aligned} \tag{1.1.20}$$

or, after solving for $\mathrm{d}U$,

$$\mathrm{d}U = T\mathrm{d}S + f \cdot \mathrm{d}X. \tag{1.1.21}$$

Effectively, then, we have identified

$$\mathrm{d}Q = T\mathrm{d}S, \tag{1.1.22}$$

which is sometimes taken as the joint definition of temperature and entropy element. As mentioned above, this identification of $\mathrm{d}Q$ with $T\mathrm{d}S$ is only possible under circumstances in which both entropy and temperature have a meaning.

1.2 Ideal gas

Before proceeding with the general considerations, let us look at a simple yet important example — that of an *ideal gas*. As the name indicates, the ideal gas is the idealization of a real gas, or an ideal model for a real gas that is approximately valid under favorable circumstances. These are usually that the gas is diluted and interactions between the gas particles (atoms or molecules or both) are of no importance. Air at room temperature and normal atmospheric pressure is quite well approximated by an ideal gas — one easily shows that the very small molecules occupy only a fraction of a percent of the volume filled by the air sample.

For an ideal gas, the work is $\mathrm{d}W = f \cdot \mathrm{d}X = -P\mathrm{d}V$, assuming for now that the amount of air is not subject to change. Then, we have

$$\mathrm{d}U = T\mathrm{d}S - P\mathrm{d}V\,, \tag{1.2.1}$$

a familiar expression, of course.

It tells us that the natural variables of the internal energy are the entropy and the volume,

$$U = U(S,V)\,. \tag{1.2.2}$$

So, what is this function for the ideal gas? One useful piece of information is the thermal equation of state, the *combined gas law*

$$PV = nRT\,, \tag{1.2.3}$$

established by the work of Mariotte,* Boyle,† Amontons,‡ Charles,§ Gay-Lussac,¶ and others. As before, P stands for pressure, V for volume, T for temperature, and R is the *universal gas constant* (of about $8\,\mathrm{J/K}$), while n specifies the amount of substance in number of moles.

Since

$$P = -\frac{\partial U}{\partial V} \quad \text{and} \quad T = \frac{\partial U}{\partial S} \tag{1.2.4}$$

(we do not indicate $(\quad)_S$ or $(\quad)_V$, respectively, as we are consistently regarding U as a function of its natural variables), the combined gas law yields the first-order linear partial differential equation

$$-V\frac{\partial U}{\partial V} = nR\frac{\partial U}{\partial S}\,. \tag{1.2.5}$$

We conclude that U must be a function of $\frac{1}{V}\,\mathrm{e}^{\frac{S}{nR}}$, symbolically

$$U(S,V) = \mathrm{func}\left(\frac{1}{V}\,\mathrm{e}^{\frac{S}{nR}}\right). \tag{1.2.6}$$

Another piece of information is provided by the observation that equilibrium states that are mutually adiabatically accessible have pressures and temperatures that obey

$$\frac{P_1}{P_2} = \left(\frac{V_2}{V_1}\right)^{\gamma}. \tag{1.2.7}$$

---

*Edme MARIOTTE (circa 1620–1684) †Robert William BOYLE (1627–1691)
‡Guillaume AMONTON (1663–1705) §Jacques Alexandre César CHARLES (1746–1823)
¶Joseph Louis GAY-LUSSAC (1778–1850)

In this *isentropic* — that is: constant entropy — equation, P_1, V_1, and P_2, V_2 are pressure, volume in the two states and γ is a material constant, the so-called *isentropic index* or *adiabatic index*.

The adiabatic equation of state,

$$\text{for fixed entropy } S: \quad PV^{\gamma} = \text{constant}, \tag{1.2.8}$$

states that $V^{\gamma}\dfrac{\partial U}{\partial V}$ does not depend on V, that is

$$\frac{\partial}{\partial V}\left(V^{\gamma}\frac{\partial}{\partial V}U(S,V)\right) = 0, \tag{1.2.9}$$

and implies that U is proportional to $V^{1-\gamma}$. We thus know the function in (1.2.6) up to a proportionality factor $c(n)$ that could depend on n, but cannot depend on S or V,

$$U(S,V,n) = c(n)\left(\frac{1}{V}\,\mathrm{e}^{\frac{S}{nR}}\right)^{\gamma-1}, \tag{1.2.10}$$

where we added n to the list of variables on which U depends.

In order to progress further, we must now recognize that the internal energy is a *homogeneous function* of its variables. This simply says that a scaling of the system leads to the corresponding scaling of U,

$$U(\lambda S,\lambda V,\lambda n) = \lambda U(S,V,n). \tag{1.2.11}$$

For illustration, consider that you remove 10% of the system, without changing anything else:

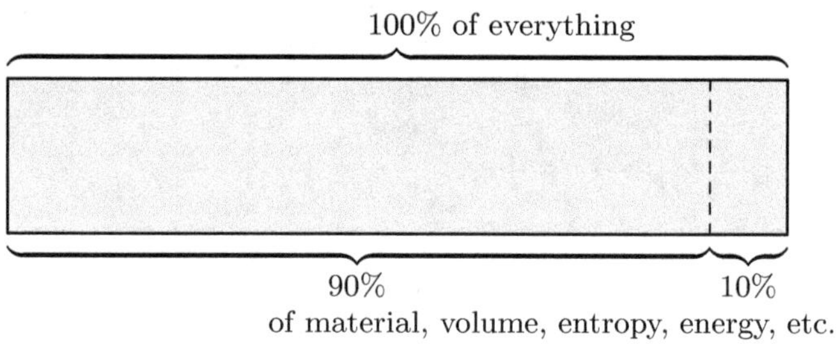

(1.2.12)

Here, with $\lambda = 0.9$, $1-\lambda = 0.1$

$$\begin{aligned} \underbrace{U(S,V,n)}_{=100\%} &= \underbrace{U(\lambda S,\lambda V,\lambda n)}_{=90\%} + \underbrace{U\big((1-\lambda)S,(1-\lambda)V,(1-\lambda)n\big)}_{=10\%} \\ &= \lambda U(S,V,n) + (1-\lambda)U(S,V,n). \end{aligned} \tag{1.2.13}$$

In other words, volume, entropy, substance, energy are *extensive* variables, their values for a composite system are the sums of the respective values of the partial systems. By contrast, temperature and pressure are *intensive*, they are the same for both parts of the split in (1.2.12).

So, let's see what this tells us. We have

$$\begin{aligned} U(\lambda S, \lambda V, \lambda n) &= c(\lambda n)\left(\frac{1}{\lambda V}\, \mathrm{e}^{\frac{\lambda S}{\lambda n R}}\right)^{\gamma - 1} \\ &= \lambda U(S, V, n) = \lambda c(n)\left(\frac{1}{V}\, \mathrm{e}^{\frac{S}{nR}}\right)^{\gamma - 1}, \end{aligned} \tag{1.2.14}$$

so that

$$\lambda^{1-\gamma} c(\lambda n) = \lambda c(n) \quad \text{or} \quad c(\lambda n) = \lambda^{\gamma} c(n) \tag{1.2.15}$$

must hold. It follows that $c(n)$ is proportional to n^{γ}. For convenience, we write the proportionality factor as $E_0 V_0^{\gamma - 1}$ with a reference energy E_0 and a reference volume V_0, and arrive at

$$U(S, V, n) = E_0\, n \left(\frac{nV_0}{V}\, \mathrm{e}^{\frac{S}{nR}}\right)^{\gamma - 1}. \tag{1.2.16}$$

The value of $E_0 V_0^{\gamma - 1}$ is not established here; it is also not relevant for any testable consequences.

To illustrate this remark, we return to the equation for $\mathrm{d}U$, but now include a term that account for changes $\mathrm{d}n$ in the amount of gas,

$$\mathrm{d}U = T\mathrm{d}S - P\mathrm{d}V + \mu \mathrm{d}n\,, \tag{1.2.17}$$

where μ is the *chemical potential*, the intensive property associated with the extensive n. Indeed, we note that each term on the right-hand side pairs an extensive variable — namely, S, V, or n — with its intensive partner — T, P, and μ, respectively. This is a generally valid pattern.

The temperature T is now understood as a function of S, V, and n,

$$T = \frac{\partial U}{\partial S} \quad \text{or} \quad T = \left(\frac{\partial U}{\partial S}\right)_{V,n} \quad \text{if we are pedantic.} \tag{1.2.18}$$

We find by differentiation that

$$T(S, V, n) = \frac{\gamma - 1}{nR} U(S, V, n)\,, \tag{1.2.19}$$

so that

$$U = \frac{nR}{\gamma - 1} T\,. \qquad (1.2.20)$$

As we shall often do, we do not specify the arguments of the various functions explicitly in the latter statement. One must, therefore, exercise care when reading these equations and not fall into the trap of regarding U as a function of T and n. That would be quite wrong. What (1.2.20) says is this: In an equilibrium state, the value of the internal energy is equal to the expression on the right, where each ingredient is measurable.

Similarly, differentiation establishes that

$$P(S, V, n) = -\frac{\partial U}{\partial V} = \frac{\gamma - 1}{V} U(S, V, n) \qquad (1.2.21)$$

or

$$U = \frac{PV}{\gamma - 1}\,, \qquad (1.2.22)$$

where the same caveat applies. Not surprisingly, we regain the combined gas law (1.2.3) from the two statements in (1.2.20) and (1.2.22) — a check of consistency, nothing more than that. The statement for μ that corresponds to (1.2.19) and (1.2.21) is the topic of Exercise 2.

1.3 Maximum property of the entropy

We return to the general considerations and look at some implications of the Second Law. A useful tool are internal constraints like the membrane in (1.1.11); these constraints could be real, or they could be imagined for the purpose of presenting an argument. Consider, in particular, the following situation, in which we have an internal wall that we can move while the system is adiabatically isolated from the bath:

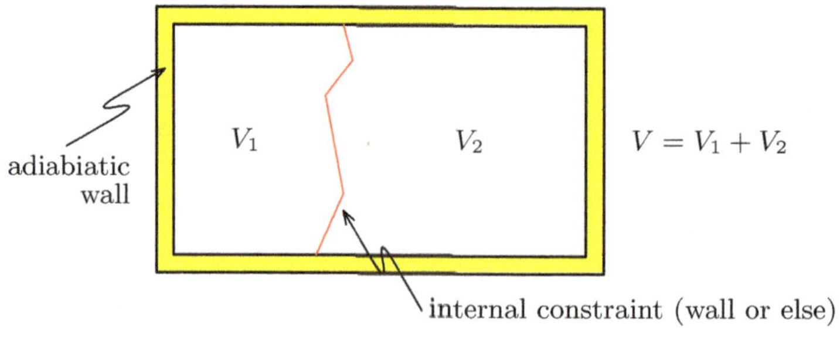

(1.3.1)

With the internal constraint, we have equilibrium state A and its entropy

$$S_{\mathsf{A}} = S(U, X; Y \neq 0)\,, \tag{1.3.2}$$

where the symbol Y stands for the constraints — two volumes with $V_1 + V_2 = V$ in (1.3.1) — and $Y = 0$ is the unconstrained case. Upon removing the internal constraint, the system evolves to equilibrium state B with entropy

$$S_{\mathsf{B}} = S(U, X; Y = 0)\,, \tag{1.3.3}$$

where $S_{\mathsf{B}} > S_{\mathsf{A}}$ as required by the Second Law. It follows that

$$S(U, X) = \underset{Y}{\mathrm{Max}}\,\{S(U, X; Y)\}\,, \tag{1.3.4}$$

that is: *The entropy is maximal in the unconstrained equilibrium state.* This extremal property of the entropy is very important.

1.4 Minimum property of the energy

There is a corresponding extremal property of the internal energy. We establish it by considering an internal wall that can conduct heat:

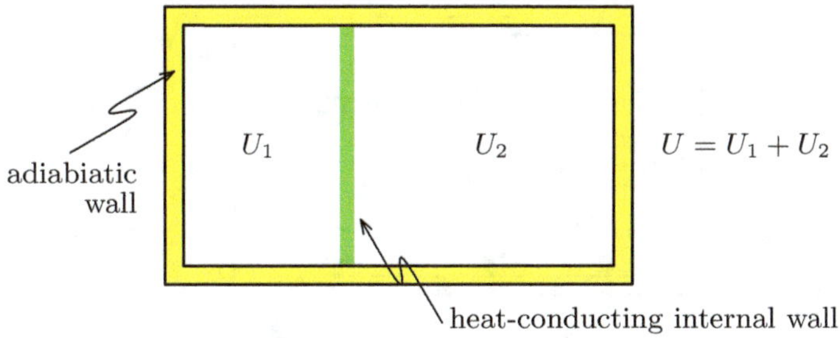

(1.4.1)

In equilibrium we have the internal energy distributed among the two parts with energy U_1 on the left and energy U_2 on the right, $U = U_1 + U_2$. Let us compare this with the out-of-equilibrium situation in which the energy on the left is $U_1 + \Delta U$ and that on the right is $U_2 - \Delta U$ with $\Delta U \neq 0$, while there is no change to the volume, material content, and so forth on either side of the internal wall. Since entropy is extensive and has the maximum property of (1.3.4), the sum of the resulting entropies must be less than the entropy for $\Delta U = 0$,

$$S(U_1 + \Delta U, \underbrace{X_1}_{\text{volume etc.}}) + S(U_2 - \Delta U, \underbrace{X_2}_{\text{volume etc.}}) < S(U_1 + U_2, \underbrace{X_1 + X_2}_{\text{sum of volumes etc.}}) . \quad (1.4.2)$$

The Second Law in (1.1.9) states that $S(U, X)$ is a monotonically growing function in the energy U and, therefore, there is an energy value $U' < U_1 + U_2$ such that

$$\underbrace{S(U_1 + \Delta U, X_1) + S(U_2 - \Delta U, X_2)}_{\text{yes constraints}} = \underbrace{S(U', X_1 + X_2)}_{\text{no constraints}} . \quad (1.4.3)$$

We conclude that by introducing internal constraints, the energy is increased ($U_1 + \Delta U + U_2 - \Delta U > U'$), or

$$U(S, X) = \operatorname*{Min}_{Y} \{U(S, X; Y)\} . \quad (1.4.4)$$

The internal energy is minimal in the unconstrained equilibrium state.

1.5 The Second Law and thermal equilibrium

Here is an application of the maximum property of the entropy. Consider two systems with a heat-conducting wall between them:

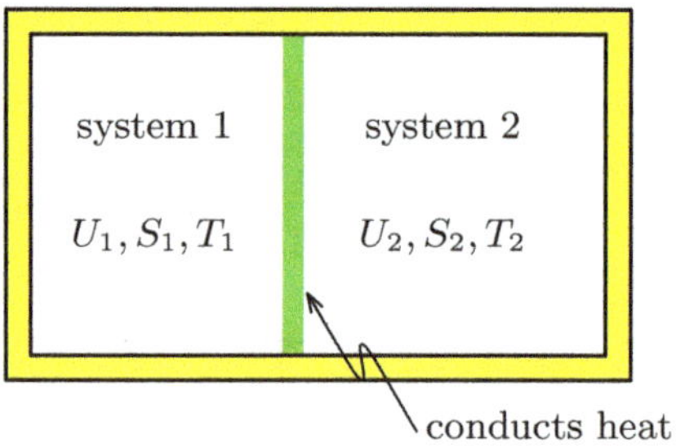

(1.5.1)

At equilibrium, system 1 has energy U_1, entropy S_1, temperature T_1 etc., and for system 2 we have U_2, S_2, T_2 and so forth. Now consider a tiny deviation from these values by a slight disturbance that changes U_1 to $U_1 + \delta U$ and U_2 to $U_2 - \delta U$. The overall change of the entropy cannot be positive

$$\delta S = \delta S_1 + \delta S_2 \le 0 . \quad (1.5.2)$$

Since only U_1 and U_2 are affected but not the configuration parameters X_1 and X_2 (volume, material content etc.), we have

$$\begin{aligned}\delta S &= \left(\frac{\partial S_1}{\partial U_1}\right)_{X_1} \delta U_1 + \left(\frac{\partial S_2}{\partial U_2}\right)_{X_2} \delta U_2 \\ &= \left(\frac{1}{T_1} - \frac{1}{T_2}\right) \delta U \leq 0 .\end{aligned} \tag{1.5.3}$$

This has to hold for all δUs, positive and negative, so that

$$T_1 = T_2 \tag{1.5.4}$$

must be true: *In equilibrium, the two systems have the same temperature.*

While this condition is not a surprise by itself, please note how we arrived at it as a simple consequence of the maximum property of the entropy, which itself is an immediate implication of the Second Law. The argument we used thereby mimics experimental procedures: You disturb the system slightly and observe what happens.

Now, imagine that we are away from equilibrium of the combined system, but have both subsystems at equilibrium with temperatures T_1 and T_2 that are different: $T_1 \neq T_2$. This is state A. State B is the equilibrium state of the combined system, which is reached from state A in adiabatic isolation, so that state B has larger entropy than state A. For the finite energy differences $\Delta U_1 = \Delta U$ and $\Delta U_2 = -\Delta U$ (small, but not infinitesimal), this means

$$\left(\frac{\partial S_1}{\partial U_1}\right)_{X_1} \Delta U_1 + \left(\frac{\partial S_2}{\partial U_2}\right)_{X_2} \Delta U_2 > 0 \tag{1.5.5}$$

for the changes that occur in the transition from state A to state B. Accordingly, we have

$$\left(\frac{1}{T_1} - \frac{1}{T_2}\right) \Delta U > 0 . \tag{1.5.6}$$

So, ΔU must be positive if $T_2 > T_1$, and negative if $T_1 > T_2$. Since ΔU is the change in the energy of system 1 during the A $\rightarrow$ B transition, this implies that the flow of heat is from the hotter system to the colder system. If two systems with different temperature are brought into contact and allowed to exchange heat, the heat flow will be against the direction of the temperature gradient.

1.6 Thermodynamical potentials: Free energy, enthalpy, free enthalpy

When we transfer a bit of heat, $\mathrm{d}Q$, to a thermodynamical system and so take it from an equilibrium state to a neighboring one, we increase the entropy in the system by

$$\mathrm{d}S = \frac{\mathrm{d}Q}{T} . \tag{1.6.1}$$

The corresponding change in temperature is related to $\mathrm{d}Q$ by the *heat capacity* C,

$$\mathrm{d}T = \frac{\mathrm{d}Q}{C} . \tag{1.6.2}$$

To achieve the same increase in temperature, we need to transfer more heat to a system with high heat capacity than to one with low heat capacity. Jointly, (1.6.1) and (1.6.2) tell us that

$$C = T\frac{\partial S}{\partial T} , \tag{1.6.3}$$

but here we have to specify the circumstances. Are we keeping the volume fixed during the process, or the pressure as in most chemical reactions? Depending on that, we distinguish between the *heat capacity for constant volume*,

$$C_V = T\left(\frac{\partial S}{\partial T}\right)_{V,n} , \tag{1.6.4}$$

and that *for constant pressure*,

$$C_P = T\left(\frac{\partial S}{\partial T}\right)_{P,n} , \tag{1.6.5}$$

and there could be others. We have also indicated that the amount of substance remains unaltered (subscript n).

Why is there a difference between C_V and C_P? If we maintain the volume while transferring heat, there is no work done ($P\,\mathrm{d}V = 0$), but if we keep the pressure constant and allow a volume change — usually $\mathrm{d}V > 0$ when $\mathrm{d}T > 0$ — the system does work on the environment, so that a part of $\mathrm{d}Q$ is handed over to the environment in the form of mechanical energy and is not available for raising the temperature. Therefore, we need to transfer more heat for the same temperature gain when $\mathrm{d}P = 0$ than when $\mathrm{d}V = 0$.

As an illustration, consider an amount of gas in a container where one wall is a movable piston, kept in place by outside pressure $P_{\rm ext}$ that balances the pressure P exerted by the gas:

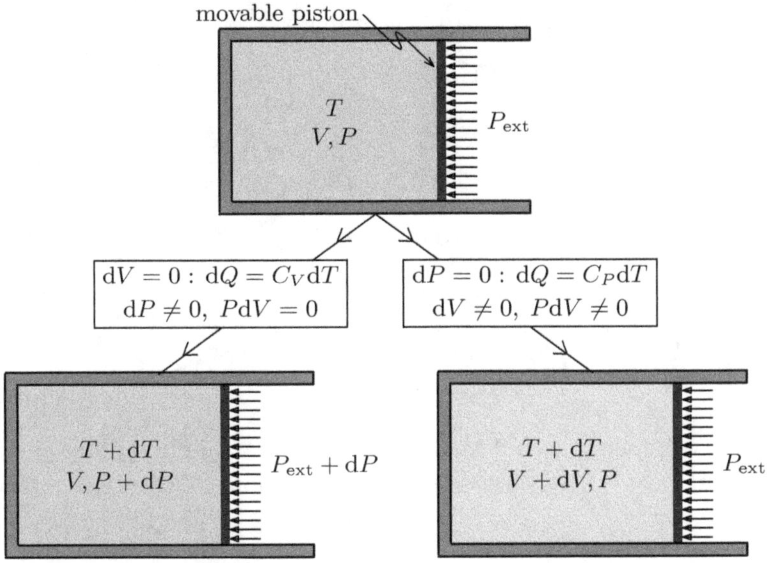

(1.6.6)

When the gas expands and the piston moves out, work is done by the gas against the external pressure. From these considerations, we infer that $C_P > C_V$ which, as discussed in Section 1.12, is indeed true quite generally; see the text after (1.12.23).

In the expression for C_V in (1.6.4), we need the entropy as a function of T, V, and n; and for C_P in (1.6.5), we need S as a function of T, P, and n. These are not supplied by (1.2.17),

$$\mathrm{d}U = T\mathrm{d}S - P\mathrm{d}V + \mu\mathrm{d}n\,, \tag{1.6.7}$$

or its multi-component generalization

$$\begin{aligned}\mathrm{d}U &= T\mathrm{d}S - P\mathrm{d}V + \sum_j \mu_j \mathrm{d}n_j \\ &= T\mathrm{d}S - P\mathrm{d}V + \boldsymbol{\mu}\cdot\mathrm{d}\boldsymbol{n}\,,\end{aligned} \tag{1.6.8}$$

which applies if we have a mixture of substances (such as air). For these, U, T, P, and $\boldsymbol{\mu}$ are function of S, V, and $\boldsymbol{n}$ as discussed in Section 1.2.

The switch of variables — from the extensive entropy S to the intensive temperature T, say — is accomplished by recalling that

$$T\mathrm{d}S = \mathrm{d}(TS) - S\mathrm{d}T\,, \tag{1.6.9}$$

the essence of integration by parts. Accordingly,

$$\mathrm{d}U = \mathrm{d}(TS) - S\mathrm{d}T - P\mathrm{d}V + \mu \cdot \delta n \tag{1.6.10}$$

or

$$\mathrm{d}(U - TS) = -S\mathrm{d}T - P\mathrm{d}V + \mu \cdot \delta n\,, \tag{1.6.11}$$

which tells us that the natural variables of

$$F = U - TS \tag{1.6.12}$$

are T, V, and n:

$$F = F(T, V, n)\,. \tag{1.6.13}$$

This quantity is called the *Helmholtz** *free energy* or simply the *free energy.* Thus, we obtain $S(T, V, n)$, needed for the calculation of C_V, as the negative T derivative of F,

$$S(T, V, n) = -\frac{\partial}{\partial T}F(T, V, n)\,, \tag{1.6.14}$$

or

$$S = -\left(\frac{\partial F}{\partial T}\right)_{V,n} \tag{1.6.15}$$

in the usual short-hand notation. Accordingly, the heat capacity for constant volume is

$$C_V = -T\left(\frac{\partial^2 F}{\partial T^2}\right)_{V,n}\,. \tag{1.6.16}$$

The transition from the internal energy U to the Helmholtz free energy F is an example of a *Legendre*† *transformation.* It helps us to switch the emphasis between an extensive variable and its intensive partner.

Another example is provided by

$$P\mathrm{d}V = \mathrm{d}(PV) - V\mathrm{d}P \tag{1.6.17}$$

---

*Hermann Ludwig Ferdinand von HELMHOLTZ (1821–1894)
†Adrien Marie LEGENDRE (1752–1833)

and its consequence

$$d(U + PV) = TdS + VdP + \mu \cdot \delta n\,, \tag{1.6.18}$$

so that the *enthalpy*, denoted by the upper-case Greek letter eta,

$$\mathrm{H} = U + PV = \mathrm{H}(S, P, n) \tag{1.6.19}$$

is a function of S, P, and n, Finally, upon switching from S to T and also from V to P, we have

$$d(U - TS + PV) = -SdT + VdP + \mu \cdot \delta n = dG\,, \tag{1.6.20}$$

which introduces

$$G = U - TS + PV = G(T, P, n)\,, \tag{1.6.21}$$

the so-called *Gibbs* free energy* or *free enthalpy*. It provides the entropy as a function of T, P, and n,

$$S(T, P, n) = -\frac{\partial}{\partial T}G(T, P, n) \tag{1.6.22}$$

or

$$S = -\left(\frac{\partial G}{\partial T}\right)_{P,n} \tag{1.6.23}$$

for short, the version of S that we use for calculating the heat capacity for constant pressure

$$C_P = -T\left(\frac{\partial^2 G}{\partial T^2}\right)_{P,n}\,. \tag{1.6.24}$$

The internal energy $U(S, V, n)$, the free energy $F(T, V, n)$, the enthalpy $\mathrm{H}(S, P, n)$, and the free enthalpy $G(T, P, n)$ are four *thermodynamical potentials*, each of them with the metrical dimension of energy. Yet others can be introduced when other pairs of intensive and extensive variables than T, S and P, V are important in the physical situation. The various thermodynamical potentials serve different purposes, and one has to use the appropriate one, as illustrated here by their roles in determining the heat capacities. We'll see more applications as we proceed.

---

*Josiah Willard GIBBS (1839–1903)

1.7 Heat capacities of the ideal gas

We use the ideal gas for illustration and recall the result in (1.2.16),

$$U(S,V,n) = E_0 n \left(\frac{nV_0}{V}\, \mathrm{e}^{\frac{S}{nR}}\right)^{\gamma - 1}, \tag{1.7.1}$$

as well as that in (1.2.19),

$$T(S,V,n) = \frac{\gamma - 1}{nR} U(S,V,n)\,. \tag{1.7.2}$$

This gives us

$$F = U - TS = \left[1 - (\gamma - 1)\frac{S}{nR}\right] U\,, \tag{1.7.3}$$

where we must remember to express the right-hand side in terms of T, V, and, n. For U, this is simple,

$$U = \frac{nRT}{\gamma - 1}\,, \tag{1.7.4}$$

and for $(\gamma - 1)\dfrac{S}{nR}$ we have

$$(\gamma - 1)\frac{S}{nR} + (\gamma - 1)\log\frac{nV_0}{V} = \log\frac{U}{E_0 n} = \log\frac{RT/E_0}{\gamma - 1}\,. \tag{1.7.5}$$

Together, they give

$$F(T,V,n) = \left[1 - \log\frac{RT/E_0}{\gamma - 1} - (\gamma - 1)\log\frac{V}{nV_0}\right]\frac{nRT}{\gamma - 1}\,. \tag{1.7.6}$$

The resulting expression for the entropy is

$$\begin{aligned} S(T,V,n) &= -\frac{\partial}{\partial T} F(T,V,n) \\ &= \frac{nR}{\gamma - 1}\log\frac{RT/E_0}{\gamma - 1} + nR\log\frac{V}{nV_0}\,, \end{aligned} \tag{1.7.7}$$

so that we find

$$C_V = T\left(\frac{\partial S}{\partial T}\right)_{V,n} = \frac{nR}{\gamma - 1} \tag{1.7.8}$$

for the heat capacity for constant volume.

We obtain the free enthalpy from the free energy by means of

$$G = F + PV \tag{1.7.9}$$

and have to express the right-hand side as a function of T, P, n. This is quite simple because the combined gas law tells us that $PV = nRT$ and

$$V = \frac{nRT}{P} \quad \text{or} \quad \frac{V}{nV_0} = \frac{RT}{PV_0}. \tag{1.7.10}$$

Accordingly,

$$\begin{aligned} G(T,P,n) &= \left[1 - \log\frac{RT/E_0}{\gamma-1} - (\gamma-1)\log\frac{RT}{PV_0}\right]\frac{nRT}{\gamma-1} + nRT \\ &= \left[\gamma - \log\frac{RT/E_0}{\gamma-1} - (\gamma-1)\log\frac{RT}{PV_0}\right]\frac{nRT}{\gamma-1}, \end{aligned} \tag{1.7.11}$$

from which first

$$S(T,P,n) = -\frac{\partial}{\partial T}G(T,P,n) = \frac{nR}{\gamma-1}\log\frac{RT/E_0}{\gamma-1} + nR\log\frac{RT}{PV_0} \tag{1.7.12}$$

and then

$$C_P = T\frac{\partial S(T,P,n)}{\partial T} = \frac{nR}{\gamma-1} + nR = \frac{\gamma nR}{\gamma-1} \tag{1.7.13}$$

follow.

Note that the final expressions for C_V and C_P make no reference to E_0 and V_0; this is as it should be because heat capacities are measurable quantities whereas E_0 and V_0 are arbitrary reference quantities that we introduced for a consistent formalism. They should only appear in intermediate steps in quantities that are not directly accessible by measurements.

We observe that the ratio of the heat capacities is the adiabatic index

$$\frac{C_P}{C_V} = \gamma, \tag{1.7.14}$$

and the difference

$$C_P - C_V = nR \tag{1.7.15}$$

is essentially the universal gas constant. We say "essentially" because the factor n here, and also in the individual expressions for C_P and C_V, just states that the heat capacity is proportional to the amount of material that is heated up. It is, therefore, common to introduce the *molar heat capacities* — also called *specific heats* — in accordance with (upper-case letters *vs* lower-case letters)

$$C_V = nc_V, \quad C_P = nc_P. \tag{1.7.16}$$

For an ideal gas, we have

$$c_V = \frac{R}{\gamma - 1}\,, \quad c_P = \frac{\gamma R}{\gamma - 1}\,, \quad c_P - c_V = R \tag{1.7.17}$$

for the molar heat capacities and their difference. The ratio is, of course, unaffected by the switch from C_V and C_P to c_V and c_P,

$$\gamma = \frac{c_P}{c_V}\,. \tag{1.7.18}$$

This relation between the adiabatic index and the ratio of the heat capacities is often regarded as a definition of γ for nonideal gases. The ratio c_P/c_V and the difference $c_P - c_V$ can be expressed by material properties more generally, and the expressions in (1.7.17) and (1.7.18) are the particular ideal-gas variants thereof. More about this in Section 1.10.

1.8 Minimum property of the free energy

We know that for an isolated system the entropy is maximal and the internal energy is minimal,

$$\mathrm{d}S = 0\,, \quad \mathrm{d}^2S < 0\,, \qquad \mathrm{d}U = 0\,, \quad \mathrm{d}^2U > 0\,, \tag{1.8.1}$$

where the differentials refer to tiny changes that move the system away from equilibrium, such as the changes introduced by internal constraints.

Let us now apply this to the situation in which the small thermodynamical system is connected to a very large bath:

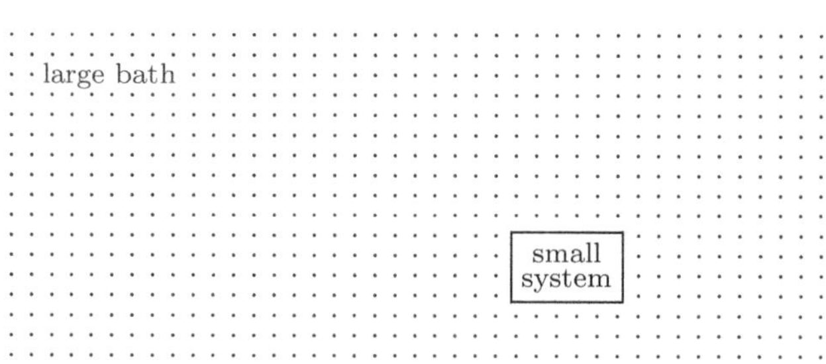

(1.8.2)

We continue to denote the entropy, energy, ... of the system by S, U, ..., and use U_{B}, S_{B}, ... for the corresponding bath quantities. Then the statements in (1.8.1) apply to the combined system,

$$\begin{aligned} \mathrm{d}(S+S_\mathrm{B}) &= 0\,, \quad \mathrm{d}^2(S+S_\mathrm{B}) < 0\,, \\ \mathrm{d}(U+U_\mathrm{B}) &= 0\,, \quad \mathrm{d}^2(U+U_\mathrm{B}) > 0\,. \end{aligned} \tag{1.8.3}$$

Now, a *bath* is huge compared with the thermodynamical system (the ocean around a bottle of water) and has, therefore, infinite heat capacity C_B and the like, that is: any finite amount of heat transferred to a temperature bath will not change the bath temperature, and any finite volume change imposed on a pressure bath will not change the bath pressure, and so forth. Accordingly, quantities such as

$$\frac{\partial^2 U_\mathrm{B}}{\partial S_\mathrm{B}^2} = \frac{\partial T_\mathrm{B}}{\partial S_\mathrm{B}} = \frac{T_\mathrm{B}}{C_\mathrm{B}} = 0 \tag{1.8.4}$$

vanish, and the second-order changes of the bath quantities in (1.8.3) do not contribute: $\mathrm{d}^2U_\mathrm{B} = 0$ and $\mathrm{d}^2S_\mathrm{B} = 0$, so that

$$\mathrm{d}(S+S_\mathrm{B}) = 0\,, \quad \mathrm{d}^2S < 0\,, \qquad \mathrm{d}(U+U_\mathrm{B}) = 0\,, \quad \mathrm{d}^2U > 0\,. \tag{1.8.5}$$

To be specific, we now consider a temperature bath, for which

$$\mathrm{d}U_\mathrm{B} = T_\mathrm{B}\,\mathrm{d}S_\mathrm{B} = -T_\mathrm{B}\,\mathrm{d}S = -\mathrm{d}(T_\mathrm{B}S) \quad \text{or} \quad \mathrm{d}(U - T_\mathrm{B}S) = 0\,, \tag{1.8.6}$$

where we exploit $\mathrm{d}S_\mathrm{B} = -\mathrm{d}S$ and $\mathrm{d}U_\mathrm{B} = -\mathrm{d}U$ as well as the constancy of the bath temperature $\mathrm{d}T_\mathrm{B} = 0$. This constancy also implies $\mathrm{d}^2(T_\mathrm{B}S) = T_\mathrm{B}\,\mathrm{d}^2S < 0$ and, together with $\mathrm{d}^2U > 0$, establishes

$$\mathrm{d}^2(U - T_\mathrm{B}S) > 0\,. \tag{1.8.7}$$

When we now restrict the states of the thermodynamical system to those which have the same temperature as the temperature bath (as they will have in equilibrium), that is: $T = T_\mathrm{B}$, we arrive at

$$\mathrm{d}(U-TS) = 0\,, \quad \mathrm{d}^2(U-TS) > 0 \quad \text{or} \quad \mathrm{d}F = 0\,, \quad \mathrm{d}^2F > 0 \tag{1.8.8}$$

with the Helmholtz free energy $F = U - TS$. This is the minimum property of the free energy for thermodynamical systems with a fixed temperature.

Similarly, by considering a pressure bath rather than a temperature bath, one establishes the minimum property of the enthalpy for systems with a fixed pressure, and by considering a temperature-and-pressure bath one establishes the minimum property of the free enthalpy for systems with given temperature and pressure.

1.9 Maxwell relations

A reasonably regular function of two variables $z(x, y)$, and almost all functions of interest in physics are like that, has equal mixed second derivatives, irrespective of the order of differentiation:

$$\frac{\partial}{\partial x}\frac{\partial z(x,y)}{\partial y}=\frac{\partial}{\partial y}\frac{\partial z(x,y)}{\partial x}\,. \tag{1.9.1}$$

In standard thermodynamical notation this appears as

$$\left(\frac{\partial}{\partial x}\left(\frac{\partial z}{\partial y}\right)_x\right)_y=\left(\frac{\partial}{\partial y}\left(\frac{\partial z}{\partial x}\right)_y\right)_x\,. \tag{1.9.2}$$

Applied to $U(S, V, n)$ with fixed n, this gives

$$\left(\frac{\partial}{\partial S}\left(\frac{\partial U}{\partial V}\right)_{S,n}\right)_{V,n}=\left(\frac{\partial}{\partial V}\left(\frac{\partial U}{\partial S}\right)_{V,n}\right)_{S,n} \tag{1.9.3}$$

or

$$-\left(\frac{\partial P}{\partial S}\right)_{V,n}=\left(\frac{\partial T}{\partial V}\right)_{S,n}\,, \tag{1.9.4}$$

which is an example of a *Maxwell* relation*. Likewise, we find

$$\left(\frac{\partial P}{\partial T}\right)_{V,n}=\left(\frac{\partial S}{\partial V}\right)_{T,n} \tag{1.9.5}$$

from the free energy $F(T, V, n)$, as well as

$$\left(\frac{\partial V}{\partial S}\right)_{P,n}=\left(\frac{\partial T}{\partial P}\right)_{S,n} \tag{1.9.6}$$

from the enthalpy $\mathrm{H}(S, P, n)$. Finally, from the free enthalpy $G(T, P, n)$ we get

$$\left(\frac{\partial V}{\partial T}\right)_{P,n}=-\left(\frac{\partial S}{\partial P}\right)_{T,n}\,. \tag{1.9.7}$$

The ratios of differentials in (1.9.6) are the reciprocals of the ratios in (1.9.5), but the two statements are not stating the same fact as the variables that are kept constant are different. An analogous remark applies to the equation pair (1.9.7) and (1.9.4).

---

*James Clerk MAXWELL (1831–1879)

In this context, let us take a look at the differential of $z(x, y)$,

$$\mathrm{d}z = \left(\frac{\partial z}{\partial x}\right)_y \mathrm{d}x + \left(\frac{\partial z}{\partial y}\right)_x \mathrm{d}y\,, \tag{1.9.8}$$

and pay due attention to the circumstances of constant x,

$$(\mathrm{d}z)_x = \left(\frac{\partial z}{\partial y}\right)_x (\mathrm{d}y)_x\,. \tag{1.9.9}$$

This implies

$$1 = \left(\frac{\partial z}{\partial y}\right)_x \left(\frac{\partial y}{\partial z}\right)_x \tag{1.9.10}$$

or

$$\left(\frac{\partial z}{\partial y}\right)_x = \frac{1}{\left(\frac{\partial y}{\partial z}\right)_x}\,, \tag{1.9.11}$$

which is a familiar statement for two dependent variables, here y and z. While we are at it, let us establish another identity by considering "iso-z" changes,

$$0 = \left(\frac{\partial z}{\partial x}\right)_y (\mathrm{d}x)_z + \left(\frac{\partial z}{\partial y}\right)_x (\mathrm{d}y)_z\,. \tag{1.9.12}$$

This gives us first

$$\left(\frac{\partial z}{\partial x}\right)_y \left(\frac{\partial x}{\partial y}\right)_z + \left(\frac{\partial z}{\partial y}\right)_x = 0 \tag{1.9.13}$$

and, equivalently,

$$\left(\frac{\partial z}{\partial x}\right)_y + \left(\frac{\partial z}{\partial y}\right)_x \left(\frac{\partial y}{\partial x}\right)_z = 0 \tag{1.9.14}$$

and finally

$$\left(\frac{\partial x}{\partial y}\right)_z \left(\frac{\partial y}{\partial z}\right)_x \left(\frac{\partial z}{\partial x}\right)_y = -1\,, \tag{1.9.15}$$

which is known as the *chain relation*. Its cyclic structure helps in memorizing this statement.

1.10 Heat capacities

Here are examples for the use of Maxwell relations. First, we ask how the heat capacity for constant volume, C_V, changes with the volume. After noting that

$$\begin{aligned}\left(\frac{\partial C_V}{\partial V}\right)_{T,n} &= \left(\frac{\partial}{\partial V}\left[T\left(\frac{\partial S}{\partial T}\right)_{V,n}\right]\right)_{T,n} \\ \text{T is constant} \quad &= T\left(\frac{\partial}{\partial V}\left(\frac{\partial S}{\partial T}\right)_{V,n}\right)_{T,n} \\ \text{differentiations interchanged} \quad &= T\left(\frac{\partial}{\partial T}\left(\frac{\partial S}{\partial V}\right)_{T,n}\right)_{V,n},\end{aligned} \tag{1.10.1}$$

we recall the Maxwell relation in (1.9.5), $\left(\frac{\partial S}{\partial V}\right)_{T,n} = \left(\frac{\partial P}{\partial T}\right)_{V,n}$, and then have

$$\left(\frac{\partial C_V}{\partial V}\right)_{T,n} = T\left(\frac{\partial^2 P}{\partial T^2}\right)_{V,n}. \tag{1.10.2}$$

Likewise we find that the P derivative of C_P is

$$\begin{aligned}\left(\frac{\partial C_P}{\partial P}\right)_{T,n} &= T\left(\frac{\partial}{\partial P}\left(\frac{\partial S}{\partial T}\right)_{P,n}\right)_{T,n} \\ &= T\left(\frac{\partial}{\partial T}\underbrace{\left(\frac{\partial S}{\partial P}\right)_{T,n}}_{=-\left(\frac{\partial V}{\partial T}\right)_{P,n}}\right)_{P,n} \\ &= -T\left(\frac{\partial^2 V}{\partial T^2}\right)_{P,n}.\end{aligned} \tag{1.10.3}$$

Note that we encounter the temperature derivative of the *thermal expansion coefficient* α here,

$$\alpha = \frac{1}{V}\left(\frac{\partial V}{\partial T}\right)_{P,n}, \tag{1.10.4}$$

that is

$$\left(\frac{\partial C_P}{\partial P}\right)_{T,n} = -T\left(\frac{\partial}{\partial T}(\alpha V)\right)_{P,n} = -T\left[\alpha \underbrace{\left(\frac{\partial V}{\partial T}\right)_{P,n}}_{=\alpha V} + V\left(\frac{\partial \alpha}{\partial T}\right)_{P,n}\right]$$

$$= -TV\left[\alpha^2 + \left(\frac{\partial \alpha}{\partial T}\right)_{P,n}\right]. \tag{1.10.5}$$

For our last example, we take another look at the constant-volume heat capacity,

$$C_V = T\left(\frac{\partial S}{\partial T}\right)_{V,n}, \tag{1.10.6}$$

with the aim of relating it to that of constant pressure.

$$C_P = T\left(\frac{\partial S}{\partial T}\right)_{P,n}. \tag{1.10.7}$$

For this purpose, we view the entropy as a function of T, P, and n, so that

$$\mathrm{d}S = \left(\frac{\partial S}{\partial T}\right)_{P,n} \mathrm{d}T + \left(\frac{\partial S}{\partial P}\right)_{T,n} \mathrm{d}P + \left(\frac{\partial S}{\partial n}\right)_{T,P} \cdot \mathrm{d}n\,. \tag{1.10.8}$$

Upon restricting to changes of T and P only, but not of n, this becomes

$$(\mathrm{d}S)_n = \left(\frac{\partial S}{\partial T}\right)_{P,n} (\mathrm{d}T)_n + \left(\frac{\partial S}{\partial P}\right)_{T,n} (\mathrm{d}P)_n\,. \tag{1.10.9}$$

Now, we consider the case where also the volume is constant and obtain

$$\left(\frac{\partial S}{\partial T}\right)_{V,n} = \left(\frac{\partial S}{\partial T}\right)_{P,n} + \left(\frac{\partial S}{\partial P}\right)_{T,n} \left(\frac{\partial P}{\partial T}\right)_{V,n}. \tag{1.10.10}$$

We multiply by T, recognize C_V and C_P and use the Maxwell relation in (1.9.7), $\left(\frac{\partial S}{\partial P}\right)_{T,n} = -\left(\frac{\partial V}{\partial T}\right)_{P,n}$, and so arrive at

$$C_V = C_P - T\left(\frac{\partial V}{\partial T}\right)_{P,n} \left(\frac{\partial P}{\partial T}\right)_{V,n}. \tag{1.10.11}$$

Here, we use the chain relation in the form

$$\left(\frac{\partial P}{\partial T}\right)_{V,n} \left(\frac{\partial T}{\partial V}\right)_{P,n} \left(\frac{\partial V}{\partial P}\right)_{T,n} = -1 \tag{1.10.12}$$

or

$$\left(\frac{\partial P}{\partial T}\right)_{V,n} = -\left(\frac{\partial V}{\partial T}\right)_{P,n}\left(\frac{\partial P}{\partial V}\right)_{T,n} \tag{1.10.13}$$

and so get to

$$C_V = C_P + T\underbrace{\left(\frac{\partial V}{\partial T}\right)_{P,n}^2}_{=(\alpha V)^2}\left(\frac{\partial P}{\partial V}\right)_{T,n}. \tag{1.10.14}$$

The thermal expansion coefficient α of (1.10.4) appears here, and so does the *isothermal compressibility* K_T,

$$K_T = -\frac{1}{V}\left(\frac{\partial V}{\partial P}\right)_{T,n}. \tag{1.10.15}$$

After putting the pieces together, we have

$$C_P - C_V = \frac{\alpha^2 T}{K_T}V\,. \tag{1.10.16}$$

All quantities appearing in this identity can be measured in suitable experiments.

For a check, we verify the identity for the ideal gas. Owing to the combined gas law $PV = nRT$, we have the thermal expansion coefficient

$$\alpha = \frac{1}{V}\left(\frac{\partial V}{\partial T}\right)_{P,n} = \frac{1}{V}\left(\frac{\partial}{\partial T}\frac{nRT}{P}\right)_{P,n} = \frac{1}{V}\frac{nR}{P} = \frac{1}{T} \tag{1.10.17}$$

and the isothermal compressibility

$$K_T = -\frac{1}{V}\left(\frac{\partial V}{\partial P}\right)_{T,n} = -\frac{1}{V}\left(\frac{\partial}{\partial P}\frac{nRT}{P}\right)_{T,n} = -\frac{1}{V}\left(-\frac{nRT}{P^2}\right) = \frac{1}{P}, \tag{1.10.18}$$

so that

$$C_P - C_V = \frac{\left(\frac{1}{T}\right)^2}{\frac{1}{P}}VT = \frac{PV}{T} = nR, \tag{1.10.19}$$

consistent with what we found earlier; see (1.7.15).

What about the ratio $\dfrac{C_P}{C_V}$? To find an expression for it, we first exploit the chain relation and the pertinent Maxwell relations to establish useful alternative expressions for the heat capacities. Namely,

$$\begin{aligned}C_V &= T\left(\frac{\partial S}{\partial T}\right)_{V,n}\\&= T\underbrace{\left(\frac{\partial S}{\partial T}\right)_{V,n}\left(\frac{\partial T}{\partial V}\right)_{S,n}\left(\frac{\partial V}{\partial S}\right)_{T,n}}_{=-1}\left(\frac{\partial V}{\partial T}\right)_{S,n}\left(\frac{\partial S}{\partial V}\right)_{T,n}\\&= -T\left(\frac{\partial V}{\partial T}\right)_{S,n}\left(\frac{\partial S}{\partial V}\right)_{T,n}\end{aligned} \qquad (1.10.20)$$

and

$$\begin{aligned}C_P &= T\left(\frac{\partial S}{\partial T}\right)_{P,n}\\&= T\underbrace{\left(\frac{\partial S}{\partial T}\right)_{P,n}\left(\frac{\partial T}{\partial P}\right)_{S,n}\left(\frac{\partial P}{\partial S}\right)_{T,n}}_{=-1}\left(\frac{\partial P}{\partial T}\right)_{S,n}\left(\frac{\partial S}{\partial P}\right)_{T,n}\\&= -T\left(\frac{\partial P}{\partial T}\right)_{S,n}\left(\frac{\partial S}{\partial P}\right)_{T,n}.\end{aligned} \qquad (1.10.21)$$

Their ratio is then

$$\begin{aligned}\frac{C_P}{C_V} &= \underbrace{\left(\frac{\partial P}{\partial T}\right)_{S,n}\left(\frac{\partial V}{\partial T}\right)_{S,n}^{-1}}_{=\left(\frac{\partial P}{\partial V}\right)_{S,n}}\underbrace{\left(\frac{\partial S}{\partial P}\right)_{T,n}\left(\frac{\partial S}{\partial V}\right)_{T,n}^{-1}}_{=\left(\frac{\partial V}{\partial P}\right)_{T,n}}\\&= \left(\frac{\partial P}{\partial V}\right)_{S,n}\left(\frac{\partial V}{\partial P}\right)_{T,n}\\&= \left(\frac{\partial V}{\partial P}\right)_{S,n}^{-1}\left(\frac{\partial V}{\partial P}\right)_{T,n}\end{aligned} \qquad (1.10.22)$$

or

$$\frac{C_P}{C_V} = \frac{-\frac{1}{V}\left(\frac{\partial V}{\partial P}\right)_{T,n}}{-\frac{1}{V}\left(\frac{\partial V}{\partial P}\right)_{S,n}} = \frac{K_T}{K_S}, \qquad (1.10.23)$$

where we recognize the isothermal compressibility K_T of (1.10.15) and encounter also K_S, the isentropic compressibility or *adiabatic compressibility*,

$$K_S = -\frac{1}{V}\left(\frac{\partial V}{\partial P}\right)_{S,n} . \tag{1.10.24}$$

Again, a quick check for the ideal gas is worth the effort. Since PV^γ does not depend on V for processes with constant entropy, we have

$$K_S = \frac{1}{\gamma P} \tag{1.10.25}$$

and with K_T from (1.10.18) we get

$$\frac{C_P}{C_V} = \frac{\frac{1}{P}}{\frac{1}{\gamma P}} = \gamma\,, \tag{1.10.26}$$

consistent with the earlier result in (1.7.18).

The results for the difference $C_P - C_V$ and the ratio $\frac{C_P}{C_V}$ can be combined to yield explicit expression for C_V and C_P themselves. We have

$$\begin{aligned} C_V &= \frac{C_P - C_V}{C_P - C_V} C_V = \frac{C_P - C_V}{\frac{C_P}{C_V} - 1} = \frac{\frac{\alpha^2}{K_T} VT}{\frac{K_T}{K_S} - 1} \\ &= \frac{K_S}{K_T} \frac{\alpha^2 T}{K_T - K_S} V \end{aligned} \tag{1.10.27}$$

and

$$C_P = C_V \frac{K_T}{K_S} = \frac{\alpha^2 T}{K_T - K_S} V\,. \tag{1.10.28}$$

1.11 Gibbs–Duhem relation

The functions $U(S, V, n)$, $F(T, V, n)$, $H(S, P, n)$, and $G(T, P, n)$ are the four thermodynamical potentials that we get from the two pairs (T, S) and (P, V) of intensive and extensive variables. What about the pair, or pairs, (μ, n)? One can also do a Legendre transformation on them, but it is not useful as often as the standard ones. In addition, there is a trap that one must avoid.

To understand it, consider a function $f(X)$ of extensive variables X with linear homogeneity,

$$f(\lambda X) = \lambda f(X)\,, \tag{1.11.1}$$

where "linear" indicates a single factor of λ, rather than λ^2 ("quadratic") or λ^3 ("cubic"), or another power. Differentiation with respect to λ at $\lambda = 1$ establishes rather immediately that

$$f(X) = \sum_j X_j \frac{\partial f(X)}{\partial X_j} = X \cdot \frac{\partial f}{\partial X}(X)\,, \tag{1.11.2}$$

where, of course, the variables X_1, X_2, X_3, ... are regarded as independent, but we do not write $\left(\frac{\partial f}{\partial X_j}\right)_{X_k \text{ with } k \neq j}$ or similar; there is no need for this pedantry.

We apply this to $U(S, V, n)$ with the linear homogeneity of (1.2.11),

$$U(\lambda S, \lambda V, \lambda n) = \lambda U(S, V, n)\,. \tag{1.11.3}$$

Accordingly, we have

$$U(S, V, n) = S\left(\frac{\partial U}{\partial S}\right)_{V,n} + V\left(\frac{\partial U}{\partial V}\right)_{S,n} + n \cdot \left(\frac{\partial U}{\partial n}\right)_{S,V} \tag{1.11.4}$$

or

$$U = ST - VP + n \cdot \mu\,. \tag{1.11.5}$$

So, has U suddenly become a function of S, T, V, P, n, and μ? Of course not. We have $T(S, V, n)$, $P(S, V, n)$, and $\mu(S, V, n)$ here, the natural variables continue to be S, V, and n.

The comparison of

$$\begin{aligned} \mathrm{d}U &= \mathrm{d}(ST - VP + n \cdot \mu) \\ &= T\mathrm{d}S - P\mathrm{d}V + \mu \cdot \mathrm{d}n + S\mathrm{d}T - V\mathrm{d}P + n \cdot \mathrm{d}\mu \end{aligned} \tag{1.11.6}$$

with

$$\mathrm{d}U = T\mathrm{d}S - P\mathrm{d}V + \mu \cdot \mathrm{d}n \tag{1.11.7}$$

now establishes the *Gibbs–Duhem* relation*

$$S\mathrm{d}T - V\mathrm{d}P + n \cdot \mathrm{d}\mu = 0\,. \tag{1.11.8}$$

*Pierre Maurice Marie DUHEM (1861–1916)

It states that adjacent equilibrium states cannot differ *only* in temperature, or *only* in pressure, or *only* in one chemical potential. An important consequence is the impossibility of having a meaningful thermodynamical potential that is a function of T, P, and μ; their infinitesimal changes are not independent, which is to say that they cannot be regarded as independent variables and treated as such in the formalism.

Upon combining (1.11.5) with $G = U - TS + PV$, we obtain

$$G = n \cdot \mu \,, \tag{1.11.9}$$

where μ is a function of T, P, and n, of course. If the system is composed of one kind of material only, so that n and μ have only one variable, $n \cdot \mu = n\mu$, then the Gibbs free energy, or free enthalpy, is essentially the chemical potential.

The Gibbs–Duhem relation links the intensive quantities T, P, μ and tells us that they are not independent of each other. Another way of arriving at the same conclusion is the following. Consider, for example, the temperature as a function of the extensive quantities S, V, n,

$$T(S, V, n) = \left(\frac{\partial U}{\partial S}\right)_{V,n} . \tag{1.11.10}$$

If we scale the system by a factor λ, the temperature does not change

$$T(\lambda S, \lambda V, \lambda n) = T(S, V, n) \,, \tag{1.11.11}$$

that is: the temperature is a zeroth-order (λ^0 on the right) homogeneous function of S, V, and $n = (n_1, n_2, \ldots, n_J)$ if there are J different kinds of material in the system. The same is true for the other intensive quantities, the pressure P and the chemical potentials $\mu = (\mu_1, \mu_2, \ldots, \mu_J)$. Now, let's choose

$$\lambda = \left(\sum_{j=1}^{J} n_j\right)^{-1} = \frac{1}{n_{\text{tot}}} \,, \tag{1.11.12}$$

with n_{tot} the total number of moles in the system. Then

$$\lambda n = \frac{1}{n_{\text{tot}}} n = \left(\frac{n_1}{n_{\text{tot}}}, \frac{n_2}{n_{\text{tot}}}, \ldots, \frac{n_J}{n_{\text{tot}}}\right) = (\nu_1, \nu_2, \ldots, \nu_J) = \nu \tag{1.11.13}$$

is the collection of molar fractions, which is intensive, since $n \to \lambda n$ gives $n_{\text{tot}} \to \lambda n_{\text{tot}}$ and $\nu \to \nu$. Likewise, the molar entropy s and the molar volume v, introduced in accordance with

$$s = \frac{S}{n_{\text{tot}}} \quad \text{and} \quad v = \frac{V}{n_{\text{tot}}}, \tag{1.11.14}$$

are intensive, so that

$$T(S, V, n) = T(s, v, \nu) \tag{1.11.15}$$

states the intensive temperature as a function of the intensive variables s, v, and ν. They are $J+1$ in number, not $J+2$, because

$$\sum_{j=1}^{J} \nu_j = 1 \tag{1.11.16}$$

constrains the ν_js to unit sum. Therefore, we need one less intensive variable to specify the value of another intensive quantity than we need extensive variables to state the value of another extensive quantity, such as $U(S, V, n)$ which is extensive and has $J+2$ extensive arguments as function variables. The one extra variable simply states the system size (such as n_{tot} does), which is needed for extensive quantities but not for intensive ones.

1.12 Stability criteria; coexisting phases

In Section 1.5 we established that two systems that can exchange heat have the same temperature when the composed total system is in an equilibrium state. Exercise 7 establishes further that the two systems have equal pressure if they can do work on each other, and they have the same chemical potentials if the systems can exchange matter (particles = atoms, molecules, ...). Let us now return to this discussion and extend it considerably by a study of the conditions that must be obeyed if the system is in a *stable* thermodynamical equilibrium. The stability in question is that with respect to small perturbations, such as the spontaneous fluctuations that are occurring all the time in a thermodynamical system. It is an unavoidable consequence of the very nature of a thermodynamical system — namely that we can characterize it sufficiently by a small number of macroscopic parameters (volume, temperature, ...) — that there are these ubiquitous fluctuations. "Stability" then means that the system returns to the equilibrium and that the fluctuations do not grow from tiny, unnoticeable bits to macroscopic proportions.

We consider a system composed of two or more phases, such as gas, liquid, and solid (steam, water, ice for H_2O), a total of K phases; here sketched for $K = 6$:

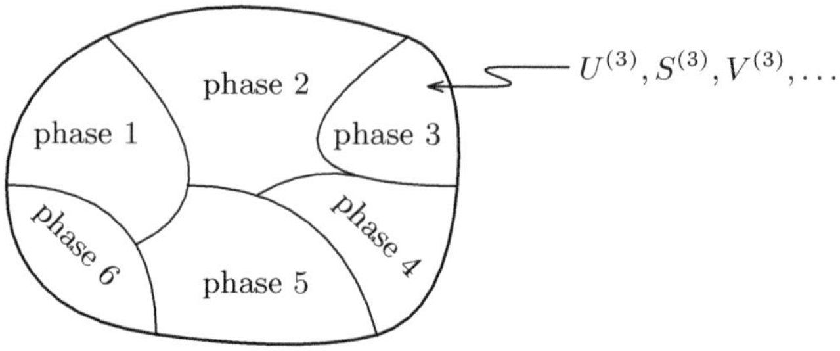

(1.12.1)

The phases are in equilibrium, with the respective equilibrium values $U^{(k)}$, $S^{(k)}$, $V^{(k)}$, $n^{(k)}$ and so forth for the kth phase. Now, small internal fluctuations of entropy, volume, and particle number will take the system a bit away from the equilibrium and this is necessarily accompanied by an increase of the total internal energy,

$$(\delta U)_{S,V,n} = \sum_{k=1}^{K} \delta U^{(k)} = \sum_{k=1}^{K} \Big[T^{(k)}\delta S^{(k)} - P^{(k)}\delta V^{(k)} + \mu^{(k)} \cdot \delta n^{(k)}\Big] \geq 0\,, \tag{1.12.2}$$

where we have written the first-order contributions to $\delta U^{(k)}$ only. In this redistribution of entropy, volume, and material content between the phases, we do not change their sums, as indicated by the subscript $_{S,V,n}$. Thus, the null sums

$$\sum_{k=0}^{K} \delta S^{(k)} = 0\,, \quad \sum_{k=0}^{K} \delta V^{(k)} = 0\,, \quad \sum_{k=0}^{K} \delta n^{(k)} = 0 \tag{1.12.3}$$

constrain the permissible $\delta S^{(k)}$s, $\delta V^{(k)}$s, and $\delta n^{(k)}$s, but other than that these fluctuations are independent. We can, therefore, look at the situation in which we have only fluctuations of the entropy, but not of the volume or the particle numbers. Then

$$(\delta U)_{S,V,n} = \sum_{k=1}^{K} T^{(k)}\delta S^{(k)} \geq 0 \quad \text{with} \quad \sum_{k=1}^{K} \delta S^{(k)} = 0\,, \tag{1.12.4}$$

which has to hold for *all* thinkable $\delta S^{(k)}$s. In particular, if it holds for certain $\delta S^{(k)}$s, it must also hold when all $\delta S^{(k)}$s change sign. Accordingly, we must have

$$\sum_{k=1}^{K} T^{(k)} \delta S^{(k)} = 0\,, \tag{1.12.5}$$

which implies that all temperatures are the same

$$T^{(1)} = T^{(2)} = T^{(3)} = \cdots = T^{(K)}\,. \tag{1.12.6}$$

For, if some temperatures are larger and others are smaller than their average value, then $\delta S^{(k)} < 0$ for the smaller ones and $\delta S^{(k)} > 0$ for the larger ones would result in $\sum_k T^{(k)} \delta S^{(k)} > 0$, which must not happen.

Likewise we conclude that all pressures are the same,

$$P^{(1)} = P^{(2)} = \cdots = P^{(K)}\,, \tag{1.12.7}$$

by considering fluctuations of only the volumes, and that all chemical potentials are the same,

$$\mu^{(1)} = \mu^{(2)} = \mu^{(3)} = \cdots = \mu^{(K)} \tag{1.12.8}$$

by considering $\delta S^{(k)} = 0$, $\delta V^{(k)} = 0$, and $\delta n^{(k)} \neq 0$.

In summary, we have thus established that all phases have the same temperature, the same pressure, the same chemical potentials — the same values of the intensive variables — when the phases are in equilibrium: There is one common temperature, one common pressure, and one common set of chemical potentials. The transitivity of the Zeroth Law is evident. Further, since we can introduce an arbitrary partitioning within each phase, and so split a phase in two, it also follows that the intensive variables are constant throughout each phase.

What about the energy contribution to the total internal energy from the interfaces between physically different phases? These surface contributions are nonzero (there could be surface tension at the liquid-gas interface, for example) but negligibly small compared with the bulk contributions that we have been considering. This is the consequence of the very small size of atoms and the correspondingly very large number of atoms in a mole. If we have N atoms in total, then $\sim N^{\frac{2}{3}}$ of them are near an interface, because the bulk volume is three-dimensional whereas a surface layer is two-dimensional. Thus, the fraction of atoms contributing to interface-related energies is $N^{-\frac{1}{3}}$. This is a tiny fraction for a mole ($N \sim 10^{24}$, $N^{-\frac{1}{3}} \sim 10^{-8}$) and still rather small for a micromole ($N \sim 10^{18}$, $N^{-\frac{1}{3}} \sim 10^{-6}$). We leave it at that for now and will take a closer at interfaces in Section 1.18.

In this reasoning, and others to come, we rely on the usual thermodynamical circumstances of systems made up by a huge number of constituents. When, however, the system consists of a rather small number of constituents, the situation is quite different and all aspects require careful reconsideration. In particular, it is possible that quantum properties of the constituents and their interactions play a significant role. This is why the study of relatively small systems with the tools of thermodynamics has been termed "quantum thermodynamics." It is still an emerging field today (2018), in which even the most basic concepts — such as heat, work, entropy — are topics of controversial debates. We shall not step onto this mine field.

Let us now look at a single phase and consider an arbitrary partitioning:

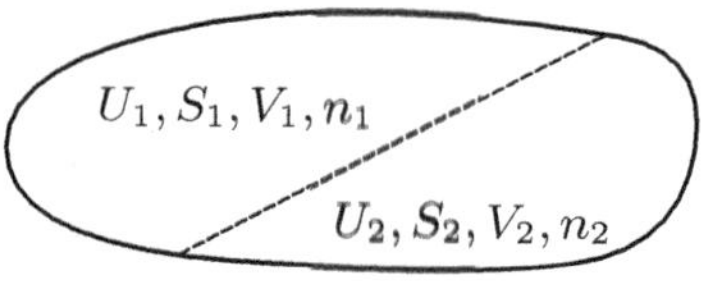

$$\begin{aligned} U &= U_1 + U_2 \\ S &= S_1 + S_2 \\ V &= V_1 + V_2 \\ n &= n_1 + n_2 \end{aligned} \tag{1.12.9}$$

The extensive variables add, and the intensive variables are the same in equilibrium,

$$T_1 = T_2\,, \quad P_1 = P_2\,, \quad \mu_1 = \mu_2\,. \tag{1.12.10}$$

A small fluctuation of the entropy, $\delta S_1 = -\delta S_2$, necessarily increases the total internal energy

$$\begin{aligned} (\delta U)_{S,V,n} &= U_1(S_1 + \delta S_1, V_1, n_1) - U_1(S_1, V_1, n_1) \\ &\quad + U_2(S_2 + \delta S_2, V_2, n_2) - U_2(S_2, V_2, n_2) \\ &= \left(\frac{\partial U_1}{\partial S_1}\right)_{V_1,n_1} \delta S_1 + \left(\frac{\partial U_2}{\partial S_2}\right)_{V_2,n_2} \delta S_2 \\ &\quad + \frac{1}{2}\left(\frac{\partial^2 U_1}{\partial S_1^2}\right)_{V_1,n_1} (\delta S_1)^2 + \frac{1}{2}\left(\frac{\partial^2 U_2}{\partial S_2^2}\right)_{V_2,n_2} (\delta S_2)^2 \\ &\quad + \text{(higher-order terms)} \\ &> 0\,. \end{aligned} \tag{1.12.11}$$

Since $\left(\frac{\partial U_1}{\partial S_1}\right)_{V_1,n_1} = T_1 = T_2 = \left(\frac{\partial U_2}{\partial S_2}\right)_{V_2,n_2}$ and $\delta S_2 = -\delta S_1$, the first-order terms take care of themselves, so that the second-order terms, which dominate for small enough fluctuations, must give a positive contribution,

$$\begin{aligned}&\left(\frac{\partial^2 U_1}{\partial S_1^2}\right)_{V_1,n_1}(\delta S_1)^2+\left(\frac{\partial^2 U_2}{\partial S_2^2}\right)_{V_2,n_2}(\delta S_2)^2\\&=\left(\frac{\partial T_1}{\partial S_1}\right)_{V_1,n_1}(\delta S_1)^2+\left(\frac{\partial T_2}{\partial S_2}\right)_{V_2,n_2}(\delta S_2)^2>0\end{aligned} \tag{1.12.12}$$

for $\delta S_1 = -\delta S_2 \neq 0$.

Since the chosen partitioning is arbitrary, it follows that

$$\left(\frac{\partial T}{\partial S}\right)_{V,n} > 0 \tag{1.12.13}$$

is necessary in a stable equilibrium. Otherwise, small fluctuations of the entropy would decrease the internal energy and so drive the system spontaneously to a state with lower energy, thereby identifying the previous equilibrium as unstable. This is quite analogous to the familiar mechanical situation where a stable point of vanishing force requires a minimum of the potential energy rather than a maximum or a saddle point.

We recall the definition of the heat capacity for constant volume in (1.6.4),

$$C_V = T\left(\frac{\partial S}{\partial T}\right)_{V,n}, \tag{1.12.14}$$

and infer

$$C_V > 0 \tag{1.12.15}$$

from the stability criterion in (1.12.13). A positive heat capacity is, indeed, necessary for stability. As we have seen in Section 1.5, heat flows from the hotter part to the colder part, and if the heat capacity of both parts is positive, this flux of heat decreases the hot temperature and increases the cold temperature leading to an eventual temperature equality. For negative heat capacities, the opposite would happen; the hotter part would get even hotter and the colder part colder, thereby increasing the temperature difference and the heat flux.

We return to the partitioning in (1.12.9) and now consider fluctuations of only the volume. The analog of the argument that took us to recognizing that the second derivative of U with respect to S must be positive, now leads to

$$\left(\frac{\partial^2 U}{\partial V^2}\right)_{S,n} > 0 \tag{1.12.16}$$

or

$$\left(\frac{\partial P}{\partial V}\right)_{S,n} < 0\,. \tag{1.12.17}$$

Here, we encounter the adiabatic compressibility of (1.10.15),

$$K_S = -\frac{1}{V}\left(\frac{\partial V}{\partial P}\right)_{S,n}, \tag{1.12.18}$$

and infer that

$$K_S > 0\,. \tag{1.12.19}$$

Again, this is as it should be: Putting additional pressure on the system decreases the volume; it does not increase the volume.

If we want to say something similar about the isothermal compressibility of (1.10.15),

$$K_T = -\frac{1}{V}\left(\frac{\partial V}{\partial P}\right)_{T,n} = -\frac{1}{V\left(\frac{\partial P}{\partial V}\right)_{T,n}}, \tag{1.12.20}$$

we have to work with the free energy $F = U - TS = F(T,V,n)$. Fluctuations in the volumes of the partitioning in (1.12.9) result in second-order changes of the free energy

$$(\delta F)_{T,V,n} = \frac{1}{2}\left(\frac{\partial^2 F_1}{\partial V_1^2}\right)_{T_1,n_1}(\delta V_1)^2 + \frac{1}{2}\left(\frac{\partial^2 F_2}{\partial V_2^2}\right)_{T_2,n_2}(\delta V_2)^2 \tag{1.12.21}$$

and the requirement $\delta F > 0$ implies first

$$\left(\frac{\partial^2 F}{\partial V^2}\right)_{T,n} = -\left(\frac{\partial P}{\partial V}\right)_{T,n} > 0 \tag{1.12.22}$$

and then

$$K_T > 0\,. \tag{1.12.23}$$

As a consequence of (1.10.23), that is: $C_P K_S = C_V K_T$, it follows that $C_P > 0$. Further, (1.10.16) implies that $C_P - C_V > 0$, which in turn requires $K_T > K_S$. It is easier to compress a system connected to a thermal reservoir than a system that is adiabatically isolated.

For the free energy considered just now, we have

$$\mathrm{d}F = -S\mathrm{d}T - P\mathrm{d}V + \mu \cdot \mathrm{d}n\,. \tag{1.12.24}$$

The partitioning in (1.12.9) assigns partial values of the extensive variables V and n to the two (fictitious) subsystems but not partial values of the intensive variables T. It makes no sense to "redistribute temperature." Therefore, the stability requirement of positive second derivatives applies only to the extensive variables.

The general situation is as follows. For a system with extensive variables $X_1, X_2, X_3, \ldots$ and corresponding intensive variables $y_1, y_2, y_3, \ldots$, so that

$$\mathrm{d}U = \sum_m y_m \mathrm{d}X_m \tag{1.12.25}$$

is the differential of the internal energy, and a thermodynamical potential Φ with

$$\mathrm{d}\Phi = \sum_{m\leq M} y_m \mathrm{d}X_m - \sum_{m>M} X_m \mathrm{d}y_m\,, \tag{1.12.26}$$

we have

$$\left(\frac{\partial y_m}{\partial X_m}\right)_{\substack{\text{all other } X_m (m \leq M)\\ \text{and all } y_m (m > M)}} > 0 \tag{1.12.27}$$

for stable equilibrium. Note the important general feature that all these criteria refer to the response of an intensive variable to a change of the corresponding extensive variable. Examples are $\frac{\partial T}{\partial S} > 0$ in (1.12.13) and $-\frac{\partial P}{\partial V} > 0$ in (1.12.17) and (1.12.22). Which other variables are kept fixed, depends on the thermodynamical potential considered.

1.13 Gibbs phase rule

We return once more to the situation with K coexisting phases in equilibrium, as depicted in (1.12.1), and regard the chemical potential as a function of the other intensive variables,

$$\mu = \mu(T, P, \nu) \quad \text{with} \quad \sum_{j=1}^{J} \nu_j = 1\,, \tag{1.13.1}$$

where $\nu_1, \nu_2, \ldots, \nu_J$ are the molar fractions if there are J different substances in the system. Equal chemical potentials for all K phases requires

$$\mu_j^{(k)}(T, P, \nu^{(k)}) = \mu_j^{(k')}(T, P, \nu^{(k')}) \tag{1.13.2}$$

for $1 \leq k < k' \leq K$ and $1 \leq j \leq J$, so that we have $(K-1)J$ equations in total obeyed by $2 + (J-1)K$ variables — temperature and pressure plus $J-1$ independent ν_j values per phase. The number of independent intensive variables — that is: the number of *thermodynamical degrees of freedom* (TDOF) — is, therefore,

$$\#(\text{TDOF}) = 2 + (J-1)K - (K-1)J = 2 + J - K\,. \tag{1.13.3}$$

This is the *Gibbs phase rule.*

It tells us that a J-component system can have at most $2+J$ coexisting phases, since the number of thermodynamical degrees of freedom cannot be negative. For values of T and P where the maximum number of phases coexist, there are no TDOF. Accordingly, these (T, P) pairs are isolated points in a T, P-diagram.

The simplest situation is that of a single-component system ($J = 1$), where $K = 1, 2, 3$ are the allowed values, so that a T, P-diagram could look like this:

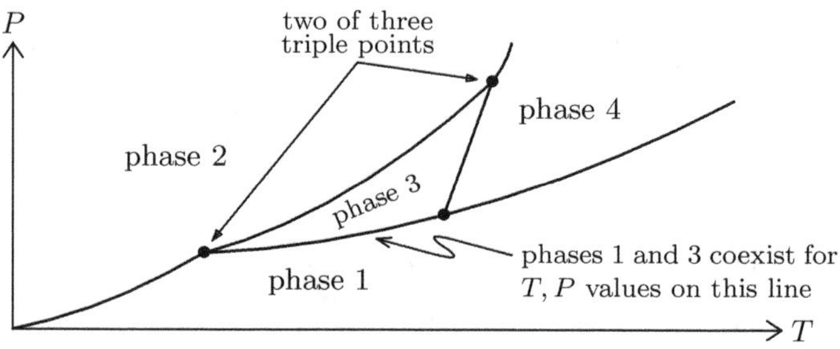

(1.13.4)

There are extended regions with a single phase in the stable equilibrium, separated from each other by the lines along which two phases coexist. In those extended single-phase regions, the temperature T and the pressure P take on values independently of one another, whereas there is only one value of P for a given value of T on one of the coexistence lines. And then there are those isolated points — the so-called *triple points*, three of them in the sketch — where one has three phases coexisting.

A well-known example is the triple point of H_2O at temperature $T = 273.16\,\text{K} = 0.01°\,\text{C}$ and pressure $P = 611.73\,\text{Pa} = 6 \times 10^{-3}\,\text{atm}$, where we have ice and water and steam in coexistence. What makes this triple point so peculiar is that very small changes of P and T move the system to a region where one has only water, or only ice, or only steam. Triple points have applications as temperature standards, simply because of the particular values for temperature and pressure that need to be realized to have all three phases at the same time in thermodynamical equilibrium — stable equilibrium in fact.

Let us now consider one of the lines along which two phases coexist. The (T, P) values on the line must be such that the two chemical potentials are the same,

$$\mu^{(1)}(T,P) = \mu^{(2)}(T,P), \tag{1.13.5}$$

where $\mu^{(1)}(T,P)$ in the function of T and P that gives the chemical potential for phase 1 and $\mu^{(2)}(T,P)$ is the analogous function for phase 2. Since we are considering a system of a single component, there is only one molar fraction and it is just equal to unity, $\nu = \nu_1 = 1$, so that we do not need to keep ν as an argument of $\mu^{(1)}$ and $\mu^{(2)}$ in the formalism.

What would we find if we evaluated $\mu^{(1)}(T,P)$ for (T,P) values in the region of phase 2? Simply that

$$\mu^{(1)}(T,P) > \mu^{(2)}(T,P) \quad \text{for phase 2,} \tag{1.13.6}$$

and likewise

$$\mu^{(2)}(T,P) > \mu^{(1)}(T,P) \quad \text{for phase 1.} \tag{1.13.7}$$

This is so because the free enthalpy, or Gibbs free energy, equals

$$G = n \cdot \mu(T,P,n) \tag{1.13.8}$$

as we found in (1.11.9) above, which gives

$$\frac{G}{n} = \mu(T,P) \tag{1.13.9}$$

for the molar free enthalpy of the single-component system. Now, since G is minimal for stable equilibrium when the temperature and the pressure have given values, the phase with smallest chemical potential is the actual phase if more than one phase are possible. Hence, $\mu^{(1)} > \mu^{(2)}$ for phase 2 and $\mu^{(2)} > \mu^{(1)}$ for phase 1.

If we picture $\mu^{(1)}(T,P)$ and $\mu^{(2)}(T,P)$ as height above the T,P-plane, then each of these functions is a so-called *Gibbs surface* and where they intersect we have a coexistence line for the two phases. This sketch is an attempt at illustrating the matter:

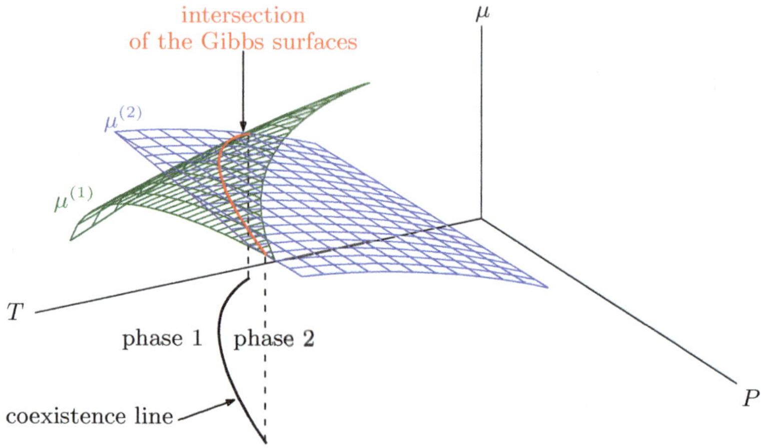

(1.13.10)

For phase 1, we have $\mu^{(1)}(T,P)$, which is smaller than $\mu^{(2)}(T,P)$ for phase 2 in the high-temperature part of the plot on the left but larger at low temperatures. The stable phase is the one with the smaller chemical potential.

Actually, the extrapolation of $\mu^{(1)}$ or $\mu^{(2)}$ beyond the coexistence line is more mathematical than physical. Physically it is rather difficult to drive the system across the coexistence line because one gets a metastable state that spontaneously, and rather rapidly, turns into the other phase. Therefore, one can really do this crossing over only to (T,P) values that are close to the coexistence line. Familiar examples for such metastable systems is water cooled below freezing temperature or heated above boiling temperature while staying in liquid form. One speaks of *supercooled* or *superheated* water.

The phase diagram of a normal substance is not as complicated as the one sketched in (1.13.4), and its typical appearance is like this:

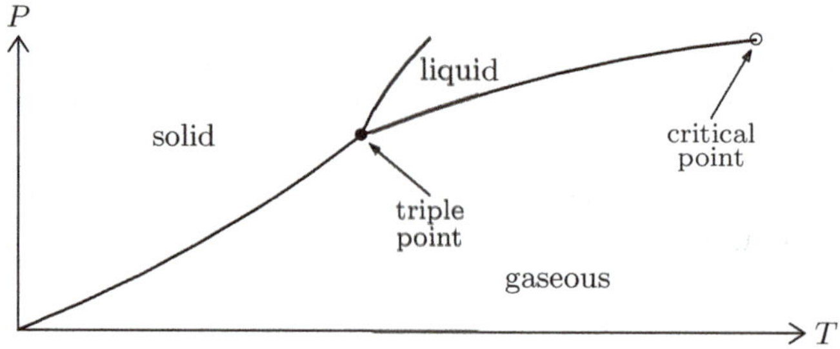

(1.13.11)

The three phases are the ones we call solid, liquid, and gaseous (ice, water, steam for H_2O). Along the coexistence lines we have the Gibbs surfaces intersecting such that the chemical potential is continuous when crossing the line but its derivative is not. Where $\mu^{(1)}(T,P)$ and $\mu^{(2)}(T,P)$ intersect, we have a cusp in the Gibbs surface of least chemical potential. This cusp can eventually disappear, and when this happens the coexistence line ends in a so-called *critical point*, beyond which the Gibbs surfaces $\mu^{(1)}(T,P)$ and $\mu^{(2)}(T,P)$ can no longer be distinguished. In (1.13.11), this also means that the distinction between the liquid phase and the gaseous phase becomes meaningless. Think of the following analogy: It is meaningful to say that the Buda part of Budapest is on the right-hand side of the Danube, and the Pest part is on the left-hand side, but it makes no sense to have such a statement about Paris or Tokyo. Similarly, for (T,P) points too far away from the liquid-gas coexistence line, it is not meaningful to say that the system is in the liquid or the gaseous phase.

1.14 Clausius–Clapeyron equation

We look at the vicinity of a coexistence line, between phase 1 and phase 2:

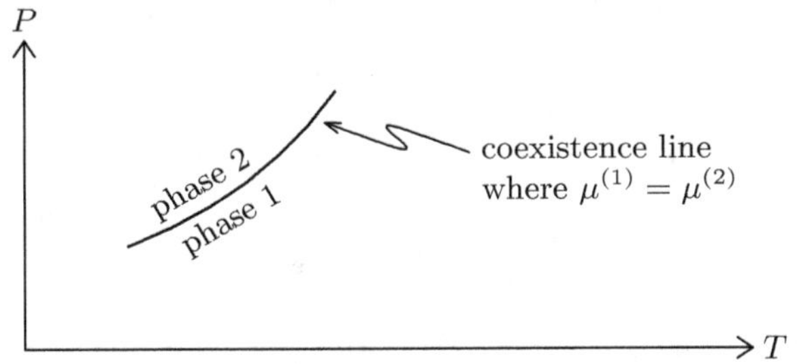

(1.14.1)

On the coexistence line, we have the (T,P) pairs for which

$$\mu^{(1)}(T,P) = \mu^{(2)}(T,P)\,, \tag{1.14.2}$$

where we continue to restrict the discussion to systems with a simple component ($\nu = \nu_1 = 1$) and a single chemical potential in each phase. This determines the coexistence pressure as a function of the temperature, which we denote by $\overline{P}(T)$, and the corresponding coexistence value of the chemical potential is

$$\overline{\mu}(T) = \mu^{(1)}(T,\overline{P}(T)) = \mu^{(2)}(T,\overline{P}(T))\,. \tag{1.14.3}$$

How do $\overline{P}(T)$ and $\overline{\mu}(T)$ respond to changes of the temperature?

One answer to this question proceeds from the Gibbs–Duhem relation (1.11.8),

$$n\,\mathrm{d}\mu = -S\,\mathrm{d}T + V\,\mathrm{d}P \tag{1.14.4}$$

or

$$\mathrm{d}\mu = -\frac{S}{n}\,\mathrm{d}T + \frac{V}{n}\,\mathrm{d}P = -s\,\mathrm{d}T + v\,\mathrm{d}P\,, \tag{1.14.5}$$

where $s = \dfrac{S}{n}$ is the molar entropy (= entropy per mole) and $v = \dfrac{V}{n}$ is the molar volume; see (1.11.14). Thus

$$\left(\frac{\partial\mu}{\partial T}\right)_P = -s \quad \text{and} \quad \left(\frac{\partial\mu}{\partial P}\right)_T = v\,, \tag{1.14.6}$$

which imply

$$\frac{\mathrm{d}}{\mathrm{d}T}\bar{\mu}(T) = -s^{(1)} + v^{(1)}\frac{\mathrm{d}}{\mathrm{d}T}\bar{P}(T) = -s^{(2)} + v^{(2)}\frac{\mathrm{d}}{\mathrm{d}T}\bar{P}(T)\,, \tag{1.14.7}$$

so that

$$\frac{\mathrm{d}}{\mathrm{d}T}\bar{P}(T) = \frac{s^{(1)} - s^{(2)}}{v^{(1)} - v^{(2)}} \tag{1.14.8}$$

and

$$\begin{aligned}\frac{\mathrm{d}}{\mathrm{d}T}\bar{\mu}(T) &= \frac{s^{(1)}v^{(2)} - s^{(2)}v^{(1)}}{v^{(1)} - v^{(2)}} \\ &= -\frac{s^{(1)}/v^{(1)} - s^{(2)}/v^{(2)}}{1/v^{(1)} - 1/v^{(2)}}\,.\end{aligned} \tag{1.14.9}$$

Here, $s^{(1)} = s^{(1)}\big(T, \bar{P}(T)\big)$ and $v^{(1)} = v^{(1)}\big(T, \bar{P}(T)\big)$ refer to phase 1 on the coexistence line in (1.14.1), and $s^{(2)}$ and $v^{(2)}$ to phase 2. It is common to write

$$\begin{aligned}\Delta s(T) &= s^{(1)}\big(T, \bar{P}(T)\big) - s^{(2)}\big(T, \bar{P}(T)\big)\,, \\ \Delta v(T) &= v^{(1)}\big(T, \bar{P}(T)\big) - v^{(2)}\big(T, \bar{P}(T)\big)\end{aligned} \tag{1.14.10}$$

for the changes in molar entropy and molar volume when crossing the coexistence line at the (T, P) point in question. These changes are functions of T, and their ratio gives us the derivative of $\bar{P}(T)$,

$$\frac{\mathrm{d}\bar{P}}{\mathrm{d}T} = \frac{\Delta s}{\Delta v}\,, \tag{1.14.11}$$

and

$$\frac{\mathrm{d}\overline{\mu}}{\mathrm{d}T} = -\frac{\Delta(s/v)}{\Delta(1/v)} \tag{1.14.12}$$

is the analogous expression for $\overline{\mu}(T)$.

The differential equation (1.14.11) is known as the *Clausius** *–Clapeyron*† *equation.* Note that it has a certain similarity with the Maxwell relation in (1.9.5): Both equate the ratio of related changes of pressure and temperature with the ratio of related changes of entropy and volume. The two equations do refer to quite different circumstances, however.

As we cross the coexistence line from phase 2 to phase 1, there is a finite change in volume (water is much denser than steam) and a finite change in entropy, while pressure and temperature change continuously or not at all, depending on how we move from one side of the coexistence line to the other. At the transition between the phases, T and $\overline{P}(T)$ have their values at the point in question on the coexistence line. Then, there is work done, $\Delta W = -Pn\,\Delta v$, and heat released (steam $\rightarrow$ water, water $\rightarrow$ ice) or absorbed (ice $\rightarrow$ water, water $\rightarrow$ steam), $\Delta Q = Tn\,\Delta s$. The intensive quantity is the so-called *latent heat* q, a term coined by Black,‡

$$q = \frac{\Delta Q}{n} = T\,\Delta s\,, \tag{1.14.13}$$

here stated as heat per mole; other conventions are also common, in particular that of reporting the latent heat as heat per kg of material. The heat associated with a phase transition is called latent (= hidden) because it is not accompanied by a change in temperature.

The latent heat q, the volume change Δv, and the coexistence pressure $\overline{P}$ are measurable quantities. Their values can be determined as functions of the temperature T by suitable experiments, so that both sides in the Clausius–Clapeyron equation (1.14.11) and are independently accessible. The correctness of the equation can thus be confirmed by actual data.

At a critical point, $(T,P) = \bigl(T_\mathrm{c}, \overline{P}(T_\mathrm{c}) = P_\mathrm{c}\bigr)$ where the two chemical-potential functions become one, there is no difference in molar volume, $\Delta v(T_\mathrm{c}) = 0$. Likewise, there is no difference in molar entropy, $\Delta s(T_\mathrm{c}) = 0$, so that the right-hand side of the Clausius–Clapeyron equation is ill-defined at the endpoint of the coexistence line, but $\frac{\mathrm{d}}{\mathrm{d}T}\overline{P}(T)$ usually has a meaningful limit for $0 < T_\mathrm{c} - T \rightarrow 0$.

---

*Rudolf Julius Emanuel (born Rudolf Gottlieb) CLAUSIUS (1822–1888)
†Benoît Paul Émile CLAPEYRON (1799–1864) ‡Joseph BLACK (1728–1799)

1.15 Phase transition; Maxwell construction

It is common to describe various equilibrium states by v, P-diagrams, where one follows curves of constant entropy (adiabatic changes) or curves of constant temperature (isothermal changes) as they correspond to the system adiabatically isolated or in contact with a heat bath. In an isothermal curve, a phase transition appears as a jump in volume, for example:

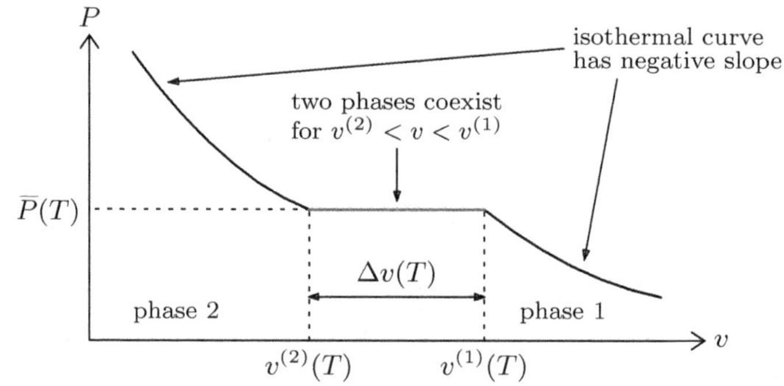

(1.15.1)

Here, $(T, \overline{P}(T))$ is a point on the coexistence line, that is: $\overline{P}(T)$ is the solution of the Clausius–Clapeyron equation; the molar volumes of the two phases are $v^{(1)}(T)$ and $v^{(2)}(T)$, and $\Delta v(T)$ denotes their difference. Where the isothermal curve is continuous within one phase, it has a negative slope since $\left(\frac{\partial P}{\partial v}\right)_T < 0$ in stable equilibrium; see (1.12.22). Since $v^{(2)} < v^{(1)}$ for the two phases in (1.15.1), one would usually think of phase 1 to be the gaseous phase and phase 2 to be the liquid phase, with gas and liquid coexisting for $v^{(2)} < v < v^{(1)}$.

If, say, we have a large molar volume $v > v^{(1)}(T)$ to begin with and then reduce the volume isothermally — while the system is kept in contact with a heat bath, that is — the pressure increases until we reach $v^{(1)}$. Further reduction of the volume is not accompanied by an increase in pressure; rather we are converting gas to liquid at the coexistence pressure $\overline{P}(T)$. This goes on until all gas is liquefied and molar volume $v^{(2)}(T)$ is reached. Any further volume reduction will then come with a pressure increase.

Note the difference between the v, P diagram (1.15.1) and the T, P diagram in (1.14.1). The whole range $v^{(2)} < v < v^{(1)}$ in (1.15.1) corresponds to a single point on the coexistence line in (1.14.1). Clearly, we have a more detailed account by the isothermal curve in (1.15.1), one such curve for each temperature T.

Pressure as derived from the free energy, $P = -\left(\frac{\partial F}{\partial V}\right)_{T,n}$, is a function of T, V, and n, and for a one-component system we can exploit the homogeneity of the intensive pressure and the extensive free energy to express both as functions of the temperature T and the molar volume $v = \frac{V}{n}$. That is

$$\begin{aligned} F(T,V,n) &= nF\left(T, \frac{V}{n}, 1\right) = nf(T,v)\,, \\ P = -\left(\frac{\partial F}{\partial V}\right)_{T,n} &= -\left(\frac{\partial f}{\partial v}\right)_T = P(T,v)\,. \end{aligned} \tag{1.15.2}$$

We have the molar free energy $f^{(1)}(T,v)$ in phase 1, and $f^{(2)}(T,v)$ in phase 2. Therefore,

$$\begin{aligned} \overline{P}(T) &= P^{(1)}(T, v^{(1)}(T)) = P^{(2)}(T, v^{(2)}(T)) \\ &= -\left(\frac{\partial f^{(1)}}{\partial v}\right)_T = -\left(\frac{\partial f^{(2)}}{\partial v}\right)_T \end{aligned} \tag{1.15.3}$$

holds for $\big(T, \overline{P}(T)\big)$ on the coexistence line for phases 1 and 2 in (1.14.1); on this line we also have equal chemical potentials,

$$\overline{\mu}(T) = \mu^{(1)}(T, v^{(1)}(T)) = \mu^{(2)}(T, v^{(2)}(T))\,. \tag{1.15.4}$$

These two equations together determine the two molar volumes $v^{(1)}(T)$ and $v^{(2)}(T)$ of the two phases at the coexistence line. The system may have any volume v between those two values: $v^{(1)}(T) > v > v^{(2)}(T)$ in (1.15.1). For simplicity in notation, we shall not indicate the T dependence of $v^{(1)}$ and $v^{(2)}$ from now on, while taking for granted that this T dependence is understood.

In a graph of an isothermal curve — an *isotherm* — of the molar free energy, the slope of $f^{(1)}$ at $v^{(1)}$ is equal to the slope of $f^{(2)}$ at $v^{(2)}$:

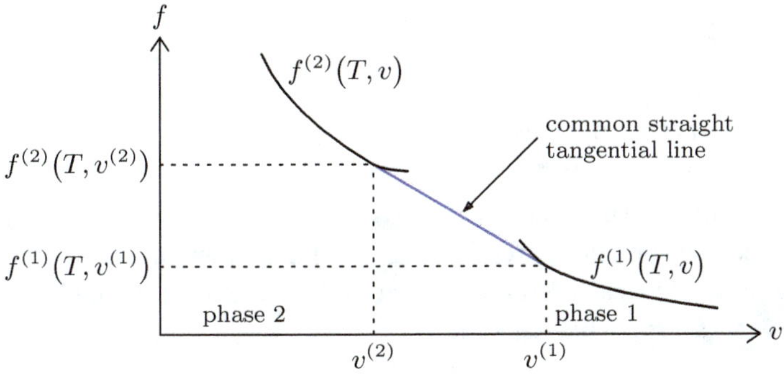

(1.15.5)

The sketch tells us that

$$f^{(2)}(T,v^{(2)}) - f^{(1)}(T,v^{(1)}) = \underbrace{-\bar{P}(T)}_{\text{common slope}}(v^{(2)} - v^{(1)}) \qquad (1.15.6)$$

or

$$(f + \bar{P}v)^{(1)} = (f + \bar{P}v)^{(2)} . \qquad (1.15.7)$$

This is just $\mu^{(1)} = \mu^{(2)}$, since

$$f + Pv = \frac{F + PV}{n} = \frac{G}{n} = \mu , \qquad (1.15.8)$$

and so confirms that this construction of the common tangential straight line is exactly what the equality of the two chemical potentials requires. Here, as already indicated in (1.15.4), the chemical potential is that associated with the free energy, a function of T and v,

$$\mu(T,v) = f(T,v) + P(T,v)v , \qquad (1.15.9)$$

in contrast to $\mu(T,P) = \frac{1}{n}G(T,P,n)$ in Section 1.14.

It follows that the free energy for molar volumes between $v^{(1)}$ and $v^{(2)}$ is

$$v^{(2)} < v < v^{(1)} : \; f(T,v) = f^{(1)}\frac{v - v^{(2)}}{v^{(1)} - v^{(2)}} + f^{(2)}\frac{v^{(1)} - v}{v^{(1)} - v^{(2)}} . \qquad (1.15.10)$$

In this region, we have a mixture of the two phases with the fractions given by the weights in this convex sum of $f^{(1)} = f^{(1)}(T,v^{(1)}(T))$ and $f^{(2)}(T,v^{(2)}(T))$,

$$\begin{aligned} \text{fraction of phase 1} &= \frac{v - v^{(2)}}{v^{(1)} - v^{(2)}} , \\ \text{fraction of phase 2} &= \frac{v^{(1)} - v}{v^{(1)} - v^{(2)}} , \end{aligned} \qquad (1.15.11)$$

which have unit sum, as they should.

Situations such as the one sketched in (1.15.5) arise also in approximate descriptions of real physical systems, where one may obtain a free energy $F = nf(T,v)$ with a graph like this one:

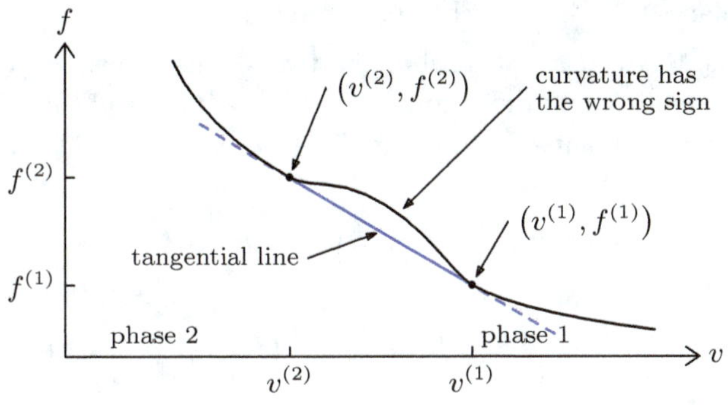

(1.15.12)

where the curvature of the isothermal curve has the wrong sign in some range, $\left(\frac{\partial^2 f}{\partial v^2}\right)_T < 0$ or $\left(\frac{\partial P}{\partial v}\right)_T > 0$, in violation of the stability condition in (1.12.22). One then replaces $f(T, v)$ in that range by the straight line that is the common tangential line for two points. These are determined by

$$\frac{\partial f}{\partial v}(T, v^{(1)}) = \frac{\partial f}{\partial v}(T, v^{(2)}) \tag{1.15.13}$$

and

$$f(T, v^{(1)}) + \overline{P}(T)v^{(1)} = f(T, v^{(2)}) + \overline{P}(T)v^{(2)}, \tag{1.15.14}$$

which state what we know, namely that the pressure $\overline{P} = -\left(\frac{\partial f}{\partial v}\right)_T$ and the chemical potential $\overline{\mu} = f + Pv$ are the same for both values of the molar volumes.

In the v, P-diagram that shows the graph of the negative v-derivative of $f(T, v)$,

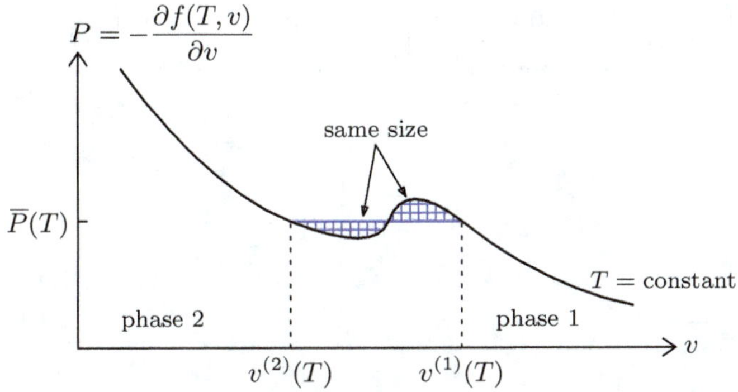

(1.15.15)

a straight line replaces $\left(-\frac{\partial f}{\partial v}\right)_T$ in the coexistence region, as we have seen in (1.15.1). In this *Maxwell construction*, the two shaded areas are of the same size, and this requirement determines $v^{(1)}(T)$, $v^{(2)}(T)$, and thus $\overline{P}(T)$. It is, in fact, the requirement of equal chemical potentials, which we can state in the form

$$\begin{aligned}\big(v^{(1)}-v^{(2)}\big)\overline{P}(T) &= f\big(T,v^{(2)}\big)-f\big(T,v^{(1)}\big)\\ &= \int\limits_{v^{(2)}}^{v^{(1)}} \mathrm{d}v\left(-\frac{\partial f}{\partial v}(T,v)\right)_T . \qquad (1.15.16)\end{aligned}$$

The first expression is the area of the rectangle defined by $v^{(2)}$, $v^{(1)}$, and $\overline{P}(T)$; the last expression is the area under the graph of $\left(-\frac{\partial f}{\partial v}\right)_T$ for the same v interval. The two areas are the same if the two shaded areas are.

1.16 Van der Waals gas

A familiar and important example is the approximate equation of state for a real gas that is known as the *van der Waals* equation*,

$$\Big(P+\frac{a}{v^2}\Big)(v-b)=RT \quad \text{with} \quad a,b>0\,, \qquad (1.16.1)$$

which differs from the ideal gas equation $Pv = RT$ by a phenomenological modification of P to $P+\frac{a}{v^2}$ and v to $v-b$. Roughly speaking, this subtraction of b accounts for the volume that the molecules occupy, and the addition of $\frac{a}{v^2}$ to the "kinetic pressure" P accounts for the resistance of the molecules against squeezing them; more about the physical significance of a and b in Section 6.5.

Isotherms for the van der Waals gas come in two kinds: the regular isotherms with $\left(\frac{\partial P}{\partial v}\right)_T < 0$, and the irregular ones for which $\left(\frac{\partial P}{\partial v}\right)_T > 0$ in some range of the molar volume v:

*Johannes Diderik VAN DER WAALS (1837–1923)

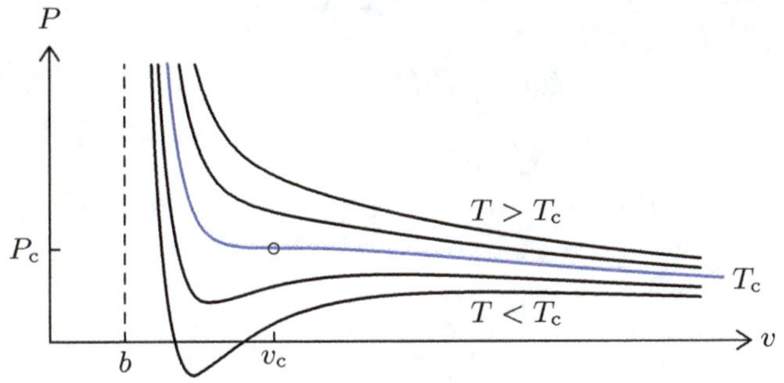

(1.16.2)

Above a certain critical temperature T_c the isotherms are regular, below T_c they are irregular, and the isotherm (in blue color) for $T = T_c$ has a saddle point (marked by ∘) at which $\left(\frac{\partial P}{\partial v}\right)_T = 0$ and also $\left(\frac{\partial^2 P}{\partial v^2}\right)_T = 0$.

With

$$P(T, v) = \frac{RT}{v - b} - \frac{a}{v^2}, \tag{1.16.3}$$

for which

$$\left(\frac{\partial P}{\partial v}\right)_T = -\frac{RT}{(v-b)^2} + \frac{2a}{v^3} \tag{1.16.4}$$

and

$$\left(\frac{\partial^2 P}{\partial v^2}\right)_T = \frac{2RT}{(v-b)^3} - \frac{6a}{v^4}, \tag{1.16.5}$$

we have

$$\begin{aligned} RT_c v_c^3 &= 2a(v_c - b)^2 \\ \text{and} \quad RT_c v_c^4 &= 3a(v_c - b)^3 \end{aligned} \tag{1.16.6}$$

for the critical temperature T_c and the critical molar volume v_c at that saddle point. This gives

$$RT_c = \frac{8}{27}\frac{a}{b} \quad \text{and} \quad v_c = 3b \tag{1.16.7}$$

and then

$$P_c = P(T_c, v_c) = \frac{1}{27}\frac{a}{b^2} \tag{1.16.8}$$

for the critical pressure. Whereas $Pv = RT$ holds for an ideal gas, we have

$$P_{\mathrm{c}}v_{\mathrm{c}} = \frac{3}{8}RT_{\mathrm{c}} \tag{1.16.9}$$

for the van der Waals gas at the critical point.

Let us now consider a subcritical temperature $T < T_{\mathrm{c}}$ and take a look at the Maxwell construction in the range of $v^{(2)} < v < v^{(1)}$. For the coexistence pressure $\overline{P}(T)$ we have

$$\begin{aligned}\overline{P}(T) &= P\bigl(T, v^{(1)}\bigr) = P\bigl(T, v^{(2)}\bigr)\\ &= \frac{RT}{v^{(1)} - b} - \frac{a}{{v^{(1)}}^2} = \frac{RT}{v^{(2)} - b} - \frac{a}{{v^{(2)}}^2}\,,\end{aligned} \tag{1.16.10}$$

which are two equations for the three quantities that we need to determine: $v^{(1)}$, $v^{(2)}$, and $\overline{P}(T)$. The third equation is provided by the Maxwell construction

$$\begin{aligned}\bigl(v^{(1)} - v^{(2)}\bigr)\overline{P}(T) &= \int_{v^{(2)}}^{v^{(1)}} \mathrm{d}v \left(\frac{RT}{v-b} - \frac{a}{v^2}\right)\\ &= RT \log\frac{v^{(1)} - b}{v^{(2)} - b} - a\frac{v^{(1)} - v^{(2)}}{v^{(1)}v^{(2)}}\,,\end{aligned} \tag{1.16.11}$$

or

$$\overline{P}(T) + \frac{a}{v^{(1)}v^{(2)}} = \frac{RT}{v^{(1)} - v^{(2)}} \log\frac{v^{(1)} - b}{v^{(2)} - b}\,. \tag{1.16.12}$$

We use (1.16.10) to express RT and $\overline{P}(T)$ in terms of $v^{(1)}$ and $v^{(2)}$,

$$RT = \frac{a}{\bigl[v^{(1)}v^{(2)}\bigr]^2}\Bigl(v^{(1)} + v^{(2)}\Bigr)\Bigl(v^{(1)} - b\Bigr)\Bigl(v^{(2)} - b\Bigr) \tag{1.16.13}$$

and

$$\overline{P}(T) = \frac{a}{\bigl[v^{(1)}v^{(2)}\bigr]^2}\Bigl[v^{(1)}v^{(2)} - \Bigl(v^{(1)} + v^{(2)}\Bigr)b\Bigr]\,, \tag{1.16.14}$$

and then use these expressions in (1.16.12) to establish that all $v^{(1)}$, $v^{(2)}$ pairs obey the equation

$$\frac{v^{(1)}}{v^{(1)} - b} + \frac{v^{(2)}}{v^{(2)} - b} = \frac{v^{(1)} + v^{(2)}}{v^{(1)} - v^{(2)}} \log\frac{v^{(1)} - b}{v^{(2)} - b}\,. \tag{1.16.15}$$

Upon finding the paired values of $v^{(1)}$ and $v^{(2)}$ from here, one then gets the coexistence pressure as a function of temperature from (1.16.13) and (1.16.14).

Each $v^{(1)}$, $v^{(2)}$ pair thus gives two points in the v, P-diagram. Together all these points trace out the boundary of the *coexistence region*, the region in the v, P-diagram where the van der Waals equation needs to be modified by the Maxwell construction for consistency with the physical constraints.

Each $v^{(1)}$, $v^{(2)}$ pair also identifies one point $\bigl(T, \overline{P}(T)\bigr)$ on the coexistence line in the T, P-diagram of (1.14.1). The change in volume $\Delta v(T) = v^{(1)}(T) - v^{(2)}(T)$ is known once the solution of (1.16.15) is found for the subcritical temperature T in question. How about the corresponding change in the molar entropy $\Delta s(T) = s^{(1)}(T) - s^{(2)}(T)$ that appears in the numerator of the Clausius–Clapeyron equation (1.14.11)? We find it from

$$s(T, v) = -\left(\frac{\partial f(T, v)}{\partial T}\right)_v \tag{1.16.16}$$

after first establishing

$$f(T, v) = f_0(T) - RT \log \frac{v - b}{b} - \frac{a}{v} \tag{1.16.17}$$

as a consequence of

$$P(T, v) = -\left(\frac{\partial f(T, v)}{\partial v}\right)_T = \frac{RT}{v - b} - \frac{a}{v^2}\,. \tag{1.16.18}$$

Here, $f_0(T)$ is a v-independent function of T that would be relevant in the entropy

$$s(T, v) = -\frac{\partial f_0(T)}{\partial T} + R \log \frac{v - b}{b} \tag{1.16.19}$$

and the constant-volume specific heat

$$c_V = T\left(\frac{\partial s(T, v)}{\partial v}\right)_v = -T\frac{\partial^2 f_0(T)}{\partial T^2} \tag{1.16.20}$$

but has no bearing on the entropy difference

$$\Delta s(T) = s\bigl(T, v^{(1)}(T)\bigr) - s\bigl(T, v^{(2)}(T)\bigr) = R \log \frac{v^{(1)} - b}{v^{(2)} - b}\,. \tag{1.16.21}$$

We thus recognize that the logarithm in (1.16.12) and (1.16.15) is $\Delta s(T)/R$, and the Clausius–Clapeyron equation for the van der Waals gas is

$$\frac{\mathrm{d}\overline{P}(T)}{\mathrm{d}T} = \frac{R}{v^{(1)} - v^{(2)}} \log \frac{v^{(1)} - b}{v^{(2)} - b}, \tag{1.16.22}$$

where we encounter the right-hand side of (1.16.12), divided by T. As it should be, the right-hand side in (1.16.22) does not change when we interchange $v^{(1)}$ and $v^{(2)}$ because it cannot matter how we assign the labels "1" and "2" to the two phases. For the same reason, both sides in (1.16.15) are invariant under $v^{(1)} \leftrightarrow v^{(2)}$.

Upon combining (1.16.22) with (1.16.12), we have

$$T\frac{\mathrm{d}\overline{P}}{\mathrm{d}T} = \overline{P} + \frac{a}{v^{(1)}v^{(2)}}; \tag{1.16.23}$$

see also Exercise 26. This tells us the value of $\dfrac{\mathrm{d}\overline{P}}{\mathrm{d}T}$ at the critical point,

$$T\frac{\mathrm{d}\overline{P}}{\mathrm{d}T}\bigg|_{0<T_\mathrm{c}-T\to 0} = P_\mathrm{c} + \frac{a}{v_\mathrm{c}^2} = \frac{4}{27}\frac{a}{b^2} = \frac{RT_\mathrm{c}}{2b} \tag{1.16.24}$$

or

$$\frac{\mathrm{d}\overline{P}}{\mathrm{d}T}\bigg|_{0<T_\mathrm{c}-T\to 0} = \frac{R}{2b} = 4\frac{P_\mathrm{c}}{T_\mathrm{c}}, \tag{1.16.25}$$

where the critical values of (1.16.7) and (1.16.8) are used.

One can find $v^{(1)}$, $v^{(2)}$ pairs that solve (1.16.15) by numerical methods. It is rather more useful to have analytical expressions that give the paired volumes in terms of a single parameter. One such parameterization is specified by the ansatz

$$\left.\begin{matrix} v^{(1)} \\ v^{(2)} \end{matrix}\right\} = b + by(x)\,\mathrm{e}^{\pm x} \quad \text{with} \quad y(\pm x) = y(x). \tag{1.16.26}$$

This ansatz is such that

$$\log\frac{v^{(1)} - b}{v^{(2)} - b} = 2x \quad \text{and} \quad v^{(1)} - v^{(2)} = 2by(x)\sinh x \tag{1.16.27}$$

are odd functions of the parameter x while all symmetric combinations of $v^{(1)}$ and $v^{(2)}$ are even functions of x. Accordingly, both sides of (1.16.15) are even in x,

$$\frac{v^{(1)}}{v^{(1)}-b}+\frac{v^{(2)}}{v^{(2)}-b}=2+\frac{2\cosh x}{y(x)}\,,$$
$$\frac{v^{(1)}+v^{(2)}}{v^{(1)}-v^{(2)}}\log\frac{v^{(1)}-b}{v^{(2)}-b}=2x\,\frac{1+y(x)\cosh x}{y(x)\sinh x}\,, \qquad (1.16.28)$$

and (1.16.15) becomes an equation that determines $y(x)$,

$$\frac{y(x)+\cosh x}{y(x)}=\frac{x}{\sinh x}\frac{1+y(x)\cosh x}{y(x)}\,, \qquad (1.16.29)$$

which is solved by

$$y(x)=\frac{\cosh x\,\sinh x-x}{x\cosh x-\sinh x}=\frac{\dfrac{1}{2x}\sinh(2x)-1}{\cosh x-\dfrac{1}{x}\sinh x}\,. \qquad (1.16.30)$$

This is an even function of x as it should be.

With $y(x)$ at hand now, we can find $v^{(1)},v^{(2)}$ pairs from (1.16.26) and then the corresponding values of RT and $\overline{P}(T)$ from (1.16.13) and (1.16.14). This establishes the coexistence line in the T,P diagram, the van der Waals version of what is sketched in (1.13.11) and (1.14.1):

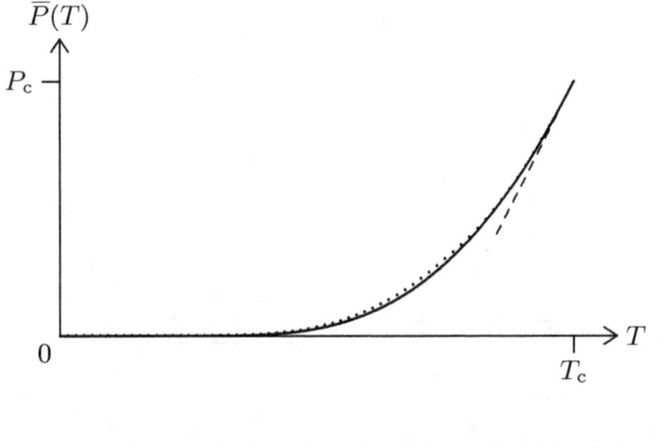

(1.16.31)

The line ends in the critical point, but there is no triple point here as we only have two phases for the van der Waals gas, not three. Regarding the dashed line, see (1.16.35) below; the dotted curve is the subject matter of Exercises 22 and 23.

Near the critical point, where $T \lesssim T_c$ and $P \lesssim P_c$, we have

$$\begin{aligned}\bar{P}(T) &\cong P_c - (T_c - T)\left.\frac{dP(T)}{dT}\right|_{T=T_c} \\ &= P_c - (T_c - T)\frac{R}{b}\left.\frac{x}{y(x)\sinh x}\right|_{x=0}\end{aligned} \tag{1.16.32}$$

upon recalling (1.16.22) and (1.16.27). Here,

$$y(0) = \left.\frac{\frac{d}{dx}(\cosh x \sinh x - x)}{\frac{d}{dx}(x\cosh x - \sinh x)}\right|_{x=0} = \left.\frac{2(\sinh x)^2}{x \sinh x}\right|_{x=0} = 2\,, \tag{1.16.33}$$

so that

$$\bar{P}(T) \cong P_c - \frac{1}{2b}(RT_c - RT) \tag{1.16.34}$$

or, with $RT_c = 8bP_c$ from (1.16.7) and (1.16.8),

$$\frac{P(T)}{P_c} \cong \frac{4T - 3T_c}{T_c} \tag{1.16.35}$$

near the critical point — consistent with (1.16.25), as it should be. This linear approximation is indicated by the dashed straight line in (1.16.31).

The isotherms in (1.16.2) are incorrect in the coexistence region, where they have to be replaced by $(v, \bar{P}(T))$ points in accordance with (1.15.5), so that the isotherms acquire the correct form that is sketched in (1.15.1). Accordingly, we must replace (1.16.2) by

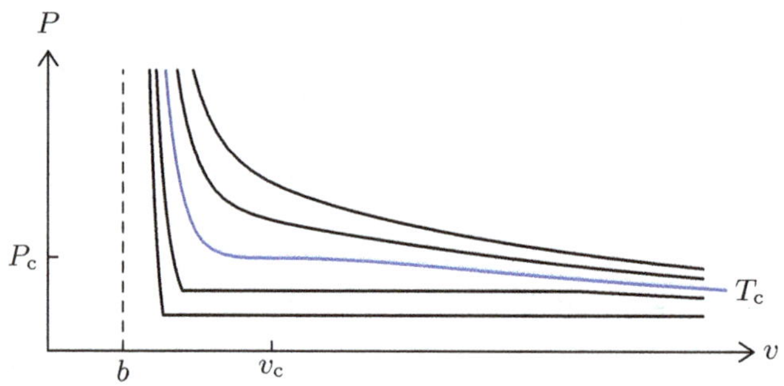

(1.16.36)

Note that the coexistence region for the lowest-temperature isotherm shown ($T = 0.8T_c$) extends beyond the v range covered.

For $x = 0$, we have $v^{(1)} = v^{(2)} = v_c = 3b$, telling us that $y(0) = 2$ and confirming the value found in (1.16.33). Near the critical point, we have

$|x| \ll 1$ and the power series

$$\begin{aligned}\frac{1}{x}\sinh x &= 1+\frac{1}{6}x^2+\frac{1}{120}x^4+\frac{1}{5040}x^6+\cdots,\\ \cosh x-\frac{1}{x}\sinh x &= x\frac{\mathrm{d}}{\mathrm{d}x}\left(\frac{1}{x}\sinh x\right)\\ &= \frac{1}{3}x^2+\frac{1}{30}x^4+\frac{1}{840}x^6+\cdots \end{aligned}\tag{1.16.37}$$

establish first

$$\begin{aligned}\frac{1}{2x}\sinh(2x)-1 &= \frac{1}{6}(2x)^2+\frac{1}{120}(2x)^4+\frac{1}{5040}(2x)^6+\cdots\\ &= \frac{2}{3}x^2\left(1+\frac{1}{5}x^2+\frac{2}{105}x^4+\cdots\right),\\ \left(\cosh x-\frac{1}{x}\sinh x\right)^{-1} &= \left[\frac{1}{3}x^2\left(1+\frac{1}{10}x^2+\frac{1}{280}x^4+\cdots\right)\right]^{-1}\\ &= \frac{3}{x^2}\left(1-\frac{1}{10}x^2+\frac{9}{1400}x^4+\cdots\right)\end{aligned}\tag{1.16.38}$$

and then

$$\begin{aligned}y(x) &= 2\left(1+\frac{1}{5}x^2+\frac{2}{105}x^4+\cdots\right)\left(1-\frac{1}{10}x^2+\frac{9}{1400}x^4+\cdots\right)\\ &= 2+\frac{1}{5}x^2+\frac{23}{2100}x^4+\cdots,\end{aligned}\tag{1.16.39}$$

where the ellipsis stands for terms of order $x^6=\left(x^2\right)^3$ and higher powers of x^2. Some applications of (1.16.26) with (1.16.30) and (1.16.39) are the topics of Exercises 22–24.

1.17 Coexisting phases: Entropy, heat capacity

The graph in (1.15.5) depicts an isotherm of the molar free energy $f(T,v)$ for two coexisting phases whereby the value of $f(T,v)$ in the coexistence region $v^{(2)}<v<v^{(1)}$ is a linear-in-v interpolation between the boundary values $f^{(1)}\bigl(T,v^{(1)}(T)\bigr)$ and $f^{(2)}\bigl(T,v^{(2)}(T)\bigr)$, as stated in (1.15.10). We introduce the symbol $\lceil X \rfloor$ for such an interpolation of a thermodynamical

quantity $X(T, v)$, that is

$$\lceil X \rfloor(T, v) = X^{(1)}(T, v^{(1)}(T)) \frac{v - v^{(2)}(T)}{v^{(1)}(T) - v^{(2)}(T)} + X^{(2)}(T, v^{(2)}(T)) \frac{v^{(1)}(T) - v}{v^{(1)}(T) - v^{(2)}(T)}, \qquad (1.17.1)$$

quite explicitly. Then

$$f(T, v) = \left\{ \begin{array}{ll} f^{(1)}(T, v) & \text{for } v^{(1)} < v \text{ (phase 1)} \\ \lceil f \rfloor(T, v) & \text{for } v^{(1)} < v < v^{(2)} \text{ (coexisting phases)} \\ f^{(2)}(T, v) & \text{for } v < v^{(2)} \text{ (phase 2)} \end{array} \right\} \qquad (1.17.2)$$

applies for all subcritical temperatures T and all molar volumes v.

The Δ symbol of (1.14.10)–(1.14.12) is convenient when we exhibit the v-independent and the proportional-to-v contributions to $\lceil X \rfloor$,

$$\lceil X \rfloor = \frac{\Delta(X/v)}{\Delta(1/v)} + \frac{\Delta X}{\Delta v} v, \qquad (1.17.3)$$

where the two ratios $\frac{\Delta(\ )}{\Delta(\ )}$ depend only on the temperature T. The verification for $X(T, v) = T$ and $X(T, v) = v$ is an immediate check. Further, we illustrate this by the free energy [see (1.15.9)],

$$f(T, v) = \mu(T, v) - P(T, v)v, \qquad (1.17.4)$$

for which

$$\begin{aligned} \Delta f &= \left(\mu^{(1)} - P^{(1)}v^{(1)}\right) - \left(\mu^{(2)} - P^{(2)}v^{(2)}\right) \\ &= \Delta\mu - \Delta(Pv) = -\overline{P}\Delta v \\ \text{and} \quad \Delta(f/v) &= \Delta(\mu/v) - \Delta P = \overline{\mu}\Delta(1/v) \end{aligned} \qquad (1.17.5)$$

since $\mu^{(1)}(T, v^{(1)}(T)) = \overline{\mu}(T) = \mu^{(2)}(T, v^{(2)}(T))$ and $P^{(1)}(T, v^{(1)}(T)) = \overline{P}(T) = P^{(2)}(T, v^{(2)}(T))$, so that

$$\lceil f \rfloor = \frac{\Delta(f/v)}{\Delta(1/v)} + \frac{\Delta f}{\Delta v} v = \overline{\mu} - \overline{P} v. \qquad (1.17.6)$$

Indeed, this is the coexistence-region version of (1.17.4).

We also get $-\left(\frac{\partial}{\partial v}\lceil f\rfloor\right)_T = \overline{P}$ as we should, and we learn something new when differentiating with respect to T,

$$-\left(\frac{\partial}{\partial T}\lceil f\rfloor\right)_v = -\frac{\mathrm{d}\overline{\mu}}{\mathrm{d}T} + \frac{\mathrm{d}\overline{P}}{\mathrm{d}T}v = \frac{\Delta(s/v)}{\Delta(1/v)} + \frac{\Delta s}{\Delta v}v = \lceil s\rfloor\,, \tag{1.17.7}$$

where the Clausius–Clapeyron equation (1.14.11) and its companion (1.14.12) enter, and the final step is (1.17.3) for $X = s$. It follows that the molar entropy for subcritical temperatures is stated analogously to the molar free energy in (1.17.2),

$$s(T,v) = \left\{\begin{array}{ll} s^{(1)}(T,v) & \text{for phase 1} \\ \lceil s\rfloor(T,v) & \text{for coexisting phases} \\ s^{(2)}(T,v) & \text{for phase 2} \end{array}\right\}. \tag{1.17.8}$$

In particular, we note that $s(T,v)$ is continuous at $v = v^{(1)}(T)$ and $v^{(2)}(T)$.

Let us see how this emerges from (1.17.2) without invoking (1.17.6). The T derivative of the right-hand side in (1.17.1) involves

$$\frac{\mathrm{d}}{\mathrm{d}T}X^{(1)} = \frac{\mathrm{d}}{\mathrm{d}T}X^{(1)}(T,v^{(1)}(T)) = \left(\frac{\partial X}{\partial T}\right)_v^{(1)} + \left(\frac{\partial X}{\partial v}\right)_T^{(1)}\frac{\mathrm{d}v^{(1)}}{\mathrm{d}T} \tag{1.17.9}$$

and the analog for $X^{(2)}$ as well as

$$\begin{aligned}\left(\frac{\partial}{\partial T}\frac{v-v^{(2)}}{v^{(1)}-v^{(2)}}\right)_v &= -\left(\frac{\partial}{\partial T}\frac{v^{(1)}-v}{v^{(1)}-v^{(2)}}\right)_v \\ &= -\frac{1}{v^{(1)}-v^{(2)}}\left(\frac{\mathrm{d}v^{(1)}}{\mathrm{d}T}\frac{v-v^{(2)}}{v^{(1)}-v^{(2)}} + \frac{\mathrm{d}v^{(2)}}{\mathrm{d}T}\frac{v^{(1)}-v}{v^{(1)}-v^{(2)}}\right) \\ &= -\frac{1}{v^{(1)}-v^{(2)}}\left\lceil\frac{\mathrm{d}v}{\mathrm{d}T}\right\rfloor,\end{aligned} \tag{1.17.10}$$

so that

$$\begin{aligned}\left(\frac{\partial}{\partial T}\lceil X\rfloor\right)_v &= \left\lceil\left(\frac{\partial X}{\partial T}\right)_v + \left(\frac{\partial X}{\partial v}\right)_T\frac{\mathrm{d}v}{\mathrm{d}T}\right\rfloor - \frac{X^{(1)}-X^{(2)}}{v^{(1)}-v^{(2)}}\left\lceil\frac{\mathrm{d}v}{\mathrm{d}T}\right\rfloor \\ &= \left\lceil\left(\frac{\partial X}{\partial T}\right)_v + \left(\left(\frac{\partial X}{\partial v}\right)_T - \frac{\Delta X}{\Delta v}\right)\frac{\mathrm{d}v}{\mathrm{d}T}\right\rfloor.\end{aligned} \tag{1.17.11}$$

For $X = f$, then, we have

$$\left(\frac{\partial f}{\partial T}\right)_v = -s \quad\text{and}\quad \left(\frac{\partial f}{\partial v}\right)_T = -P = -\overline{P} = \frac{\Delta f}{\Delta v}\,, \tag{1.17.12}$$

and (1.17.7) follows.

While $\left(\frac{\partial}{\partial T}\lceil f \rfloor\right)_v = \left\lceil\left(\frac{\partial f}{\partial T}\right)_v\right\rfloor$, this is not the generic situation, as there is usually a difference between $\left(\frac{\partial}{\partial T}\lceil X \rfloor\right)_v$ and $\left\lceil\left(\frac{\partial X}{\partial T}\right)_v\right\rfloor$, given by the additional term in (1.17.11),

$$\left(\frac{\partial}{\partial T}\lceil X \rfloor\right)_v - \left\lceil\left(\frac{\partial X}{\partial T}\right)_v\right\rfloor = \left\lceil\left(\left(\frac{\partial X}{\partial v}\right)_T - \frac{\Delta X}{\Delta v}\right)\frac{\mathrm{d}v}{\mathrm{d}T}\right\rfloor, \qquad (1.17.13)$$

which has two contributions, one from (1.17.9), the other from (1.17.10). Both result from the change in $v^{(1)}(T)$ and $v^{(2)}(T)$ when $T \to T + \mathrm{d}T$. The $\left(\frac{\partial X}{\partial v}\right)_T$ term accounts for the response of $X(T, v)$ when $v \to v + \mathrm{d}v$, and the $\frac{\Delta X}{\Delta v}$ term accounts for the change in the fractions of the phases 1 and 2. The work associated with $\mathrm{d}v^{(1)}$ and $\mathrm{d}v^{(2)}$ and the latent heat required for the conversion of one phase into the other compensate for each other in the sense that there is no net work, as the molar volume is not changed. That, however, does not mean that the right-hand side in (1.17.13) is always zero.

As we have seen, the two contributions to the extra term take care of themselves in $-\left(\frac{\partial}{\partial T}\lceil f \rfloor\right)_v = \lceil s \rfloor$, but that is not the case for $\left(\frac{\partial}{\partial T}\lceil s \rfloor\right)_v$. There really *is* an extra contribution to the constant-volume specific heat in the coexistence region,

$$\begin{aligned} c_V\Big|_{v^{(2)} < v < v^{(1)}} &= T\left(\frac{\partial}{\partial T}\lceil s \rfloor\right)_v = \left\lceil c_V + T\left(\left(\frac{\partial s}{\partial v}\right)_T - \frac{\Delta s}{\Delta v}\right)\frac{\mathrm{d}v}{\mathrm{d}T}\right\rfloor \\ &= \lceil c_V \rfloor + T\left\lceil\left(\left(\frac{\partial P}{\partial T}\right)_v - \frac{\mathrm{d}\bar{P}}{\mathrm{d}T}\right)\frac{\mathrm{d}v}{\mathrm{d}T}\right\rfloor, \end{aligned} \qquad (1.17.14)$$

where the Maxwell relation (1.9.5) and the Clausius–Clapeyron equation (1.14.11) are used. The specific heat in the coexistence region is not equal to what the linear interpolation of the specific heats in phases 1 and 2 would give, the values of $c_V^{(1)}$ and $c_V^{(2)}$ are modified when entering the coexistence region through the borders at at $v = v^{(1)}$ and $v = v^{(2)}$, and then these modified values are interpolated. We have here an example of a discontinuous change of a thermodynamical quantity at a phase transition.

We need the T derivatives of $v^{(1)}(T)$ and $v^{(2)}(T)$ for the evaluation of the extra term in (1.17.14). We get them from the Clausius–Clapeyron equation,

$$\begin{aligned}\frac{\Delta s}{\Delta v} = \frac{\mathrm{d}}{\mathrm{d}T}\overline{P}(T) &= \frac{\mathrm{d}}{\mathrm{d}T}P^{(1)}(T, v^{(1)}(T)) \\ &= \left(\frac{\partial P}{\partial T}\right)_v^{(1)} + \left(\frac{\partial P}{\partial v}\right)_T^{(1)} \frac{\mathrm{d}v^{(1)}}{\mathrm{d}T}\end{aligned} \tag{1.17.15}$$

and likewise for phase 2, so that

$$\frac{\mathrm{d}v^{(1,2)}}{\mathrm{d}T} = -\left(\frac{\partial v}{\partial P}\right)_T^{(1,2)} \left(\left(\frac{\partial P}{\partial T}\right)_v^{(1,2)} - \frac{\mathrm{d}\overline{P}}{\mathrm{d}T}\right). \tag{1.17.16}$$

Accordingly,

$$c_V\Big|_{v^{(2)}<v<v^{(1)}} = \lceil c_V \rfloor - T\left\lceil \left(\frac{\partial v}{\partial P}\right)_T \left(\left(\frac{\partial P}{\partial T}\right)_v - \frac{\mathrm{d}\overline{P}}{\mathrm{d}T}\right)^2 \right\rfloor, \tag{1.17.17}$$

and we note that there is never a decrease of the specific heat because

$$\left(\frac{\partial v}{\partial P}\right)_T = -K_T v < 0\,, \tag{1.17.18}$$

where K_T is the isothermal compressibility of (1.10.15); Exercise 70 deals with an important example in which the additional term in (1.17.17) vanishes as a consequence of $\left(\frac{\partial P}{\partial T}\right)_v = \frac{\mathrm{d}\overline{P}}{\mathrm{d}T}$ at the boundary. We can also relate $\left(\frac{\partial P}{\partial T}\right)_v$ to material constants,

$$\left(\frac{\partial P}{\partial T}\right)_v = -\left(\frac{\partial P}{\partial v}\right)_T \left(\frac{\partial v}{\partial T}\right)_P = \frac{1}{K_T v}\alpha v = \frac{\alpha}{K_T} \tag{1.17.19}$$

with an application of the chain relation (1.9.15) and the thermal expansion coefficient α of (1.10.4); note the implicit reference to Exercise 17. In each phase, then,

$$\begin{aligned}&-T\left(\frac{\partial v}{\partial P}\right)_T \left(\left(\frac{\partial P}{\partial T}\right)_v - \frac{\mathrm{d}\overline{P}}{\mathrm{d}T}\right)^2 \\ &= TK_T v\left(\frac{\alpha}{K_T} - \frac{\Delta s}{\Delta v}\right)^2 = T\frac{\alpha^2 v}{K_T}\left(1 - \frac{K_T}{\alpha}\frac{\Delta s}{\Delta v}\right)^2 \\ &= (c_P - c_V)\left(1 - \frac{K_T}{\alpha}\frac{\Delta s}{\Delta v}\right)^2,\end{aligned} \tag{1.17.20}$$

after recalling (1.10.16) in the last step. In summary, we have

$$c_V(T,v) = \begin{cases} c_V^{(1)}(T,v) & \text{for phase 1} \\ \left[c_V + (c_P - c_V)\left(1 - \frac{K_T}{\alpha}\frac{\Delta s}{\Delta v}\right)^2\right](T,v) & \text{for coexisting phases} \\ c_V^{(2)}(T,v) & \text{for phase 2} \end{cases} \tag{1.17.21}$$

or

$$c_V(T,v) = \begin{cases} c_V^{(1)}(T,v) & \text{for phase 1} \\ \left[c_P - 2T\alpha v\frac{\mathrm{d}\bar{P}}{\mathrm{d}T} + TK_T v\left(\frac{\mathrm{d}\bar{P}}{\mathrm{d}T}\right)^2\right](T,v) & \text{for coexisting phases} \\ c_V^{(2)}(T,v) & \text{for phase 2} \end{cases} \tag{1.17.22}$$

for the constant-volume specific heat. All the material constants that appear here — c_V, c_P, α, K_T — can be measured in both phases, and so can the latent heat $T\Delta s$ and the volume change Δv of the phase transition. The right-hand side of (1.17.21) or (1.17.22) in the coexistence region is, therefore, predictable and can be compared with measured specific heats for intermediate values of v.

For an illustration of the discontinuity in the constant-volume specific heat, we consider the van der Waals gas of Section 1.16 and evaluate the extra term in (1.17.17) at the critical point. We combine $\left(\frac{\partial P}{\partial T}\right)_v$ from (1.16.3) with (1.16.23) to

$$\begin{aligned} T\left(\left(\frac{\partial P}{\partial T}\right)_v - \frac{\mathrm{d}\bar{P}}{\mathrm{d}T}\right)^{(1)} &= \frac{RT}{v^{(1)} - b} - \left(P(T,v^{(1)}) + \frac{a}{v^{(1)}v^{(2)}}\right) \\ &= -\frac{a\Delta v}{v^{(1)}v^{(2)}}\frac{1}{v^{(1)}} \end{aligned} \tag{1.17.23}$$

and

$$T\left(\left(\frac{\partial P}{\partial T}\right)_v - \frac{\mathrm{d}\bar{P}}{\mathrm{d}T}\right)^{(2)} = \frac{a\Delta v}{v^{(1)}v^{(2)}}\frac{1}{v^{(2)}}, \tag{1.17.24}$$

so that they contribute a factor $\propto (\Delta v)^2$ in

$$
\begin{aligned}
c_V - \lceil c_V \rfloor &= -T \left\lceil \left(\frac{\partial v}{\partial P}\right)_T \left(\left(\frac{\partial P}{\partial T}\right)_v - \frac{\mathrm{d}\overline{P}}{\mathrm{d}T} \right)^2 \right\rfloor \\
&= \frac{1}{T} \left(\frac{a\Delta v}{v^{(1)} v^{(2)}} \right)^2 \left\lceil \left(-v^2 \left(\frac{\partial P}{\partial v}\right)_T \right)^{-1} \right\rfloor , \qquad (1.17.25)
\end{aligned}
$$

which matches the vanishing of both $\left(\frac{\partial P}{\partial v}\right)_T$ and $\left(\frac{\partial^2 P}{\partial v^2}\right)_T$ at the critical point, where $\Delta v = 0$. Further, we make use of (1.16.13) in

$$
\begin{aligned}
\left(-v^2 \left(\frac{\partial P}{\partial v}\right)_T \right)^{(1)} &= \frac{RT {v^{(1)}}^2}{\left(v^{(1)} - b\right)^2} - \frac{2a}{v^{(1)}} \\
&= \frac{a}{v^{(1)} v^{(2)}} \left[\left(v^{(1)} + v^{(2)}\right) \frac{v^{(1)}}{v^{(2)}} \frac{v^{(2)} - b}{v^{(1)} - b} - 2v^{(2)} \right] \\
&= \frac{a\Delta v}{v^{(1)} v^{(2)}} \left[\frac{v^{(1)}}{v^{(2)}} \frac{v^{(2)} - b}{v^{(1)} - b} - \frac{2b}{v^{(1)} - b} \right] \\
&= \frac{a(\Delta v)^2}{4v_\mathrm{c}^3} \left[1 + \{\text{terms of order } \Delta v\} \right] . \qquad (1.17.26)
\end{aligned}
$$

The last step exploits

$$
\left.\begin{matrix} v^{(1)} \\ v^{(2)} \end{matrix}\right\} = v_\mathrm{c} \pm \frac{1}{2} \Delta v + \cdots , \qquad (1.17.27)
$$

where $v_\mathrm{c} = 3b$ is the critical volume of (1.16.7) and the ellipsis stands for terms proportional to $(\Delta v)^2$ and higher powers of $(\Delta v)^2$. The leading term stated in (1.17.26) is invariant under exchanging $v^{(1)}$ and $v^{(2)}$ and, therefore, we have

$$
\left\lceil \left(-v^2 \left(\frac{\partial P}{\partial v}\right)_T \right)^{-1} \right\rfloor = \frac{4v_\mathrm{c}^3}{a(\Delta v)^2} \left[1 + \{\text{terms of order } (\Delta v)^2\} \right] \qquad (1.17.28)
$$

and find,

$$
\left(c_V - \lceil c_V \rfloor \right)_\mathrm{c} = \frac{4a}{T_\mathrm{c} v_\mathrm{c}} = \frac{9}{2} R . \qquad (1.17.29)
$$

The discontinuous change in the constant-volume specific heat is quite substantial, it is three times the value of a mono-atomic ideal-gas, $c_V = \frac{3}{2}R$, obtained for $\gamma = \frac{5}{3}$ in (1.7.17). This is also the value of c_V for the van der Waals gas outside the coexistence region; see Exercise 120, where $f_0(T)$ of (1.16.17) and (1.16.20) is found.

1.18 Interfaces

The final subject matter of this review of basic thermodynamical concepts concerns interfaces between phases. For simplicity, we assume that the interfaces are planar, as depicted here:

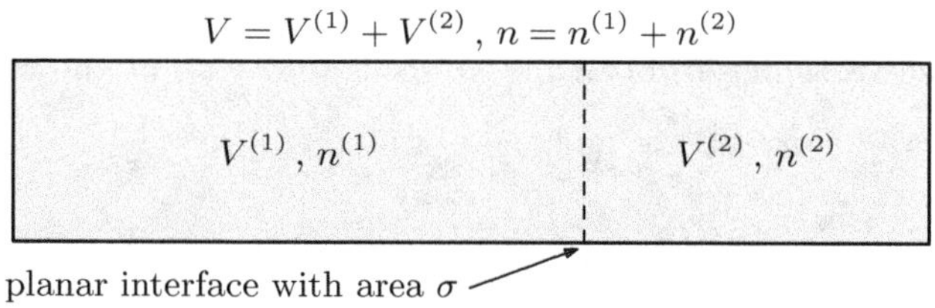

(1.18.1)

We associate an energy contribution with the interface and this energy should be proportional to the interface area σ. Accordingly, we write $\tau\,\mathrm{d}\sigma$ for the work associated with a change of area σ, where τ is the *surface tension*. In the pair σ, τ, we have the extensive variable σ and the intensive variable τ.

Then

$$\mathrm{d}U = T\,\mathrm{d}S - P\,\mathrm{d}V + \mu \cdot \mathrm{d}n + \tau\,\mathrm{d}\sigma \tag{1.18.2}$$

is the differential of the internal energy for small changes that take us from an equilibrium state to an adjacent one. If the surface is not a simple plane but has a more complicated geometry, the $\tau\,\mathrm{d}\sigma$ term is replaced by a integral over the surface; we do not deal with these complications here.

The surface tension

$$\tau = \left(\frac{\partial U}{\partial \sigma}\right)_{S,V,n} \geq 0 \tag{1.18.3}$$

cannot be negative in a stable equilibrium state. If it were, an increase in surface area would decrease the energy, so that a rougher surface (perhaps folded over) would be energetically favorable. We would then keep folding until the whole volume is filled with surface. This is *not* what the phenomenology tells us.

For a real interface, we need two phases as depicted in (1.18.1). With $K = 2$ in the Gibbs phase rule (1.13.3), we then have $\#(\mathrm{TDOF}) = J$, that is: τ is determined by J intensive variables, where J is the number of components in $\nu = (\nu_1, \nu_2, \ldots, \nu_J)$. If there is only one component, a single variable matters, the temperature T, say, and since $U(S, V, n, \sigma)$ is homogeneous in the extensive variables S, V, n, σ and is itself extensive, we have

$$U = TS - PV + \mu n + \tau\sigma \,. \tag{1.18.4}$$

As already mentioned earlier, the contributions TS, PV, μn, and also U itself, are proportional to the total number of molecules, N, whereas $\tau\sigma$ is a small contribution $\sim N^{\frac{2}{3}} = N \times N^{-\frac{1}{3}}$, so that the surface energy has a negligible influence on bulk properties.

The interface sketched in (1.18.1) has an extension in the direction normal to the surface of a certain width w. A density profile looks like this:

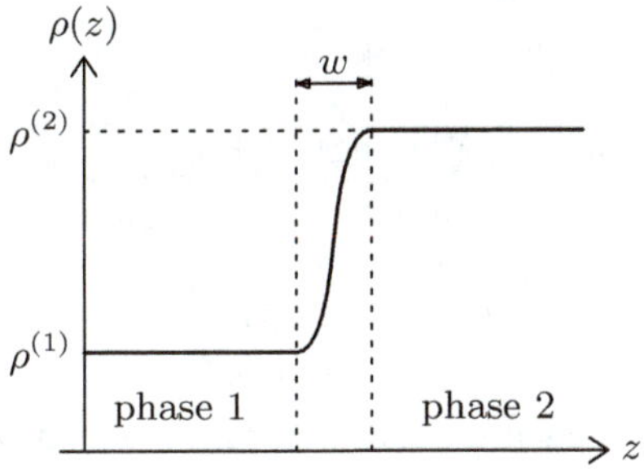

(1.18.5)

where $\mathrm{d}z\,\rho(z)$ is the number of moles between z and $z + \mathrm{d}z$. The density (per unit length) $\rho(z)$ grows from $\rho^{(1)}$ to $\rho^{(2)}$ over the distance of a few molecular diameters — the width w.

Let us compare the realistic boundary of width w with a fictitious boundary described by a step function at $z = z_\mathrm{d}$ (subscript $_\mathrm{d}$ stands for "discontinuous"):

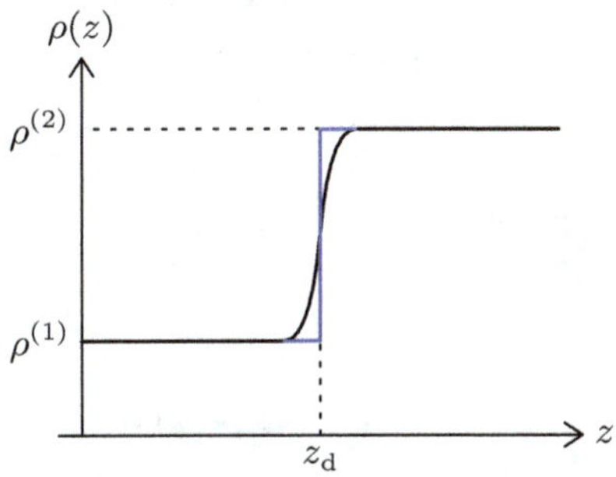

(1.18.6)

We then have the fictitious, hypothetical situation in which phase 1 reaches up to z_d, and we have phase 2 for all $z > z_d$, with

$$\begin{aligned} \mathrm{d}U^{(1)} &= T\mathrm{d}S^{(1)} - P\mathrm{d}V^{(1)} + \mu\,\mathrm{d}n^{(1)}\,, \\ \mathrm{d}U^{(2)} &= T\mathrm{d}S^{(2)} - P\mathrm{d}V^{(2)} + \mu\,\mathrm{d}n^{(2)} \end{aligned} \tag{1.18.7}$$

for these two phases as well as

$$\begin{aligned} U^{(1)} &= TS^{(1)} - PV^{(1)} + \mu n^{(1)}\,, \\ U^{(2)} &= TS^{(2)} - PV^{(2)} + \mu n^{(2)}\,, \end{aligned} \tag{1.18.8}$$

where $V^{(1)} + V^{(2)} = V$ is the volume of the system. For other extensive properties, there is a difference between the sum of their values for these fictitious phases and the true value of the system,

$$\underbrace{X}_{\text{total system}} = \underbrace{X^{(1)} + X^{(2)}}_{\text{fictitious phases}} + \underbrace{X^{(\mathrm{IF})}}_{\text{interface contribution}} \tag{1.18.9}$$

with $X = S$ or $X = n$ or $X = U$. The interface contribution $X^{(\mathrm{IF})}$ depends on the location z_d, where we put the step function.

A choice of z_d that suggests itself is that for which $n = n^{(1)} + n^{(2)}$ and $n^{(\mathrm{IF})} = 0$. We have

$$n^{(1)} = \int_{-\infty}^{z_d} \mathrm{d}z\,\rho_d(z)\,, \quad n^{(2)} = \int_{z_d}^{\infty} \mathrm{d}z\,\rho_d(z) \tag{1.18.10}$$

with the discontinuous density

$$\rho_d(z) = \begin{cases} \rho^{(1)} & \text{for } z < z_d \\ \rho^{(2)} & \text{for } z > z_d \end{cases} \tag{1.18.11}$$

and "$-\infty$" and "∞" simply mean z values that correspond to the extension of the system in z direction. The difference between $n = \int_{-\infty}^{\infty} \mathrm{d}z\,\rho(z)$ and $n^{(1)} + n^{(2)}$ is

$$n^{(\mathrm{IF})} = \int_{-\infty}^{\infty} \mathrm{d}z\,\bigl[\rho(z) - \rho_d(z)\bigr]\,, \tag{1.18.12}$$

and the plot in (1.18.6) tells us that we can indeed choose z_d such that

$n^{(\mathrm{IF})} = 0$, or

$$\int_{-\infty}^{\infty} \mathrm{d}z\, \rho_{\mathrm{d}}(z) = \int_{-\infty}^{\infty} \mathrm{d}z\, \rho(z)\,. \tag{1.18.13}$$

This equation then determines z_{d}, the location of the hypothetical zero-width interface, the so-called *Gibbs dividing surface.* In these integrals, it is, of course, sufficient to integrate over a z range $z^{(1)} < z < z^{(2)}$ such that $\rho(z^{(1)}) = \rho^{(1)}$ and $\rho(z^{(2)}) = \rho^{(2)}$, that is: $z^{(1)}$ and $z^{(2)}$ are in the bulk, far enough away from the interface that the difference of the two integrals does not change if we move $z^{(1)}$ further to the left or $z^{(2)}$ further to the right.

Regarding the internal energy, the differential for $U^{(\mathrm{IF})}$ is

$$\begin{aligned} \mathrm{d}U^{(\mathrm{IF})} &= \mathrm{d}U - \mathrm{d}U^{(1)} - \mathrm{d}U^{(2)} \\ &= T\mathrm{d}\, S^{(\mathrm{IF})} + \mu\, \mathrm{d}n^{(\mathrm{IF})} + \tau\, \mathrm{d}\sigma\,, \end{aligned} \tag{1.18.14}$$

where $\mathrm{d}V = \mathrm{d}\left(V^{(1)} + V^{(2)}\right) = 0$ is taken into account. If we now agree on the z_{d} value of the Gibbs dividing surface, then $\mathrm{d}n^{(\mathrm{IF})} = 0$ as well, so that

$$\mathrm{d}U^{(\mathrm{IF})} = T\,\mathrm{d}S^{(\mathrm{IF})} + \tau\, \mathrm{d}\sigma\,. \tag{1.18.15}$$

The homogeneity of the internal energy then implies

$$U^{(\mathrm{IF})} = TS^{(\mathrm{IF})} + \tau\sigma\,, \tag{1.18.16}$$

and we conclude that

$$\tau = \frac{U^{(\mathrm{IF})} - TS^{(\mathrm{IF})}}{\sigma} = \frac{F^{(\mathrm{IF})}}{\sigma}\,. \tag{1.18.17}$$

The surface tension is the Helmholtz free energy associated with the interface per unit area.

Matters are more complicated if there are several components, and we cannot choose z_{d} such that $n_j^{(\mathrm{IF})} = 0$ for all j. Usually, however, the various components are not quite on equal footing — in the physics, that is, not in the formalism. Take, for example, the two-component case of a solvent (water) and a solute (salt) with molar amounts n_1 and n_2, respectively. We then choose the Gibbs dividing surface such that $n_1^{(\mathrm{IF})} = 0$, and have

$$\mathrm{d}U^{(\mathrm{IF})} = T\,\mathrm{d}S^{(\mathrm{IF})} + \mu_2\, \mathrm{d}n_2^{(\mathrm{IF})} + \tau\, \mathrm{d}\sigma\,, \tag{1.18.18}$$

for which the corresponding Gibbs–Duhem relation is

$$\sigma\, \mathrm{d}\tau = -S^{(\mathrm{IF})}\, \mathrm{d}T - n_2^{(\mathrm{IF})}\, \mathrm{d}\mu_2\,. \tag{1.18.19}$$

At constant temperature ($\mathrm{d}T = 0$), this yields

$$\frac{n_2^{(\mathrm{IF})}}{\sigma} = -\left(\frac{\partial\tau}{\partial\mu_2}\right)_T = -\left(\frac{\partial\tau}{\partial\rho_2}\right)_T\left(\frac{\partial\rho_2}{\partial\mu_2}\right)_T , \tag{1.18.20}$$

known as the *Gibbs adsorption isotherm.* Since σ is positive and stability requires $\left(\frac{\partial\rho_2}{\partial\mu_2}\right)_T > 0$, it follows that $n_2^{(\mathrm{IF})}$ and $\left(\frac{\partial\tau}{\partial\rho_2}\right)_T$ have opposite signs. The surface tension decreases when the solute concentration is increased if there is an excess of solute at the interface ($n_2^{(\mathrm{IF})} > 0$), and *vice versa.* This observation is important for surface physics and has many technical applications.

If there are several solutes, matters can get complicated, and the possibility of chemical reactions can add much to the complexity of the situation. We leave it at that.

Chapter 2

Statistical ensembles

After this review of some aspects of thermodynamics, we now turn to statistical mechanics proper. It deals with the molecular, microscopic foundation of thermodynamics, with the goal of understanding the rather simple properties of large systems — simple in the sense that we can characterize them with a small number of parameter values — as a consequence of the physical laws obeyed by the very many ($\sim 10^{20}$ for a millimole) atomic particles that compose the system.

Clearly, there is something very remarkable here: As we know from experience, it is much harder to investigate systems with three, four, five, ... constituents than with one or two — the classical mechanics of one sun plus one planet is a tractable two-body problem, whereas the three-body problem of sun plus planet with moon is quite complex. Yet, we get an essential simplification when the number of constituents is large. This is, of course, the consequence of a shift of emphasis: For sun plus planet with moon it is useful and desirable to know the three position vectors as a function of time with great precision, but for 10^{20} constituents such detailed information is not only unavailable, it is also not useful. Stretch your imagination (beyond the elastic limit, I'm afraid) and assume that you are given the initial state of a system with 10^{20} point-like constituents (6×10^{20} coordinates and velocity components for a classical-mechanics description) and then you solve the equation of motion to find the positions and velocities at a later time. It appears that such a description would give a deterministic account of what is the case. This is, however, an illusion. Those 6×10^{20} initial values have unavoidable imprecisions, be it for the sole reason that only a finite number of decimals can be specified. As a consequence of the large number of variables, then, these tiny errors in the initial values grow rapidly into sizeable inaccuracies for any realistic interacting system, and whatever we calculate is of little or no value.

2.1 The benefit of large numbers

Statistical mechanics does not fall into this trap. The ambition is, contrary to such a detailed description, to arrive at the few-parameter description that catches the essential features in the same manner as it does in thermodynamics. We do need, however, to deal with those large numbers of constituents. This is where statistics enters. It enables us to focus on the properties that are *typical* for such a large system of 10^{20} constituents, the properties that (almost) all such systems have in common, irrespective of the microscopic differences between them. Despite the use of statistical methods, our conclusions will not be vague or ambiguous but quite precise. There, the large numbers come to the rescue because they lead to fluctuations (and thus to imprecision of our predictions) that are very small compared with the typical values.

Here is an elementary example for an illustration of this point. Consider N atoms in a volume V that is divided in two equal halves, and each atom has an equal chance of being in the first half or the second. Accordingly, the probability of having N_1 atoms in the first half and $N_2 = N - N_1$ atoms in the second half is

$$p(N_1) = \frac{1}{2^N}\binom{N}{N_1} = \frac{1}{2^N}\frac{N!}{N_1!(N-N_1)!}\,. \tag{2.1.1}$$

The expected value of a function of N_1 is then

$$\langle f(N_1)\rangle = \sum_{N_1=0}^{\infty} p(N_1)f(N_1)\,, \tag{2.1.2}$$

where no terms with $N_1 > N$ contribute to the sum. For example, we have

$$\begin{aligned}\left\langle\binom{N_1}{k}\right\rangle &= \sum_{N_1=0}^{\infty}\frac{1}{2^N}\underbrace{\binom{N}{N_1}\binom{N_1}{k}}_{=\binom{N}{k}\binom{N-k}{N_1-k}}\\ &= \frac{1}{2^N}\binom{N}{k}\underbrace{\sum_{N_1=k}^{\infty}\binom{N-k}{N_1-k}}_{=2^{N-k}}\\ &= \frac{1}{2^k}\binom{N}{k}\,,\end{aligned} \tag{2.1.3}$$

(no contribution from $N_1 < k$ — annotation pointing to the second equality sign) with the particular cases of $k = 0$, $k = 1$, and $k = 2$:

$$\begin{aligned}\langle 1\rangle &= \sum_{N_1=0}^{\infty} p(N_1) = 1\,,\\ \langle N_1\rangle &= \frac{1}{2}N\,,\\ \left\langle \frac{1}{2}N_1(N_1-1)\right\rangle &= \frac{1}{8}N(N-1)\,,\end{aligned} \tag{2.1.4}$$

the first of which is checking the normalization of the $p(N_1)$s to unit sum. From these we obtain

$$\langle N_1^2\rangle = \underbrace{\langle N_1(N_1-1)\rangle}_{\frac{1}{4}N(N-1)=} + \underbrace{\langle N_1\rangle}_{=\frac{2}{4}N} = \frac{1}{4}N(N+1)\,. \tag{2.1.5}$$

It follows that the average number of atoms in the first half-volume is

$$\langle N_1\rangle = \frac{1}{2}N\,, \tag{2.1.6}$$

consistent with anybody's expectations. The spread in N_1, that is: the square root of the variance

$$\begin{aligned}\left\langle (N_1 - \langle N_1\rangle)^2\right\rangle &= \left\langle N_1^2\right\rangle - \langle N_1\rangle^2\\ &= \frac{1}{4}N(N+1) - \left(\frac{1}{2}N\right)^2 = \frac{1}{4}N\,,\end{aligned} \tag{2.1.7}$$

is

$$\delta N_1 = \frac{1}{2}\sqrt{N}\,. \tag{2.1.8}$$

Now, for $N = 10^{20}$, this spread is $\frac{1}{2}\times 10^{10}$, quite a large number in absolute size, yet tiny on the scale set by the average $\langle N_1\rangle = \frac{1}{2}N$,

$$\frac{\delta N_1}{\langle N_1\rangle} = \frac{1}{\sqrt{N}} = 10^{-10}\,. \tag{2.1.9}$$

We observe — and this is an important lesson — that all typical values of N_1 are so close to the average value that deviations can often be ignored completely. Usually, there is no need to pay attention to corrections of the relative order of 10^{-10}.

The averages $\langle f(N_1)\rangle$ require a comment. We can think of them in two different ways. First, we can imagine a *large ensemble of such systems*, each with a definite value of N_1 and the number of systems with the same

N_1 value in proportion of the probability $p(N_1)$. This imagined ensemble is the so-called *Gibbs ensemble* for this situation; it has many systems with $N_1 \simeq \langle N_1 \rangle = \frac{1}{2}N$ and very few with $|N_1 - \langle N_1 \rangle| > \sqrt{N}$. The average $\langle f(N_1) \rangle$ is then the average value for this ensemble.

Second, we can think of $\langle f(N_1) \rangle$ as the value that we get by measuring $f(N_1)$ for *one system at different times*. As the atoms move around the volume, they enter or leave the first half-volume and so make N_1 fluctuate in time. This motion is often so rapid — a gas atom has a speed of hundreds of m/s at room temperature — that the number N_1 fluctuates even while we are measuring $f(N_1)$, so that a single measurement yields a time-averaged number.

2.2 Ergodic principle

The equivalence of time averages for one system and the ensemble averages of the fictitious, imagined Gibbs ensemble of systems is very plausible. When it is true, the system is called *ergodic*. It is a basic assumption of statistical mechanics that realistic systems possess this ergodicity. Whether or not one can derive the ergodic property from a microscopic description is an open question, but it is generally believed to be true for all real systems. While it is indeed possible to invent model systems that are not ergodic (see Exercise 34) these models are so obviously over-simplified that such examples have no bearing on our understanding of real systems. Numerical simulations of systems with rather few molecules show ergodic behavior even then. We shall, therefore, not hesitate to accept the *ergodic principle*, that is: the assumption that realistic systems are ergodic, as a building block of statistical mechanics.

If, for the sake of the argument, we employ the language of classical mechanics, then the configuration of the joint center-of-mass state of N atoms is a point in the $6N$-dimensional phase space. Given that there are limits in resolution, let us tile the phase space with small cells, so small that there is no noticeable difference between the properties of the system for the points within one cell, but perhaps between phase space points from neighboring cells. Mindful of the lessons of quantum mechanics, we count one *microstate* per phase space volume of $(2\pi\hbar)^{3N}$, where $2\pi\hbar$ is Planck's* constant:

---

*Max Karl Ernst Ludwig PLANCK (1858–1947)

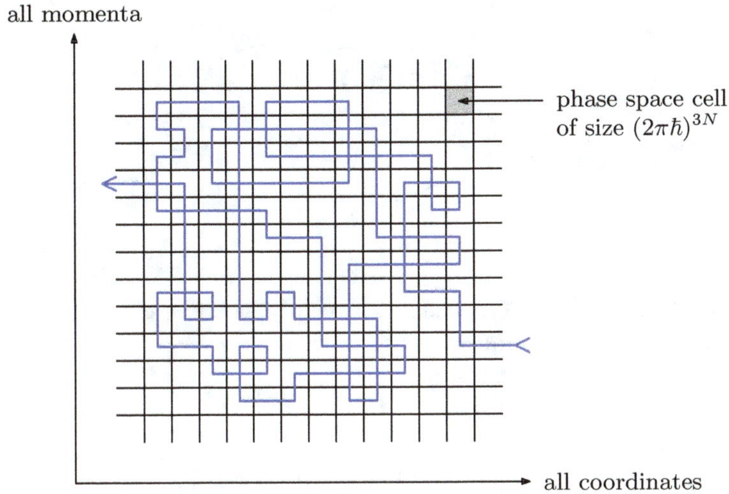

(2.2.1)

Then, according to the ergodic principle, the trajectory in phase space visits all accessible phase space cells equally often. Here, "accessible" refers to the conditions that specify the *macrostate*: particle number N, volume V, energy E, and possibly others. Macroscopic quantities, thereby, have natural uncertainties that are normally quite small on the macroscopic scale while being large on the microscopic scale. In particular, when we say that the system as a whole has a known energy E, there really is an energy range $E - \delta E \cdots E$. If we like, we can once more refer to quantum mechanics and recall that $\delta E \gtrsim \dfrac{\hbar}{\delta t}$ if δt is the duration of an energy measurement. In fact, the imprecision of our knowledge of the energy E is usually much larger than this quantum bound.

We note that the two-dimensional sketch in (2.2.1) is grossly misleading, and so are other sketches below. For a single particle, we have a six-dimensional phase space, and each cell is six-dimensional as well and has five-dimensional faces. Clearly, such a drawing is a caricature rather than a portrait already for a single particle, the more so when the particle number is huge. No such drawing, therefore, must ever be read as resembling the correct picture.

2.3 Microcanonical ensemble

For given E, there is a $(6N - 1)$-dimensional *energy surface* in the $6N$-dimensional phase space:

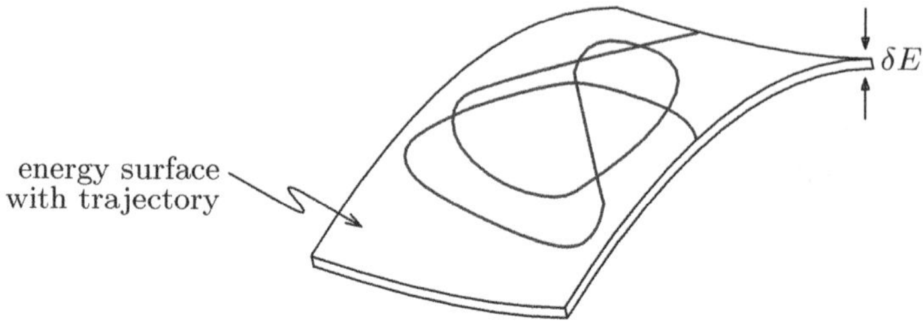

(2.3.1)

but it is actually a thin $6N$-dimensional layer of thickness δE. This layer occupies the phase-space volume

$$\begin{aligned}&\int (\mathrm{d}\boldsymbol{r}_1)(\mathrm{d}\boldsymbol{p}_1)\cdots(\mathrm{d}\boldsymbol{r}_N)(\mathrm{d}\boldsymbol{p}_N)\,\chi\big(E-\delta E<H(\boldsymbol{r},\boldsymbol{p})<E\big)\\ &=\int (\mathrm{d}\boldsymbol{r}_1)(\mathrm{d}\boldsymbol{p}_1)\cdots(\mathrm{d}\boldsymbol{r}_N)(\mathrm{d}\boldsymbol{p}_N)\,\Big[\chi\big(H(\boldsymbol{r},\boldsymbol{p})<E\big)\\ &\qquad\qquad -\chi\big(H(\boldsymbol{r},\boldsymbol{p})<E-\delta E\big)\Big]\,, \qquad (2.3.2)\end{aligned}$$

where $H(\boldsymbol{r},\boldsymbol{p})$ is the Hamilton* function for the N particles (atoms, molecules, ...) with position vectors $\boldsymbol{r}_1$, $\boldsymbol{r}_2$, ..., $\boldsymbol{r}_N$ and momentum vectors $\boldsymbol{p}_1$, $\boldsymbol{p}_2$, ..., $\boldsymbol{p}_N$, and

$$\chi(x)=\begin{cases}1 & \text{if } x \text{ is true}\\ 0 & \text{if } x \text{ is false}\end{cases} \qquad (2.3.3)$$

is the characteristic function for the statement x.

The phase-space volume occupied by all $\boldsymbol{r},\boldsymbol{p}$ such that $H(\boldsymbol{r},\boldsymbol{p})<E$ is usually roughly proportional to a power of E, $\propto E^M$, with M a multiple of the particle number N, such as $M=\frac{3}{2}N$ in the example in (2.3.15) below. Then the layer with $E-\delta E<H(\boldsymbol{r},\boldsymbol{p})<E$ occupies the fraction

$$\frac{E^M-(E-\delta E)^M}{E^M}=1-\left(1-\frac{\delta E}{E}\right)^M \qquad (2.3.4)$$

of this volume. For $\delta E \ll E$, we have

$$\text{fraction}=1-\mathrm{e}^{-M\frac{\delta E}{E}}\,. \qquad (2.3.5)$$

Now, let us be unrealistically ambitious and assume that we can know the energy with a relative precision of $\dfrac{\delta E}{E}=10^{-8}$ for a micromole of substance

*Sir William Rowan HAMILTON (1805–1865)

($M \sim N \sim 10^{18}$), so that the exponent $M\frac{\delta E}{E}$ is (roughly) of the order $\sim 10^{10}$, and

$$\text{fraction} = 1 \tag{2.3.6}$$

for all practical purposes. We thus observe that a very thin energy shell $E - \delta E < H(\boldsymbol{r}, \boldsymbol{p}) < E$ contains essentially all the phase-space volume enclosed by the energy surface $E = H(\boldsymbol{r}, \boldsymbol{p})$. We can, therefore, replace the characteristic function $\chi(E - \delta E < H(\boldsymbol{r}, \boldsymbol{p}) < E)$ by $\chi(H(\boldsymbol{r}, \boldsymbol{p}) < E)$ without introducing a noticeable mistake.

Therefore, still sticking to the language of classical mechanics, we count as many as

$$\Omega(E, V, N) = \frac{1}{(2\pi\hbar)^{3N}} \int (\mathrm{d}\boldsymbol{r}_1) \cdots (\mathrm{d}\boldsymbol{r}_N)(\mathrm{d}\boldsymbol{p}_1) \cdots (\mathrm{d}\boldsymbol{p}_N)\, \chi\big(H(\boldsymbol{r}, \boldsymbol{p}) < E\big) \tag{2.3.7}$$

microstates, one per phase-space cell of volume $(2\pi\hbar)^{3N}$, for a system with N particles, volume V, and energy E. We should remember, however, that this is essentially the number of cells in a thin energy layer, but the count does not depend on the thickness δE of the layer in any relevant way and, therefore, we do not even specify this thickness.

Each of these many microstates is consistent with the macroscopic variables E, V, N and all the microstates together compose the so-called *microcanonical ensemble*. According to Boltzmann* and others, the entropy of such a thermodynamical system is

$$S = k_{\mathrm{B}} \log\big(\Omega(E, V, N)\big)\,, \tag{2.3.8}$$

where Boltzmann's constant k_{B} is the ratio of the universal gas constant R and Avogadro's† number N_{A}, also known as Loschmidt's number,

$$k_{\mathrm{B}} = \frac{R}{N_{\mathrm{A}}} = \frac{8.314\,\mathrm{J/K}}{6.022 \times 10^{23}} = 1.38 \times 10^{-23}\,\mathrm{J/K}\,. \tag{2.3.9}$$

We justify this fundamental relation between the entropy and the count of microstates by its consequences.

First, Boltzmann's entropy is properly extensive. A system composed of subsystems A and B, with respective counts Ω_A and Ω_B for their mi-

*Ludwig Eduard BOLTZMANN (1844–1906)
†Lorenzo Romano Amedeo Carlo AVOGADRO, Conte di Quaregna e Cerreto (1776–1856)

crostates, has as many as $\Omega_{A\&B} = \Omega_A \Omega_B$ microstates, and

$$\begin{aligned} S_{A\&B} &= k_{\mathrm{B}} \log(\Omega_A \Omega_B) \\ &= k_{\mathrm{B}} \log \Omega_A + k_{\mathrm{B}} \log \Omega_B \\ &= S_A + S_B \end{aligned} \tag{2.3.10}$$

follows.

Second, Boltzmann's entropy is consistent with the Second Law of thermodynamics of Section 1.1.4, as we see (i) by noting that $\Omega(V, N, E)$ increases if E increases, so that $\left(\frac{\partial S}{\partial E}\right)_{V,N} > 0$, and (ii) by considering the maximum property of the entropy of Section 1.3. When we introduce constraints ($Y \neq 0$), we restrict the system additionally, and this will usually reduce the number of accessible microstates and will never increase it,

$$\Omega(E, V, N; \underbrace{Y \neq 0}_{\text{yes constraints}}) \leq \Omega(E, V, N; \underbrace{Y = 0}_{\text{no constraints}}) = \Omega(E, V, N)\,. \tag{2.3.11}$$

As an immediate consequence, we have

$$S(Y \neq 0) \leq S(Y = 0) \tag{2.3.12}$$

or

$$S(E, V, N) = \operatorname*{Max}_{Y} \left\{ S(E, V, N; Y) \right\}\,. \tag{2.3.13}$$

Earlier, we kept emphasising that the thermodynamical system must be in an equilibrium state (or, possibly, quite close to one) for properties such as temperature and entropy to be meaningful. The same emphasis is due here. We can phrase it by saying that the system must have reached a state of ergodicity, that is: each phase-space cell is visited equally often by the system trajectory in time, as sketched in (2.2.1), or, equivalently, we have an ensemble of trajectories, which uniformly cover the accessible phase space volume, the thin energy layer in (2.3.1).

As a further check, we consider an ideal gas of N structureless atoms in volume V. The Hamilton function is then just the kinetic energy

$$H(\boldsymbol{r}, \boldsymbol{p}) = \sum_{j=1}^{N} \frac{1}{2m} \boldsymbol{p}_j^2\,. \tag{2.3.14}$$

We have, with Exercise 35,

$$\Omega(E,V,N) = \frac{1}{(2\pi\hbar)^{3N}} \int \underbrace{(\mathrm{d}\boldsymbol{r}_1)\cdots(\mathrm{d}\boldsymbol{r}_N)}_{=V^N} \underbrace{(\mathrm{d}\boldsymbol{p}_1)\cdots(\mathrm{d}\boldsymbol{p}_N)\chi\Big(\sum_j \boldsymbol{p}_j^2 < 2mE\Big)}_{\substack{\text{volume of a } 3N\text{-dimensional} \\ \text{ball with radius } \sqrt{2mE}}}$$

$$= \frac{V^N}{(2\pi\hbar)^{3N}} \frac{\pi^{\frac{3N}{2}}}{\left(\frac{3N}{2}\right)!} \sqrt{2mE}^{\,3N}$$

$$= \left(\frac{\pi}{(2\pi\hbar)^2}\left[\left(\frac{3N}{2}\right)!\right]^{-\frac{2}{3N}} 2mEV^{\frac{2}{3}}\right)^{\frac{3N}{2}}, \tag{2.3.15}$$

and so get

$$S = k_\mathrm{B} \log(\Omega(E,V,N))$$
$$= \frac{3}{2} N k_\mathrm{B} \left[\log\left(\frac{m}{2\pi\hbar^2} EV^{\frac{2}{3}}\right) - \frac{2}{3N}\log\left(\left(\frac{3}{2}N\right)!\right)\right]. \tag{2.3.16}$$

Remembering that N is a very large number, we employ Stirling's* approximation for the logarithm of the factorial, see Exercises 36 and 37,

$$\log(x!) = x\log x - x\,, \tag{2.3.17}$$

in

$$\frac{2}{3N}\log\left(\left(\frac{3}{2}N\right)!\right) = \log\frac{3N}{2} - 1\,, \tag{2.3.18}$$

and arrive at

$$\frac{2}{3}\frac{S}{Nk_\mathrm{B}} = \log\left(\frac{m}{3\pi\hbar^2}\frac{EV^{\frac{2}{3}}}{N}\right) + 1\,. \tag{2.3.19}$$

This, however, cannot be right!

2.4 Correct Boltzmann counting

When we scale the system by $N \to \lambda N$, $V \to \lambda V$, $E \to \lambda E$, the entropy has to scale alike, $S \to \lambda S$, because all four quantities are extensive variables. For the relation above, this is not so. The solution to this problem (or puzzle or paradox — pick your choice) was guessed correctly by Boltzmann

*James STIRLING (1692–1770)

and Gibbs, although they could not actually justify their answer except by its correct consequences.

The solution they offered consists of a correction to the expression for $\Omega(E,V,N)$ in (2.3.7), where a factor $\frac{1}{N!}$ is introduced:

$$\Omega(E,V,N) = \frac{1}{N!}\int \frac{(\mathrm{d}\boldsymbol{r}_1)(\mathrm{d}\boldsymbol{p}_1)}{(2\pi\hbar)^3}\cdots\frac{(\mathrm{d}\boldsymbol{r}_N)(\mathrm{d}\boldsymbol{p}_N)}{(2\pi\hbar)^3}\chi\big(H(\boldsymbol{r},\boldsymbol{p})<E\big)\,. \quad (2.4.1)$$

This, effectively, states that we have been over-counting the microstates greatly ($N!$ is a gigantic number), for which the ultimate explanation is provided by quantum mechanics and its recognition of the absolute indistinguishability of particles of the same kind. If a microstate has the N particles in some arrangement, a permutation of the particles (or a permutation of the labels we use for distinguishing them) does not give another microstate, as it would for the fully distinguishable particles of classical mechanics, but is the same microstate again. Since there are $N!$ permutations, we over-count by this huge factor if we fail to account for the indistinguishability.

With this corrected expression for $\Omega(E,V,N)$ — the jargon is that "it has the correct Boltzmann counting" of the microstates — we have the replacements

$$\left(\tfrac{3}{2}N\right)! \to \left(\tfrac{3}{2}N\right)!\,N! \quad (2.4.2)$$

and

$$\begin{aligned}\frac{2}{3N}\log\left(\left(\tfrac{3}{2}N\right)!\right) &\to \frac{2}{3N}\left[\log\left(\left(\tfrac{3}{2}N\right)!\right)+\log(N!)\right]\\ &= \frac{2}{3N}\left(\frac{3N}{2}\log\frac{3N}{2}-\frac{3N}{2}+N\log N-N\right)\\ &= \log\left(\frac{3}{2}N^{\frac{5}{3}}\right)-\frac{5}{3} \end{aligned} \quad (2.4.3)$$

in (2.3.16) and (2.3.18). The corrected version of (2.3.19) is then

$$\frac{2}{3}\frac{S}{Nk_{\mathrm{B}}} = \log\left(\frac{m}{3\pi\hbar^2}\frac{EV^{\frac{2}{3}}}{N^{\frac{5}{3}}}\right)+\frac{5}{3}\,. \quad (2.4.4)$$

This so-called Sackur*–Tetrode† equation agrees well with experimental data for gases at high enough temperatures.

---

*Otto SACKUR (1880–1914) †Hugo Martin TETRODE (1895–1931)

We make contact with our earlier results for the ideal gas by solving for E and recognizing that this is the internal energy $U(S,V,N)$,

$$U(S,V,N) = \frac{3\pi\hbar^2}{m} N \left(\frac{N}{V}\right)^{\frac{2}{3}} \mathrm{e}^{\frac{2}{3}\frac{S}{Nk_\mathrm{B}} - \frac{5}{3}} . \tag{2.4.5}$$

When comparing with the result in (1.2.16),

$$U(S,V,n) = E_0\, n \left(\frac{nV_0}{V}\, \mathrm{e}^{\frac{S}{nR}}\right)^{\gamma - 1} , \tag{2.4.6}$$

where N and n are related by $N = nN_\mathrm{A}$, we conclude first that $\gamma = \frac{5}{3}$ is the value of the adiabatic index for this ideal gas of structureless atoms, and then that

$$E_0 V_0^{\frac{2}{3}} = \frac{3\pi\hbar^2}{m} N_\mathrm{A}^{\frac{5}{3}}\, \mathrm{e}^{-\frac{5}{3}} = \frac{0.051\ \mathrm{Jm}^2}{(\text{molar mass in gram})} . \tag{2.4.7}$$

2.5 Canonical ensemble

The microcanonical ensemble is made up of all microstates consistent with the macroscopic thermodynamical system having known volume, known particle number, and known energy (actually, energy in a small range $E - \delta E \cdots E$). We know that it is useful to change the perspective if the question of interest is more easily addressed from the altered point of view. For example, we could have well-controlled temperature rather than known energy. The situation, then, is that of (1.8.2): a (small) system embedded in a (large) bath,

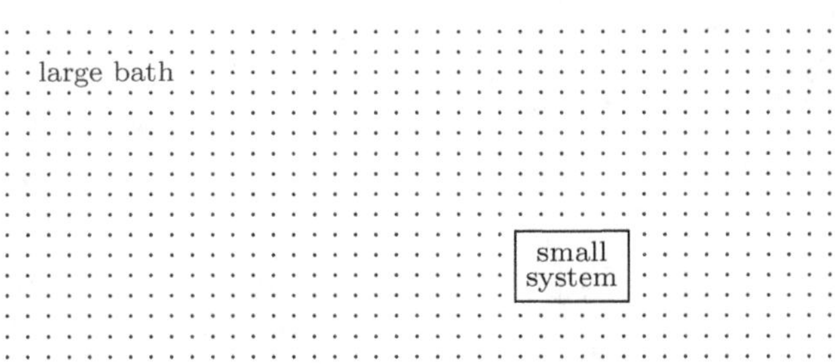

(2.5.1)

with system and bath making up a larger thermodynamical system that is isolated, and thus can be thought of as having a fixed energy E. The energy of the small system is not fixed, it can take on various values. If the small system is in its kth microstate with energy E_k, the bath has energy $E - E_k$, and there are $\Omega(E - E_k)$ microstates available. Now, the probability p_k of finding the small system in its kth state is proportional to the number of microstates available, which are here microstates of the bath. Accordingly,

$$\begin{aligned} p_k \propto \Omega(E - E_k) &= \mathrm{e}^{\log(\Omega(E - E_k))} \\ &= \mathrm{e}^{\log(\Omega(E)) - E_k \frac{\partial}{\partial E} \log(\Omega(E)) + \cdots} \end{aligned} \tag{2.5.2}$$

where the ellipses stand for higher-order terms that we can safely ignore. We have chosen to exhibit the small corrections associated with $E_k \ll E$ in the exponent because we know that $\log\bigl(\Omega(E)\bigr)$ is a relatively smooth function ($\sim$ entropy), whereas $\Omega(E)$ itself changes rapidly with E. We recall that $\dfrac{\partial}{\partial E} S(E, V, N) = \dfrac{1}{T(E, V, N)}$, that is

$$\frac{1}{T} = \left(\frac{\partial S}{\partial E}\right)_{V,N} = k_{\mathrm{B}} \left(\frac{\partial}{\partial E} \log \Omega\right)_{V,N}, \tag{2.5.3}$$

or

$$\left(\frac{\partial}{\partial E} \log \Omega\right)_{V,N} = \frac{1}{k_{\mathrm{B}} T} \equiv \beta\,. \tag{2.5.4}$$

It follows that

$$p_k \propto \mathrm{e}^{-\beta E_k} \tag{2.5.5}$$

and for the normalisation to unit sum,

$$\sum_k p_k = 1\,, \tag{2.5.6}$$

we divide by

$$Q = \sum_k \mathrm{e}^{-\beta E_k}\,, \tag{2.5.7}$$

so that

$$p_k = \frac{1}{Q} \mathrm{e}^{-\beta E_k}\,. \tag{2.5.8}$$

The function $Q(\beta, V, N)$ — which depends on V and N through the energies $E_k = E_k(V, N)$ — is the *canonical partition function*, and the ensemble specified by temperature T, volume V, and particle number N, is the *canonical ensemble.*

Energy does not have a fixed value for the canonical ensemble, but fluctuates with an average of

$$\begin{aligned}\langle E\rangle &= \sum_k p_k E_k = \frac{1}{Q}\sum_k E_k\,\mathrm{e}^{-\beta E_k}\\ &= -\frac{1}{Q}\frac{\partial}{\partial\beta}\sum_k \mathrm{e}^{-\beta E_k} = -\frac{1}{Q}\frac{\partial}{\partial\beta}Q \end{aligned}\tag{2.5.9}$$

or

$$\langle E\rangle_{V,N} = -\left(\frac{\partial}{\partial\beta}\log Q\right)_{V,N}\tag{2.5.10}$$

if we indicate that V and N are kept constant. This relation is key for recognizing that $-\frac{1}{\beta}\log Q$ is the Helmholtz free energy F.

Here is how this goes. Proceeding from

$$F = U - TS \quad\text{and}\quad \mathrm{d}U = T\mathrm{d}S - P\mathrm{d}V + \mu\cdot\mathrm{d}n\tag{2.5.11}$$

with $n = \frac{N}{N_\mathrm{A}}$, we establish

$$\begin{aligned}\mathrm{d}(\beta F) &= \mathrm{d}\left(\beta U - \frac{1}{k_\mathrm{B}}S\right)\\ &= U\mathrm{d}\beta + \beta(T\mathrm{d}S - P\mathrm{d}V + \mu\cdot\mathrm{d}n) - \frac{1}{k_\mathrm{B}}\mathrm{d}S\\ &= U\mathrm{d}\beta - \beta P\mathrm{d}V + \beta\mu\cdot\mathrm{d}n\end{aligned}\tag{2.5.12}$$

and conclude that

$$\left(\frac{\partial(\beta F)}{\partial\beta}\right)_{V,N} = U\,.\tag{2.5.13}$$

Upon identifying the internal energy here with the average value of the fluctuating energy, $U = \langle E\rangle$, we thus infer that $\beta F = -\log Q$, or

$$F(T,V,N) = -\frac{1}{\beta}\log\bigl(Q(\beta,V,N)\bigr)\,,\tag{2.5.14}$$

where $\beta = \frac{1}{k_\mathrm{B}T}$ as always. Accordingly,

$$p_k = \mathrm{e}^{-\beta(E_k - F)}\tag{2.5.15}$$

is an alternative way of writing (2.5.8).

This identification of U with $\langle E\rangle$ is only meaningful if there are relatively small fluctuations of the energy. We quantify them by their spread, the square root of the variance

$$\begin{aligned}\langle \delta E^2\rangle &= \left\langle \left(E-\langle E\rangle\right)^2\right\rangle = \langle E^2\rangle - \langle E\rangle^2 \\ &= \sum_k p_k E_k^2 - \langle E\rangle^2 \\ &= \frac{1}{Q}\left(\frac{\partial}{\partial\beta}\right)^2 Q - \left(\frac{1}{Q}\frac{\partial Q}{\partial\beta}\right)^2 \\ &= \frac{\partial}{\partial\beta}\left(\frac{1}{Q}\frac{\partial Q}{\partial\beta}\right)\end{aligned} \tag{2.5.16}$$

or, with more precise notation,

$$\langle \delta E^2\rangle_{V,N} = \left(\frac{\partial}{\partial\beta}\left(\frac{1}{Q}\frac{\partial Q}{\partial\beta}\right)\right)_{V,N} = -\left(\frac{\partial\langle E\rangle}{\partial\beta}\right)_{V,N}. \tag{2.5.17}$$

Returning once more to $\mathrm{d}U = T\mathrm{d}S - P\mathrm{d}V + \mu\cdot\mathrm{d}n$, we note that

$$\left(\frac{\partial U}{\partial T}\right)_{V,N} = T\left(\frac{\partial S}{\partial T}\right)_{V,N} = C_V \tag{2.5.18}$$

with the constant-volume heat capacity C_V of (1.6.4), so that $U = \langle E\rangle$ and $\frac{\partial}{\partial\beta} = -k_{\mathrm{B}}T^2\frac{\partial}{\partial T}$ give

$$\langle \delta E^2\rangle_{V,N} = k_{\mathrm{B}}T^2 C_V\,. \tag{2.5.19}$$

Rather remarkably, this relates the energy fluctuations in the canonical ensemble to the temperature and the heat capacity for constant volume, both measurable. On the right-hand side, T is intensive, and C_V is extensive, so that

$$\langle \delta E^2\rangle_{V,N} \propto N \tag{2.5.20}$$

and the spread $\sqrt{\langle\delta E^2\rangle_{V,N}}$ is proportional to $\sqrt{N}$. Its relative size is, therefore, of order $\frac{1}{\sqrt{N}}$,

$$\frac{\sqrt{\langle \delta E^2\rangle_{V,N}}}{\langle E\rangle_{V,N}} \propto \frac{1}{\sqrt{N}} \ll 1\,. \tag{2.5.21}$$

That is: Although the energy is fluctuating in the canonical ensemble, the fluctuations are of so small relative size that only a small range of energies around the average value $\langle E\rangle$ are important. As we recall, also in the microcanonical ensemble, we have a small range δE of energy values, so that — in the end — the canonical ensemble and the microcanonical ensemble are equivalent for practical purposes.

For illustration, we consider a model system of two-state constituents. There are N constituents with possible energy values $\pm\frac{1}{2}E_0$ for each of them:

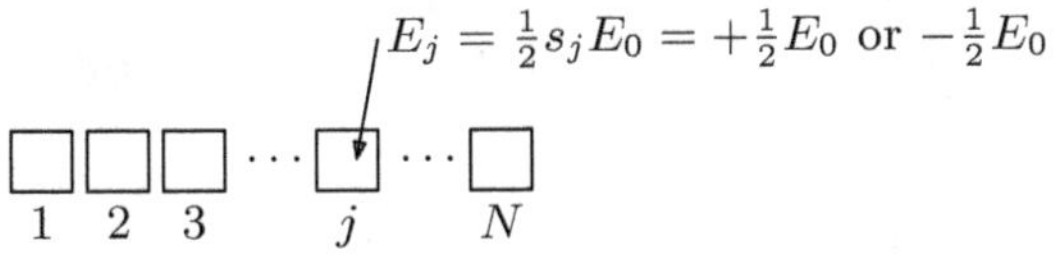

(2.5.22)

and (for now) no interaction between different constituents. In the microcanonical ensemble, the possible energy values are

$$-\frac{1}{2}NE_0,\ -\frac{1}{2}(N-2)E_0,\ \ldots,\ \frac{1}{2}(N-2)E_0,\ \frac{1}{2}NE_0\,, \tag{2.5.23}$$

and we have $\dfrac{N}{2}+\dfrac{E}{E_0}$ s_j values $+1$ and $\dfrac{N}{2}-\dfrac{E}{E_0}$ s_j values -1 for energy E. Accordingly, there are

$$\Omega(E) = \frac{N!}{\left(\frac{N}{2}+\frac{E}{E_0}\right)!\,\left(\frac{N}{2}-\frac{E}{E_0}\right)!} \tag{2.5.24}$$

states for energy E, so that we have the entropy

$$\begin{aligned} S &= k_{\mathrm{B}}\log\bigl(\Omega(E)\bigr) \\ &= k_{\mathrm{B}}\left[\log(N!) - \log\left(\left(\frac{N}{2}+\frac{E}{E_0}\right)!\right) - \log\left(\left(\frac{N}{2}-\frac{E}{E_0}\right)!\right)\right] \\ &= k_{\mathrm{B}}\left[N\log N - \left(\frac{N}{2}+\frac{E}{E_0}\right)\log\left(\frac{N}{2}+\frac{E}{E_0}\right) - \left(\frac{N}{2}-\frac{E}{E_0}\right)\log\left(\frac{N}{2}-\frac{E}{E_0}\right)\right] \end{aligned} \tag{2.5.25}$$

after using Stirling's approximation (2.3.17) three times. Since now $\dfrac{1}{T} = \dfrac{\partial}{\partial E}S$, we find

$$\beta = \frac{1}{k_{\mathrm{B}}T} = \frac{\partial}{\partial E}\log\bigl(\Omega(E)\bigr) = \frac{1}{E_0}\log\frac{NE_0-2E}{NE_0+2E}\,. \tag{2.5.26}$$

We identify the internal energy U with the energy above the minimal energy $-\frac{1}{2}NE_0$, that is: $E = -\frac{1}{2}NE_0 + U$, and solve for U, proceeding from

$$\mathrm{e}^{\beta E_0} = \frac{NE_0 - U}{U} \tag{2.5.27}$$

and arriving at

$$U = \frac{NE_0}{1 + \mathrm{e}^{\beta E_0}} \,. \tag{2.5.28}$$

Here, N is the number of constituents and E_0 is the excitation energy available for each constituent.

In the canonical ensemble, the configuration with given values for $s_1, s_2, \ldots, s_N$ has energy $\frac{1}{2}E_0 \sum_{j=1}^{N} s_j$, so that the partition function is

$$\begin{aligned}
Q &= \sum_{s_1=\pm 1} \sum_{s_2=\pm 1} \cdots \sum_{s_N=\pm 1} \underbrace{\mathrm{e}^{-\frac{1}{2}\beta E_0 \sum_{j=1}^{N} s_j}}_{=\prod_{j=1}^{N} \mathrm{e}^{-\frac{1}{2}\beta E_0 s_j}} \\
&= \sum_{s_1=\pm 1} \mathrm{e}^{-\frac{1}{2}\beta E_0 s_1} \sum_{s_2=\pm 1} \mathrm{e}^{-\frac{1}{2}\beta E_0 s_2} \ldots \sum_{s_N=\pm 1} \mathrm{e}^{-\frac{1}{2}\beta E_0 s_N} \\
&= \left(\sum_{s=\pm 1} \mathrm{e}^{-\frac{1}{2}\beta E_0 s} \right)^N = \left(\mathrm{e}^{\frac{1}{2}\beta E_0} + \mathrm{e}^{-\frac{1}{2}\beta E_0} \right)^N
\end{aligned} \tag{2.5.29}$$

or

$$Q = \left[2\cosh\left(\tfrac{1}{2}\beta E_0\right) \right]^N . \tag{2.5.30}$$

The average value of the energy is then

$$\langle E \rangle = -\frac{1}{Q}\frac{\partial}{\partial\beta} Q = -\frac{\partial}{\partial\beta} \log Q = -\frac{1}{2} N E_0 \tanh\left(\tfrac{1}{2}\beta E_0\right) , \tag{2.5.31}$$

which is negative; less than half of the constituents are in the excited state. For the comparison with the microcanonical ensemble, we look at

$$\begin{aligned}
U = \frac{1}{2} N E_0 + \langle E \rangle &= \frac{1}{2} N E_0 \left[1 - \tanh\left(\tfrac{1}{2}\beta E_0\right) \right] \\
&= N E_0 \frac{\cosh\left(\frac{1}{2}\beta E_0\right) - \sinh\left(\frac{1}{2}\beta E_0\right)}{2\cosh\left(\frac{1}{2}\beta E_0\right)} \\
&= N E_0 \frac{\mathrm{e}^{-\frac{1}{2}\beta E_0}}{\mathrm{e}^{\frac{1}{2}\beta E_0} + \mathrm{e}^{-\frac{1}{2}\beta E_0}} = \frac{N E_0}{1 + \mathrm{e}^{\beta E_0}} \,,
\end{aligned} \tag{2.5.32}$$

which is *exactly* the result in (2.5.28) for the microcanonical ensemble.

Further, we confirm that the energy fluctuations in the canonical ensemble are proportional to $\sqrt{N}$,

$$\langle \delta E^2 \rangle \underset{\substack{\uparrow \\ (2.5.17)}}{=} -\frac{\partial}{\partial \beta}\langle E \rangle = \frac{N E_0^2}{4\cosh\left(\frac{1}{2}\beta E_0\right)^2} = \left(\sqrt{N}\frac{E_0}{2\cosh\left(\frac{1}{2}\beta E_0\right)}\right)^2 , \quad (2.5.33)$$

and their relative size is proportional to $\frac{1}{\sqrt{N}}$,

$$\frac{\sqrt{\langle \delta E^2 \rangle}}{-\langle E \rangle} = \sqrt{N}\frac{E_0}{2\cosh\left(\frac{1}{2}\beta E_0\right)} \frac{2}{N E_0 \tanh\left(\frac{1}{2}\beta E_0\right)} = \frac{1}{\sqrt{N}} \frac{1}{\sinh\left(\frac{1}{2}\beta E_0\right)} . \quad (2.5.34)$$

Indeed, there is no substantial difference between the microcanonical and the canonical ensemble in this simple example.

2.6 Other ensembles

Let us now consider a generalisation of the system and situation depicted in (2.5.1). More generally than before, we now do not just allow for energy exchange between system and bath but also for other quantities that we associate with extensive variables, collectively called X, as in

$$\mathrm{d}U = T\mathrm{d}S + y \cdot \mathrm{d}X , \quad (2.6.1)$$

and their intensive partners are symbolised by y. For example, if $X = (V, N) =$ (volume, particle number), then $y = (-P, \mu) =$ (negative pressure, chemical potential). In passing, we note that we wrote $\mu\,\mathrm{d}n$ earlier and write $\mu\,\mathrm{d}N$ now, so that the chemical potential referred to a change in the number of moles earlier ($\mathrm{d}n$) while it refers to a change in particle number now ($\mathrm{d}N$). The numerical values of the two versions of μ are, therefore, related to each other by Avogadro's number, but other than that there is no difference. We trust that the context specifies which version of the chemical potential is meant, and we shall continue to use the same symbol μ in both the "number of moles" and the "number of particles" context.

We have E_k, X_k for the kth state of the system and $E - E_k, X - X_k$ for the energy of the bath and its X values if the system is in its kth state. In analogy with (2.5.2), the probability of finding the system in this state is proportional to the number of microstates available to the bath,

$$\begin{aligned} p_k \propto \Omega(E - E_k, X - X_k) &= \mathrm{e}^{\log(\Omega(E - E_k, X - X_k))} \\ &= \mathrm{e}^{\log(\Omega(E, X)) - E_k \left(\frac{\partial \log \Omega}{\partial E}\right)_X - X_k \cdot \left(\frac{\partial \log \Omega}{\partial X}\right)_E} , \end{aligned} \quad (2.6.2)$$

where we can safely neglect second-order and higher-order terms since the bath is assumed to be huge in comparison with the system.

With $S = k_\mathrm{B} \log \Omega$ and

$$\mathrm{d}S = \frac{1}{T}(\mathrm{d}U - y \cdot \mathrm{d}X) \underset{\substack{\uparrow \\ U = E \text{ here}}}{=} \frac{1}{T}(\mathrm{d}E - y \cdot \mathrm{d}X) \tag{2.6.3}$$

we have

$$\mathrm{d}\log \Omega = \beta\, \mathrm{d}E - \beta y \cdot \mathrm{d}X \tag{2.6.4}$$

and

$$\left(\frac{\partial \log \Omega}{\partial E}\right)_X = \beta\,, \quad \left(\frac{\partial \log \Omega}{\partial X}\right)_E = -\beta y \tag{2.6.5}$$

so that

$$p_k \propto \mathrm{e}^{-\beta(E_k - y \cdot X_k)} \tag{2.6.6}$$

or, with proper normalization,

$$p_k = \frac{1}{Z}\, \mathrm{e}^{-\beta(E_k - y \cdot X_k)} \tag{2.6.7}$$

where

$$Z(\beta, \beta y) = \sum_k \mathrm{e}^{-\beta E_k + \beta y \cdot X_k} \tag{2.6.8}$$

is the appropriate partition function, corresponding to the current choice of extensive quantities that can be exchanged between system and bath.

The intensive variables β and y have the same values for the bath and the system. Therefore, we did not need to note that the right-hand sides in (2.6.5) refer to the bath although the left-hand sides do.

As indicated in (2.6.8), we regard Z as a function of β and all the βy variables. The respective thermodynamical potential Λ is identified by the analog of (2.5.15),

$$p_k = \mathrm{e}^{-\beta(E_k - y \cdot X_k - \Lambda)}\,, \tag{2.6.9}$$

that is:

$$\Lambda = -\frac{1}{\beta} \log Z\,, \tag{2.6.10}$$

which is the analog of $F = -\dfrac{1}{\beta} \log Q$ in (2.5.14).

For the expected values of the energy and the pertinent X variables, we have

$$\begin{aligned}\langle E\rangle &= \sum_k p_k E_k = \frac{1}{Z}\sum_k \mathrm{e}^{-\beta(E_k - y\cdot X_k)}E_k \\ &= \frac{1}{Z}\sum_k\left(-\frac{\partial}{\partial\beta}\mathrm{e}^{-\beta(E_k - y\cdot X_k)}\right)_{\beta y} \\ &= -\frac{1}{Z}\left(\frac{\partial Z}{\partial\beta}\right)_{\beta y} = -\left(\frac{\partial \log Z}{\partial\beta}\right)_{\beta y}\end{aligned} \tag{2.6.11}$$

and likewise

$$\langle X\rangle = \left(\frac{\partial\log Z}{\partial(\beta y)}\right)_{\beta}. \tag{2.6.12}$$

These are derivatives of $\beta\Lambda = -\log Z$.

2.7 Gibbs's entropy formula

The main ingredient of the Legendre transformation that switches from the dependence on β and βy to the dependence on $\langle E\rangle$ and $\langle X\rangle$ is the observation that

$$\begin{aligned}\mathrm{d}\big(\log Z + \beta\langle E\rangle - \beta y\cdot\langle X\rangle\big) &= \mathrm{d}\beta\underbrace{\left(\frac{\partial\log Z}{\partial\beta}\right)_{\beta y}}_{=-\langle E\rangle} + \mathrm{d}(\beta y)\cdot\underbrace{\left(\frac{\partial\log Z}{\partial(\beta y)}\right)_{\beta}}_{=\langle X\rangle} \\ &\quad + \mathrm{d}\beta\,\langle E\rangle - \mathrm{d}(\beta y)\cdot\langle X\rangle \\ &\quad + \beta\,\mathrm{d}\langle E\rangle - \beta y\cdot\mathrm{d}\langle X\rangle \\ &= \beta\,\mathrm{d}\langle E\rangle - \beta y\cdot\mathrm{d}\langle X\rangle\,.\end{aligned} \tag{2.7.1}$$

We compare this with $\mathrm{d}S = \frac{1}{T}(\mathrm{d}E - y\cdot\mathrm{d}X) = k_{\mathrm{B}}\beta(\mathrm{d}E - y\cdot\mathrm{d}X)$ in (2.6.3) and so conclude that

$$S = k_{\mathrm{B}}\big(\log Z + \beta\langle E\rangle - \beta y\cdot\langle X\rangle\big)\,, \tag{2.7.2}$$

mindful, as always, that the fluctuations are of relative size so small that the expected values $\langle E\rangle$ and $\langle X\rangle$ can be identified by their values proper.

Then,

$$\begin{aligned} S &= k_{\mathrm{B}} \sum_k p_k (\log Z + \beta E_k - \beta y \cdot X_k) \\ &= k_{\mathrm{B}} \sum_k p_k \underbrace{\log\left(Z\, \mathrm{e}^{\beta(E_k - y \cdot X_k)}\right)}_{= \log \frac{1}{p_k} = -\log p_k} \end{aligned} \tag{2.7.3}$$

or

$$S = -k_{\mathrm{B}} \sum_k p_k \log p_k \,. \tag{2.7.4}$$

This is Gibbs's celebrated formula for the entropy in terms of the probabilities for the microstates of the system.

It covers as well the microcanonical ensemble, where we have a total of Ω microstates, each occurring with the same probability $p_k = \dfrac{1}{\Omega}$, and $\sum_k 1 = \Omega$ ensures correct normalization. Thus,

$$\begin{aligned} S &= -k_{\mathrm{B}} \sum_k \frac{1}{\Omega} \log \frac{1}{\Omega} = (k_{\mathrm{B}} \log \Omega) \underbrace{\sum_k \frac{1}{\Omega}}_{=1} \\ &= k_{\mathrm{B}} \log \Omega \,, \end{aligned} \tag{2.7.5}$$

indeed. We shall return to discussing (2.7.4) in Section 2.10.

2.8 Grand canonical ensemble

A particular ensemble of considerable practical importance is the *grand canonical ensemble* of an open system with given volume V, so that the system exchanges energy and particles with the bath. For a single-component system — particles of one kind only; $\beta y \cdot X_k \to \beta\mu N_k$ in (2.6.8) — we have the partition function

$$Z(\beta, V, \beta\mu) = \sum_k \mathrm{e}^{-\beta E_k + \beta\mu N_k} \,, \tag{2.8.1}$$

where the dependence on the extensive volume is indicated, and the entropy is

$$S = k_{\mathrm{B}} \bigl(\log Z + \beta \langle E \rangle - \beta\mu \langle N \rangle\bigr) \,. \tag{2.8.2}$$

The volume dependence of Z arises from the volume dependence of the system energies E_k.

Upon comparing the entropy expression with

$$U = TS - PV + \mu N \tag{2.8.3}$$

or

$$\frac{S}{k_{\mathrm{B}}} = \beta(U + PV - \mu N) = \beta\big(\langle E\rangle + PV - \mu\langle N\rangle\big) \tag{2.8.4}$$

we infer that

$$\log Z = \beta PV \tag{2.8.5}$$

in the grand canonical ensemble, with $V, \beta, \beta\mu$ as the variables of Z. This is consistent with

$$\Lambda = -\frac{1}{\beta}\log Z = \langle E\rangle - TS - \mu\langle N\rangle = U - TS - \mu N\,, \tag{2.8.6}$$

telling us that Legendre transformations $S \to T$ and $N \to \mu$ turned $U(S, V, N)$ into $\Lambda(T, V, \mu)$. Some authors call $\Lambda(T, V, \mu)$, the thermodynamical potential associated with the grand canonical ensemble, the *Landau* free energy.*

2.9 Fluctuating particle numbers

Similar to the energy fluctuation considered above for the canonical ensemble, see (2.5.17), we take a look at the fluctuations in the particle number N here. First, we note that

$$\langle N\rangle = \left(\frac{\partial}{\partial(\beta\mu)}\log Z\right)_{\beta,V} = \sum_k p_k N_k\,, \tag{2.9.1}$$

and then

$$\begin{aligned}\left\langle\delta N^2\right\rangle &= \left\langle(N - \langle N\rangle)^2\right\rangle = \left\langle N^2\right\rangle - \langle N\rangle^2 \\ &= \frac{1}{Z}\left(\frac{\partial^2}{\partial(\beta\mu)^2}Z\right)_{\beta,V} - \frac{1}{Z^2}\left(\frac{\partial Z}{\partial(\beta\mu)}\right)^2_{\beta,V} \\ &= \left(\frac{\partial}{\partial(\beta\mu)}\left(\frac{1}{Z}\frac{\partial Z}{\partial(\beta\mu)}\right)\right)_{\beta.V} = \left(\frac{\partial\langle N\rangle}{\partial(\beta\mu)}\right)_{\beta,V}\,,\end{aligned} \tag{2.9.2}$$

the obvious analog of $\left\langle\delta E^2\right\rangle = -\left(\frac{\partial\langle E\rangle}{\partial\beta}\right)_{V,N}$ in (2.5.17).

*Lev Davidovich LANDAU (1908–1968)

Let us apply these relations to the simple situation, in which we have relatively few particles in a relatively large volume:

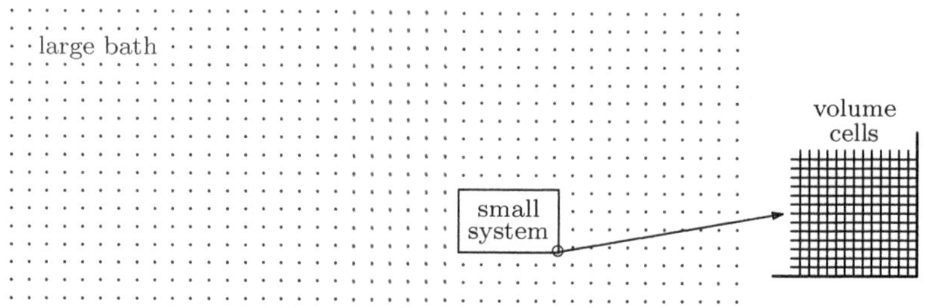

(2.9.3)

We break up the volume into many small volume cells — not to be confused with the high-dimensional phase space cells in (2.2.1) — such that we have at most one particle in a cell, never two, while most cells are empty. The count of particles in the jth cell is $n_j = 0$ or $n_j = 1$, and

$$N = \sum_{j=1}^{N_c} n_j \tag{2.9.4}$$

is the current total count of particles in the volume, where $N_c \gg N$ is the number of volume cells. All cells are equal in the sense that

$$\langle n_j \rangle = \langle n_{j'} \rangle = \langle n_1 \rangle = \frac{\langle N \rangle}{N_c} \ll 1 \tag{2.9.5}$$

is the same expected value, whether it is the jth cell or the j'th, or the first ($j = 1$).

The variance of the particle number is

$$\begin{aligned} \langle \delta N^2 \rangle &= \langle N^2 \rangle - \langle N \rangle^2 \\ &= \sum_{j,j'=1}^{N_c} \left(\langle n_j n_{j'} \rangle - \langle n_j \rangle \langle n_{j'} \rangle \right) . \end{aligned} \tag{2.9.6}$$

This is a general expression for $\langle \delta N^2 \rangle$, irrespective of circumstances, but we now apply it to the situation of a sufficiently dilute gas, with the particles far enough from each other that it can be fairly assumed that there are no statistical correlations among them. We express this by the factorization

$$\langle n_j n_{j'} \rangle = \langle n_j \rangle \langle n_{j'} \rangle \quad \text{for } j \neq j' . \tag{2.9.7}$$

In the double sum for $\langle\delta N^2\rangle$, then, all terms with $j \neq j'$ vanish, and we get

$$\langle\delta N^2\rangle = \sum_{j=1}^{N_c}\Bigl(\langle n_j^2\rangle - \langle n_j\rangle^2\Bigr). \tag{2.9.8}$$

Next, we note that $n_j^2 = n_j$ since $n_j = 0$ or $n_j = 1$ and, therefore,

$$\langle n_j^2\rangle = \langle n_j\rangle = \frac{\langle N\rangle}{N_c} \tag{2.9.9}$$

so that

$$\langle\delta N^2\rangle = \sum_{j=1}^{N_c}\left[\frac{\langle N\rangle}{N_c} - \left(\frac{\langle N\rangle}{N_c}\right)^2\right] = \langle N\rangle - \frac{\langle N\rangle^2}{N_c} \tag{2.9.10}$$

or

$$\langle\delta N^2\rangle = \langle N\rangle \tag{2.9.11}$$

in view of $\langle N\rangle \ll N_c$.

Harking back to (2.9.2), we thus have

$$\langle\delta N^2\rangle = \left(\frac{\partial\langle N\rangle}{\partial(\beta\mu)}\right)_{V,\beta} = \langle N\rangle \tag{2.9.12}$$

or, after introducing the particle density $\rho = \dfrac{\langle N\rangle}{V}$,

$$\left(\frac{\partial\rho}{\partial(\beta\mu)}\right)_{\beta} = \rho\,. \tag{2.9.13}$$

To proceed further, we recall the Gibbs–Duhem relation

$$0 = S\mathrm{d}T - V\mathrm{d}P + N\mathrm{d}\mu \tag{2.9.14}$$

or

$$\rho\,\mathrm{d}\mu = \mathrm{d}P - \frac{S}{V}\mathrm{d}T \tag{2.9.15}$$

after dividing by the volume V. It follows that

$$\rho\left(\frac{\partial\mu}{\partial\rho}\right)_T = \left(\frac{\partial P}{\partial\rho}\right)_T \tag{2.9.16}$$

or, equivalently,

$$\rho\left(\frac{\partial(\beta\mu)}{\partial\rho}\right)_\beta = \left(\frac{\partial(\beta P)}{\partial\rho}\right)_\beta . \tag{2.9.17}$$

In the context of a dilute gas of uncorrelated particles, (2.9.13) tells us that the left-hand side is equal to 1, so that

$$\left(\frac{\partial(\beta P)}{\partial\rho}\right)_\beta = 1 , \tag{2.9.18}$$

implying

$$\beta P = \rho + \text{constant} , \tag{2.9.19}$$

where the constant must vanish to ensure $P = 0$ when $\rho = 0$ (no pressure if there is no gas). With $\beta = \dfrac{1}{k_{\mathrm{B}}T}$ and $\rho = \dfrac{N}{V}$, this is

$$\beta PV = N \quad \text{or} \quad PV = k_{\mathrm{B}}NT = nRT , \tag{2.9.20}$$

the combined gas law, the thermal equation of state of an ideal gas.

This result is rather remarkable because it originates in such a simple and plausible assumption, namely that the density of particles is low and they are not correlated as they are rather far apart. These conditions are met by dilute real gases but also by dilute solutes in a solvent.

In a more general situation than that of a dilute gas, we have the pair of equations

$$\beta PV = \log\bigl(Z(\beta, V, \beta\mu)\bigr) , \quad N = \frac{\partial \log\bigl(Z(\beta, V, \beta\mu)\bigr)}{\partial(\beta\mu)} , \tag{2.9.21}$$

which jointly determine the analog of the ideal-gas equation in state in (2.9.20). As an illustration, consider the ideal-gas example of (2.3.14)–(2.4.7) and Exercise 40, for which the grand canonical partition function is

$$Z(\beta, V, \beta\mu) = \mathrm{e}^{\mathrm{e}^{\beta\mu}(2\pi\hbar^2\beta/m)^{-3/2}V} ; \tag{2.9.22}$$

show this as another exercise. Here we get $N = \log Z$ and then (2.9.20) follows from (2.9.21) immediately.

2.10 Gibbs's entropy formula as a basic statement

We shall say more about the ideal-gas scenario of noninteracting constituents before we turn to the much more complicated situation of interacting constituents, but let us first take some time to explore Gibbs's famous expression for the entropy in (2.7.4),

$$S = -k_{\mathrm{B}} \sum_k p_k \log p_k \,. \tag{2.10.1}$$

It is often taken as a fundamental statement from which most, if not all, of statistical mechanics follows. Let us, therefore, see how the Gibbs relation gives us the basic facts about the standard ensembles, when combined with

$$\langle E \rangle = \sum_k p_k E_k \tag{2.10.2}$$

and

$$\langle N \rangle = \sum_k p_k N_k \,, \tag{2.10.3}$$

and due attention is paid to the normalization of the probabilities p_k to unit sum,

$$\sum_k p_k = 1 \,. \tag{2.10.4}$$

The question is this: How do we have to choose these probabilities to ensure that the entropy is maximal under the constraints that specify the respective ensemble?

The maximum property of the entropy demands that there are no first-order changes in response to small variations of the probabilities p_k,

$$\delta p_k : \quad \mathrm{d}S = -k_{\mathrm{B}} \sum_k \delta p_k \,(\log p_k + 1) = 0 \,. \tag{2.10.5}$$

Thereby, the variations δp_k are constrained by the unit sum of the p_ks,

$$\sum_k \delta p_k = 0 \,. \tag{2.10.6}$$

In the microcanonical ensemble, when $E_k = E$ and $N_k = N$ for all permissible microstates, this is the only constraint and, therefore, we conclude that

$$\log p_k = \log p_0 \,, \tag{2.10.7}$$

all p_ks have the same value. The normalization then implies that

$$\frac{1}{p_0} = (\text{count of permissible microstates}) = \Omega(E, N, V)\,, \qquad (2.10.8)$$

which takes us back to (2.7.5).

In the canonical ensemble, we have $N_k = N$ for all permissible microstates, and the energy constraint

$$\delta\langle E\rangle = \sum_k \delta p_k\, E_k = 0\,. \qquad (2.10.9)$$

It follows that

$$\log p_k = \log p_0 - \beta E_k \qquad (2.10.10)$$

with two Lagrange* multipliers $\log p_0$ and β, whose values are determined by the constraints,

$$1 = \sum_k p_k = p_0 \sum_k \mathrm{e}^{-\beta E_k} \qquad (2.10.11)$$

and

$$\langle E\rangle = \sum_k p_k E_k = -p_0 \frac{\partial}{\partial\beta} \sum_k \mathrm{e}^{-\beta E_k}\,. \qquad (2.10.12)$$

We conclude that

$$p_k = \frac{1}{Q}\,\mathrm{e}^{-\beta E_k} \qquad (2.10.13)$$

with the canonical partition function

$$Q(\beta, V, N) = \sum_k \mathrm{e}^{-\beta E_k}\,, \qquad (2.10.14)$$

where — for given $\langle E\rangle$ — β is such that

$$\langle E\rangle = -\frac{\partial}{\partial\beta} \log Q\,. \qquad (2.10.15)$$

We are back at (2.5.8) and (2.5.10).

Finally, in the grand canonical ensemble, we have the additional constraint that enforces the prechosen value of $\langle N\rangle$,

$$\delta\langle N\rangle = 0: \quad \sum_k \delta p_k\, N_k = 0\,. \qquad (2.10.16)$$

---

*Joseph Louis de LAGRANGE (1736–1813)

Now,

$$\log p_k = \log p_0 - \beta E_k + \xi N_k \tag{2.10.17}$$

has three Lagrange multipliers, one for each constraint, namely $\log p_0$, β, and ξ, with their values determined by the constraints,

$$\begin{aligned}
1 &= \sum_k p_k = p_0 \sum_k \mathrm{e}^{-\beta E_k + \xi N_k}\,, \\
\langle E \rangle &= \sum_k p_k E_k = -p_0 \frac{\partial}{\partial \beta} \sum_k \mathrm{e}^{-\beta E_k + \xi N_k}\,, \\
\langle N \rangle &= \sum_k p_k N_k = p_0 \frac{\partial}{\partial \xi} \sum_k \mathrm{e}^{-\beta E_k + \xi N_k}\,.
\end{aligned} \tag{2.10.18}$$

Accordingly, we have

$$p_k = \frac{1}{Z}\,\mathrm{e}^{-\beta E_k + \xi N_k} \tag{2.10.19}$$

with the grand canonical partition function

$$Z(\beta, V, \xi) = \sum_k \mathrm{e}^{-\beta E_k + \xi N_k} \tag{2.10.20}$$

of (2.8.1), here writing ξ for $\beta\mu$, and

$$\begin{aligned}
\langle E \rangle &= -\frac{\partial}{\partial \beta} \log\bigl(Z(\beta, V, \xi)\bigr)\,, \\
\langle N \rangle &= \frac{\partial}{\partial \xi} \log\bigl(Z(\beta, V, \xi)\bigr)\,;
\end{aligned} \tag{2.10.21}$$

the latter repeats (2.9.1).

From the earlier discussion we know that the respective partition function contains all information about the thermodynamical system. Therefore, the Gibbs entropy expression in (2.10.1) is a valid starting point for statistical mechanics, when combined with the maximum principle for the entropy and the constraints that specify the various ensembles.

Chapter 3

Ideal gases

3.1 Getting started

We now consider systems of noninteracting particles — ideal gases. The energy is then a sum of single-particle energies

$$E_k = E^{(1)}_{k_1} + E^{(2)}_{k_2} + \cdots + E^{(N)}_{k_N} = \sum_{j=1}^{N} E^{(j)}_{k_j}\,, \tag{3.1.1}$$

where $k = (k_1, k_2, \ldots, k_N)$ labels the N-particle microstate, and $E^{(j)}_{k_j}$ is the energy of the jth particle in its k_jth state. The canonical partition function

$$Q(\beta, V, N) = \sum_k \mathrm{e}^{-\beta E_k} = \sum_{k_1}\sum_{k_2}\cdots\sum_{k_N} \mathrm{e}^{-\beta\sum_{j=1}^{N} E^{(j)}_{k_j}} \tag{3.1.2}$$

then factorizes,

$$Q(\beta, V, N) = \prod_{j=1}^{N}\sum_{k_j} \mathrm{e}^{-\beta E^{(j)}_{k_j}} = \prod_{j=1}^{N} Q^{(j)}(\beta, V)\,. \tag{3.1.3}$$

Once more, we note that the volume dependence of the partition function originates in the V dependence of the energies $E^{(j)}_{k_j}$.

If all particles are of the same kind, the product in (3.1.3) has the same single-particle factor N times,

$$Q(\beta, V, N) = \frac{1}{N!} q(\beta, V)^N \tag{3.1.4}$$

with

$$q(\beta, V) = \sum_k \mathrm{e}^{-\beta\varepsilon_k} \tag{3.1.5}$$

for the single-particle energies $\varepsilon_k = E_k^{(1)} = E_k^{(2)} = \cdots = E_k^{(N)}$. We include a factor $\dfrac{1}{N!}$ in $Q(\beta, V, N)$ for the correct Boltzmann counting in order to avoid the massive over-counting of microstates that we encountered in (2.3.19).

What about the partition function of the grand canonical ensemble? For that, we have

$$\begin{aligned} Z(\beta, V, \beta\mu) &= \sum_k \mathrm{e}^{-\beta E_k + \beta\mu N_k} = \sum_{N=0}^{\infty} \sum_k \mathrm{e}^{-\beta E_k + \beta\mu N_k} \delta_{N, N_k} \\ &= \sum_{N=0}^{\infty} \mathrm{e}^{\beta\mu N} \underbrace{\sum_k \mathrm{e}^{-\beta E_k} \delta_{N, N_k}}_{= Q(\beta, V, N)} = \sum_{N=0}^{\infty} \mathrm{e}^{\beta\mu N} Q(\beta, V, N) , \end{aligned} \tag{3.1.6}$$

where the Kronecker* delta symbol enforces the count of particles, and we recognize that the summation over k with a fixed value of N_k yields the canonical partition function. For a system of identical particles, this gives

$$Z(\beta, V, \beta\mu) = \sum_{N=0}^{\infty} \mathrm{e}^{\beta\mu N} \frac{q(\beta, V)^N}{N!} = \mathrm{e}^{\mathrm{e}^{\beta\mu} q(\beta, V)} . \tag{3.1.7}$$

As a consequence of the relations in (2.9.21), we here obtain

$$\langle N \rangle = \left(\frac{\partial}{\partial(\beta\mu)} \log Z \right)_{\beta, V} = \log Z = \beta P V \tag{3.1.8}$$

and conclude that the combined gas law $PV = N k_{\mathrm{B}} T = nRT$ holds (with, as usual, $\langle N \rangle \to N$), as it should for an ideal gas.

All of this is alright for sufficiently dilute gases at high enough temperatures when quantum effects can be ignored. A more precise treatment, however, requires that quantum aspects are properly taken into account.

3.2 Occupation numbers; bosons; fermions

A microstate has definite energy and, therefore, a quantum microstate is an eigenstate of the Hamilton operator and the eigenvalue is the energy of the microstate. For noninteracting particles, the Hamilton operator is a sum of single-particle contributions, N times the same term (with different labels) if the particles are all of the same kind. The total energy is then also a sum

*Leopold KRONECKER (1823–1891)

of single-particle energies. This much is exactly as we have it for classical particles. What is different, however, is the way we characterize states and count them: We use *occupation numbers* for this purpose. For instance, suppose that we have four particles and several single-particle states for the single-particle Hamilton operator, the ground state with energy ε_0, the first excited state with energy ε_1, the second excited state with energy ε_2, and so forth. Then, indicating the occupied states with a fat dot, here is a possible microstate:

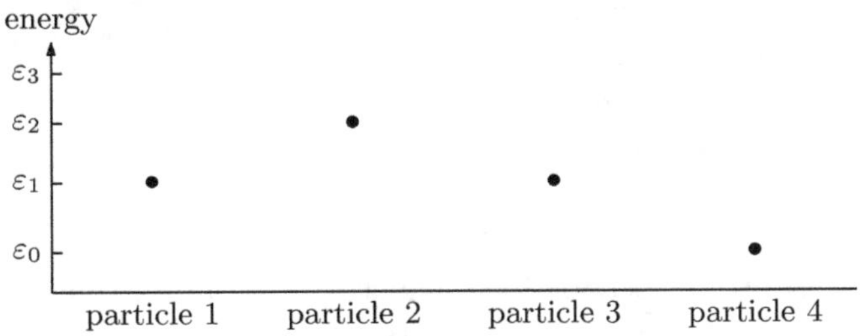

(3.2.1)

Here is another:

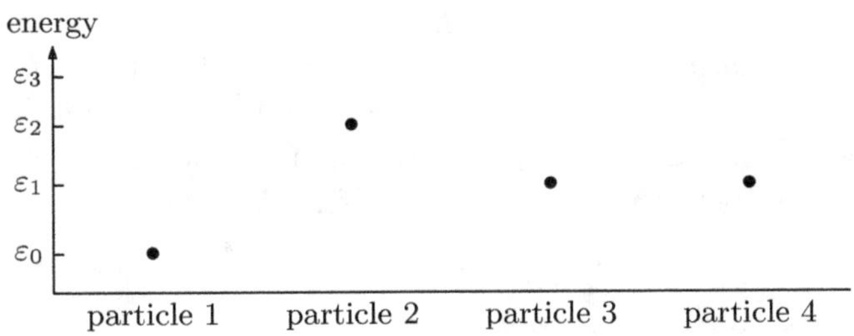

(3.2.2)

Both have the same energy $E = \varepsilon_0 + 2\varepsilon_1 + \varepsilon_2$ and there are ten more configurations with the same energy $(4!/2! = 12)$, but these are not twelve different states, it is the same state twelve times. This is so because the identical particles are fundamentally indistinguishable and, therefore, the labels "1", "2", "3", "4" for the particles have no physical significance. They only serve the purpose of facilitating the mathematical description.

It is therefore fitting to focus on what actually characterizes this four-particle state: There is one particle in the ground state, two in the first excited state, and one in the second excited state. Which particle is in which state? This is a meaningless question because of that fundamental indistinguishability. We thus *completely characterize* the state by specifying how many (not which) particles are in the various single-particle states, and we do this by a list of occupation numbers, such as

$$(n_0, n_1, n_2, n_3, n_4, \dots) = (1, 2, 1, 0, 0, 0, \dots) \tag{3.2.3}$$

for the example in (3.2.1) and (3.2.2), which is then more appropriately depicted in this way:

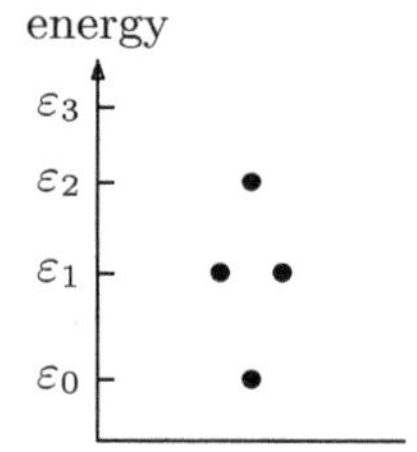

(3.2.4)

When labeling the states by occupation numbers, we ensure that each state is counted once, and only once. The correct Boltzmann counting is thereby enforced systematically.

As we recall from the courses on quantum mechanics, atomic particles are either *bosons* or *fermions*. For bosons, there is no restriction on the occupation numbers, each of them can take any integer value. For fermions, only the values $n_j = 0$ or $n_j = 1$ are allowed. At high densities or low temperatures, when the quantum aspects are important, we should expect noticeable differences between boson gases and fermion gases.

3.3 Average occupation numbers

3.3.1 *Bosons*

Let us now take a first brief look at bosonic particles, such as most neutral atoms (equal numbers of electrons and protons plus even number of neutrons give an even number of spin-$\frac{1}{2}$ constituents and thus an overall spin that is an integer, not half of an odd integer). The canonical partition function is

$$\begin{aligned} Q &= \sum_{\{n\}} \mathrm{e}^{-\beta \sum_j \varepsilon_j n_j} \delta_{N, \sum_j n_j} \\ &= \sum_{n_1=0}^{\infty} \sum_{n_2=0}^{\infty} \cdots \sum_{n_j=0}^{\infty} \cdots \mathrm{e}^{-\beta \sum_j n_j \varepsilon_j} \delta_{N, \sum_j n_j}\,, \end{aligned} \tag{3.3.1}$$

where, as it did in (3.1.6), the Kronecker delta symbol enforces the correct count of particles, namely n_1 with energy ε_1, n_2 with energy ε_2, ..., n_j with energy ε_j, for a total of

$$N = \sum_j n_j \tag{3.3.2}$$

particles. This constraint makes the evaluation of Q a bit awkward and, therefore, we turn to the grand canonical ensemble with the partition function,

$$\begin{aligned} Z &= \sum_{n_1=0}^{\infty} \cdots \sum_{n_j=0}^{\infty} \cdots \mathrm{e}^{-\beta \sum_j n_j(\varepsilon_j - \mu)} \\ &= \prod_j \sum_{n_j=0}^{\infty} \mathrm{e}^{-n_j\beta(\varepsilon_j - \mu)} \\ &= \prod_j \frac{1}{1 - \mathrm{e}^{-\beta(\varepsilon_j - \mu)}}, \end{aligned} \tag{3.3.3}$$

after recognizing that the sum over n_j is a geometric series. Upon recalling from (2.8.5) that $Z = \mathrm{e}^{\beta PV}$, we thus have

$$\beta PV = \log Z = -\sum_j \log\left(1 - \mathrm{e}^{-\beta(\varepsilon_j - \mu)}\right) \tag{3.3.4}$$

for the ideal boson gas.

The average occupation number $\langle n_j \rangle$, that is: the average number of atoms with single-particle energy ε_j, is

$$\begin{aligned} \langle n_j \rangle &= -\frac{1}{Z}\frac{\partial}{\partial(\beta\varepsilon_j)} Z = -\frac{\partial}{\partial(\beta\varepsilon_j)} \log Z \\ &\underset{\text{other terms drop out}}{=} \frac{\partial}{\partial(\beta\varepsilon_j)} \log\left(1 - \mathrm{e}^{-\beta(\varepsilon_j - \mu)}\right) \\ &= \frac{1}{1 - \mathrm{e}^{-\beta(\varepsilon_j - \mu)}} \mathrm{e}^{-\beta(\varepsilon_j - \mu)} \\ &= \frac{1}{\mathrm{e}^{\beta(\varepsilon_j - \mu)} - 1}. \end{aligned} \tag{3.3.5}$$

Since $\langle n_j \rangle$ is an average of nonnegative integers, it has to be nonnegative itself. Therefore, the chemical potential μ must be less than the ground-state energy $\varepsilon_0 = \mathrm{Min}_j \{\varepsilon_j\}$. In the limit $\varepsilon_0 > \mu \to \varepsilon_0$, the ground-state occupation $\langle n_0 \rangle$ becomes singular. Physically speaking, this means that there is a macroscopic, rather than microscopic, contribution of the $j = 0$ term to the grand canonical partition function and all other sums over j derived from it. While all other energy levels have occupation numbers of the order of one or a few atoms, there is a fraction of N atoms in the $j = 0$ ground state. This is the phenomenon of *Bose*–Einstein† condensation*, more historically precisely: Einstein condensation of a Bose gas. We leave it at that right now and shall return to this matter in Section 3.6.

3.3.2 *Fermions*

In the case of fermions, the occupation numbers n_j can only take on the values 0 and 1, so that the analog of (3.3.3) reads

$$\begin{aligned} Z = \mathrm{e}^{\beta PV} &= \sum_{n_1=0}^{1} \sum_{n_2=0}^{1} \cdots \sum_{n_j=0}^{1} \cdots \mathrm{e}^{-\beta \sum_j n_j(\varepsilon_j - \mu)} \\ &= \prod_j \sum_{n_j=0}^{1} \mathrm{e}^{-n_j \beta(\varepsilon_j - \mu)} \\ &= \prod_j \left(1 + \mathrm{e}^{-\beta(\varepsilon_j - \mu)}\right), \end{aligned} \tag{3.3.6}$$

and the logarithm of the grand canonical partition function is

$$\beta PV = \log Z = \sum_j \log\left(1 + \mathrm{e}^{-\beta(\varepsilon_j - \mu)}\right). \tag{3.3.7}$$

The average occupation number is

$$\begin{aligned} \langle n_j \rangle &= -\frac{\partial}{\partial(\beta\varepsilon_j)} \log Z \\ &= -\frac{\partial}{\partial(\beta\varepsilon_j)} \log\left(1 + \mathrm{e}^{-\beta(\varepsilon_j - \mu)}\right) \\ &= \frac{1}{\mathrm{e}^{\beta(\varepsilon_j - \mu)} + 1}. \end{aligned} \tag{3.3.8}$$

In contrast to the analogous boson expression in (3.3.5), we always have $0 < \langle n_j \rangle < 1$ here; the fermion expression is regular for all values of the chemical potential μ.

---

*Satyendranath BOSE (1894–1974) †Albert EINSTEIN (1879–1955)

3.3.3 *Dilute-gas limit*

For bosons and fermions, we have

$$\langle N\rangle = \sum_j \langle n_j\rangle = \sum_j \frac{1}{\mathrm{e}^{\beta(\varepsilon_j - \mu)} \mp 1}, \tag{3.3.9}$$

with the upper sign for bosons and the lower sign for fermions. Now, when the temperature is high and the particle density $\frac{1}{V}\langle N\rangle$ is very low, there are many more single-particle states available than there are particles, which is the situation of a dilute gas that we considered in the context of (2.9.3). Accordingly, each $\langle n_j\rangle$ must be tiny, which in turn implies that

$$\mathrm{e}^{\beta(\varepsilon_j - \mu)} \gg 1, \tag{3.3.10}$$

so that

$$\langle n_j\rangle = \mathrm{e}^{-\beta(\varepsilon_j - \mu)} \tag{3.3.11}$$

under these circumstances, both for bosons and for fermions. From

$$\langle N\rangle = \sum_j \langle n_j\rangle = \mathrm{e}^{\beta\mu} \sum_j \mathrm{e}^{-\beta\varepsilon_j}, \tag{3.3.12}$$

we then get

$$\frac{\langle n_j\rangle}{\langle N\rangle} = \frac{\mathrm{e}^{-\beta\varepsilon_j}}{\sum_{j'} \mathrm{e}^{-\beta\varepsilon_{j'}}}, \tag{3.3.13}$$

the classical Maxwell–Boltzmann factor expression for the probability that the jth state is occupied.

As way of summary, we note that

$$\langle n_j\rangle = \frac{1}{\mathrm{e}^{\beta(\varepsilon_j - \mu)} + \zeta} \quad \text{with} \begin{cases} \zeta = -1 & \text{for bosons (BE)}, \\ \zeta = +1 & \text{for fermions (FD)}, \\ \zeta = 0 & \text{for classical particles (MB)}. \end{cases} \tag{3.3.14}$$

This unifies the expressions for Bose–Einstein statistics in (3.3.5), Fermi*–Dirac† statistics in (3.3.8), and Maxwell–Boltzmann statistics in (3.3.11) — BE, FD, MB, respectively.

For the canonical and the grand canonical ensembles we have

$$-\beta F = \log Q, \quad \beta PV = \log Z \tag{3.3.15}$$

---

*Enrico FERMI (1901–1954) †Paul Adrien Maurice DIRAC (1902–1984)

and we recall that ($\langle E\rangle = U$)

$$F = \langle E\rangle - TS \quad \text{and} \quad G = \mu\langle N\rangle = \langle E\rangle - TS + PV\,. \tag{3.3.16}$$

Therefore,

$$\log Q = -\beta\mu\langle N\rangle + \beta PV = -\beta\mu\langle N\rangle + \log Z\,, \tag{3.3.17}$$

or

$$\log Q = -\beta\mu\langle N\rangle \mp \sum_j \log\left(1 \mp \mathrm{e}^{-\beta(\varepsilon_j - \mu)}\right) \tag{3.3.18}$$

after making use of the expression for $\log Z$ for the BE and FD statistics in (3.3.4) and (3.3.7), respectively. For $\mathrm{e}^{-\beta(\varepsilon_j - \mu)} \ll 1$, this is

$$\log Q = -\beta\mu\langle N\rangle + \sum_j \mathrm{e}^{-\beta(\varepsilon_j - \mu)}\,, \tag{3.3.19}$$

where we now use (3.3.12) twice,

$$\sum_j \mathrm{e}^{-\beta(\varepsilon_j - \mu)} = \langle N\rangle \quad \text{and} \quad \beta\mu = \log(\langle N\rangle) - \log\left(\sum_j \mathrm{e}^{-\beta\varepsilon_j}\right), \tag{3.3.20}$$

and get

$$\begin{aligned}\log Q &= -\langle N\rangle\log(\langle N\rangle) + \langle N\rangle\log\sum_j \mathrm{e}^{-\beta\varepsilon_j} + \langle N\rangle \\ &= \langle N\rangle\log\left(\sum_j \mathrm{e}^{-\beta\varepsilon_j}\right) - \log(\langle N\rangle!)\,,\end{aligned} \tag{3.3.21}$$

the last step employing Stirling's approximation in (2.3.17) for the logarithm of the factorial of a large number, here: $\langle N\rangle$. Upon writing N for $\langle N\rangle$, we find

$$Q = \frac{1}{N!}\left(\sum_j \mathrm{e}^{-\beta\varepsilon_j}\right)^N, \tag{3.3.22}$$

which is exactly the expression in (3.1.4) with the $\dfrac{1}{N!}$ factor of the correct Boltzmann counting. As remarked in Section 2.4, this factor was guessed by Boltzmann and Gibbs as demanded by consistency, although they could not justify it by a more basic argument. As we see now, it is ultimately a consequence of the quantum nature of the constituent particles and their fundamental and absolute indistinguishability.

3.4 Photon gas

A particular case of a boson gas is that of the *photon gas*. Photons have negligible interactions among themselves (yes, there is photon-photon scattering, but with such a small cross section that it is utterly irrelevant unless there are a lot of photons of very large energy — circumstances of no concern to us) and they are easily emitted and absorbed by the walls that confine the photon gas, so that the occupation numbers are not restricted. We assign energy $\hbar\omega_j$ to each photon of the jth kind, and so have the canonical partition function

$$Q = \sum_{\{n\}} \mathrm{e}^{-\beta \sum_j \hbar\omega_j n_j} , \tag{3.4.1}$$

where $\{n\} = \{n_1, n_2, \ldots, n_j, \ldots\}$ is the collection of occupation numbers, each taking on the values $0, 1, 2, 3, 4, \ldots$. Since the photon number has no fixed value, we do not have the Kronecker-delta factor of (3.3.1) here. Accordingly, we get

$$\begin{aligned} Q &= \prod_j \sum_{n_j=0}^{\infty} \mathrm{e}^{-\beta\hbar\omega_j n_j} = \prod_j \frac{1}{1-\mathrm{e}^{-\beta\hbar\omega_j}} , \\ \log Q &= -\sum_j \log\left(1-\mathrm{e}^{-\beta\hbar\omega_j}\right) . \end{aligned} \tag{3.4.2}$$

Essentially, this canonical partition function is the grand canonical partition function of (3.3.3) and (3.3.4) for $\varepsilon_j = \hbar\omega_j$ and $\mu = 0$.

An important, and immediate, consequence is Planck's result for the mean occupation number of the jth mode, or the jth kind of photon. It is

$$\begin{aligned} \langle n_j \rangle &= \frac{1}{Q} \sum_{\{n\}} n_j \, \mathrm{e}^{-\beta \sum_{j'} \hbar\omega_{j'} n_{j'}} = -\frac{\partial}{\partial(\beta\hbar\omega_j)} \log Q \\ &= \frac{\partial}{\partial(\beta\hbar\omega_j)} \log\left(1-\mathrm{e}^{-\beta\hbar\omega_j}\right) \end{aligned} \tag{3.4.3}$$

or

$$\langle n_j \rangle = \frac{1}{\mathrm{e}^{\beta\hbar\omega_j}-1} , \tag{3.4.4}$$

the so-called Planck distribution — the boson result of (3.3.5) for $\varepsilon_j = \hbar\omega_j$ and $\mu = 0$.

To proceed further, we need information about the possible energy values $\hbar\omega_j$ for the photon modes. For this purpose, we consider a volume of the shape of a cube of linear extension $V^{\frac{1}{3}} = L$, which we regard as very large on the scale set by the relevant photon wavelengths. It is then permissible to associate a plane wave

$$\boldsymbol{E}(\boldsymbol{r},t) = \mathrm{Re}\left(\boldsymbol{\epsilon}\,\mathrm{e}^{\mathrm{i}(\boldsymbol{k}\cdot\boldsymbol{r}-\omega t)}\right) \tag{3.4.5}$$

with a photon mode where $c|\boldsymbol{k}| = \omega$ relates the wave vector $\boldsymbol{k}$ to the (circular) frequency ω of the electromagnetic wave that propagates at the speed of light c, and $\boldsymbol{\epsilon}$ is the polarization vector. Since the electromagnetic waves are transverse ($\boldsymbol{k}\cdot\boldsymbol{\epsilon} = 0$), there are two possible polarizations for each $\boldsymbol{k}$. We thus specify a photon mode by its wave vector and its polarization.

Owing to the boundary conditions imposed by the walls that enclose the volume V, the possible wave vectors have cartesian* coordinates

$$\boldsymbol{k} \mathrel{\widehat{=}} \frac{2\pi}{L}\begin{pmatrix} k_1 \\ k_2 \\ k_3 \end{pmatrix} \tag{3.4.6}$$

with each k_j taking on the integers $0, \pm1, \pm2, \pm3, \ldots$, excluding $k_1 = k_2 = k_3 = 0$ because there are no zero-frequency photons, but this exclusion is of no further concern here. In the $\boldsymbol{k}$ space, then, we count one state per volume $\left(\frac{2\pi}{L}\right)^3 = \frac{(2\pi)^3}{V}$, as the possible $\boldsymbol{k}$s form a cubic lattice with cells of this size. In the spherical shell specified by $\omega < c|\boldsymbol{k}| < \omega + \mathrm{d}\omega$, we have as many as

$$\underset{\text{polarization}}{2} \times \underbrace{4\pi\left(\frac{\omega}{c}\right)^2\frac{\mathrm{d}\omega}{c}}_{\text{volume of the shell}} \Big/ \underbrace{\frac{(2\pi)^3}{V}}_{\text{volume occupied by one } \boldsymbol{k}} \tag{3.4.7}$$

photon modes; after simplifying, this is

$$\frac{V}{\pi^2c^3}\omega^2\mathrm{d}\omega\,. \tag{3.4.8}$$

Now, there are $\langle n_j\rangle = \left(\mathrm{e}^{\beta\hbar\omega_j} - 1\right)^{-1}$ photons in the jth mode, each contributing energy $\hbar\omega_j$, and we have $\mathrm{d}\omega\,\omega^2V/(\pi^2c^3)$ modes in the range $\omega\cdots\omega+\mathrm{d}\omega$, so that the internal energy of the photon gas is

$$U = \langle E\rangle = \frac{V}{\pi^2c^3}\int_0^\infty \mathrm{d}\omega\,\omega^2\frac{\hbar\omega}{\mathrm{e}^{\beta\hbar\omega}-1}\,, \tag{3.4.9}$$

*René DESCARTES (1596–1650)

where the sum over the mode index j is realized by the integration over ω. The energy density is

$$\frac{U}{V} = u(T) = \frac{1}{\pi^2 c^3} \frac{1}{\hbar^3 \beta^4} \underbrace{\int_0^\infty \mathrm{d}x \, \frac{x^3}{\mathrm{e}^x - 1}}_{=\pi^4/15} \tag{3.4.10}$$

or

$$u(T) = \frac{\pi^2}{15} \frac{(k_\mathrm{B} T)^4}{(\hbar c)^3} , \tag{3.4.11}$$

where we note the T^4 dependence. This is the *Stefan*–Boltzmann law*.

As a consequence, the heat capacity per volume is proportional to T^3,

$$\frac{1}{V} C_V = \frac{4\pi^2}{15} k_\mathrm{B} \left(\frac{k_\mathrm{B} T}{\hbar c} \right)^3 . \tag{3.4.12}$$

It can be so large for high temperatures because there is no limit to the number of photons we can put inside the volume.

The pressure exerted by the photon gas is available as the negative volume-derivative of the free energy $F = -\dfrac{1}{\beta} \log Q$,

$$P = \frac{1}{\beta} \left(\frac{\partial}{\partial V} \log Q \right)_\beta = -\frac{1}{\beta} \frac{\partial}{\partial V} \sum_j \log \left(1 - \mathrm{e}^{-\beta \hbar \omega_j} \right) , \tag{3.4.13}$$

where $\omega_j = c|\boldsymbol{k}| \propto V^{-\frac{1}{3}}$ since $\boldsymbol{k} \propto L^{-1} = V^{-\frac{1}{3}}$. Therefore, we have

$$\frac{\partial}{\partial V} \omega_j = -\frac{1}{3V} \omega_j \tag{3.4.14}$$

and

$$\begin{aligned} -\frac{1}{\beta} \frac{\partial}{\partial V} \log \left(1 - \mathrm{e}^{-\beta \hbar \omega_j} \right) &= \frac{1}{1 - \mathrm{e}^{-\beta \hbar \omega_j}} \frac{1}{\beta} \frac{\partial}{\partial V} \mathrm{e}^{-\beta \hbar \omega_j} \\ &= \frac{\mathrm{e}^{-\beta \hbar \omega_j}}{1 - \mathrm{e}^{-\beta \hbar \omega_j}} \frac{1}{3V} \hbar \omega_j = \frac{1}{3V} \langle n_j \rangle \hbar \omega_j , \end{aligned} \tag{3.4.15}$$

so that

$$P = \frac{1}{3V} \sum_j \langle n_j \rangle \hbar \omega_j = \frac{1}{3V} \langle E \rangle = \frac{1}{3} u . \tag{3.4.16}$$

The pressure is one-third of the energy density.

*Josef STEFAN (1835–1893)

We can understand this also from purely classical, that is: nonquantum, considerations. We recall that the energy density of the electromagnetic field is

$$u = \frac{1}{8\pi}\left(\boldsymbol{E}^2 + \boldsymbol{B}^2\right) \tag{3.4.17}$$

and the momentum current density dyadic is

$$\mathsf{T} = u\mathbf{1} - \frac{1}{4\pi}(\boldsymbol{EE} + \boldsymbol{BB})\,. \tag{3.4.18}$$

In the case of an isotropic radiation field, the momentum current density dyadic is a multiple of the unit dyadic, namely $\mathsf{T} = P\mathbf{1}$ with the pressure P, and we have $\mathrm{tr}\{\mathsf{T}\} = 3P$. On the other hand, there is the general fact that $\mathrm{tr}\{\mathsf{T}\} = u$, and $P = \frac{1}{3}u$ follows, perfectly consistent with (3.4.16).

It is, of course, also consistent with thermodynamics. For instance, we note that

$$\left(\frac{\partial U}{\partial V}\right)_T = T\left(\frac{\partial S}{\partial V}\right)_T - P = T\left(\frac{\partial P}{\partial T}\right)_V - P\,, \tag{3.4.19}$$

where the Maxwell relation in (1.9.5) has been used. Now, $U = uV = 3PV$ here, and $P = \frac{1}{3}u \propto T^4$ does not depend on V. Therefore we have

$$\left(\frac{\partial U}{\partial V}\right)_T = 3P \quad \text{and} \quad T\left(\frac{\partial P}{\partial T}\right)_V - P = 4P - P = 3P \tag{3.4.20}$$

for the expressions in (3.4.19), and so confirm that (3.4.19) is obeyed.

3.5 Phonon gas

The phonons of a solid are similar to photons inasmuch as they are also excitations of harmonic oscillators, but in the phonon case this is a low-excitation approximation. Consider the lattice of atoms that form a solid, with N atoms and coordinates q_j ($j = 1, 2, \ldots, 3N$) for the displacements from the equilibrium positions. To second order in these coordinates, we then have the potential energy

$$V(q) = V_0 + \frac{1}{2}\sum_{j,j'=1}^{3N} q_j \left(\frac{\partial^2 V}{\partial q_j \partial q_{j'}}\right)_{q=0} q_{j'} \tag{3.5.1}$$

and the kinetic energy $\frac{1}{2}\sum_{j,j'} \dot{q}_j m_{jj'} \dot{q}_{j'}$. The mass matrix with matrix elements $m_{jj'}$ and the matrix of second derivatives of $V(q)$ are symmetric and positive . Therefore, they can be diagonalized simultaneously and the new coordinates $\xi = (\xi_1, \xi_2, \ldots, \xi_{3N})$ parameterize the normal modes, for which the energy is

$$E = V_0 + \sum_{j=1}^{3N} \frac{1}{2} \overline{m}_j (\dot{\xi}_j^2 + \omega_j^2 \xi_j^2) \tag{3.5.2}$$

with $\overline{m}_j$ the effective mass of the jth normal mode and ω_j its circular frequency. In this low-excitation approximation, then, we have a collection of harmonic oscillators and their excitations are the phonons.

In the quantum-mechanical version, a state with n_j phonons of the jth kind has the excitation energy

$$E_{\{n\}} = \sum_{j=1}^{3N} \hbar\omega_j n_j \,. \tag{3.5.3}$$

It follows that

$$\begin{aligned} Q = \sum_{\{n\}} \mathrm{e}^{-\beta E_{\{n\}}} &= \sum_{n_1=0}^{\infty} \sum_{n_2=0}^{\infty} \cdots \sum_{n_{3N}=0}^{\infty} \mathrm{e}^{-\beta \sum_j \hbar\omega_j n_j} \\ &= \prod_{j=1}^{3N} \sum_{n_j=0}^{\infty} \mathrm{e}^{-\beta\hbar\omega_j n_j} \end{aligned} \tag{3.5.4}$$

or

$$Q = \prod_{j=1}^{3N} \frac{1}{1 - \mathrm{e}^{-\beta\hbar\omega_j}} \tag{3.5.5}$$

is the canonical partition function. At this point, the main difference between the photon gas of Section 3.4 and the current phonon gas is the finite number of phonon modes, whereas there are infinitely many photon modes.

Another difference is the density of modes that we need when converting sums over j into integrals over ω, as in

$$\beta F = -\log Q = \sum_{j=1}^{3N} \log(1 - \mathrm{e}^{-\beta\hbar\omega_j}) \to \int_0^{\infty} \mathrm{d}\omega \, g(\omega) \log(1 - \mathrm{e}^{-\beta\hbar\omega}) \,, \tag{3.5.6}$$

where $\mathrm{d}\omega\, g(\omega)$ is the count of phonon modes in the range $\omega \cdots \omega + \mathrm{d}\omega$. For the photon gas, we had $g(\omega) \propto \omega^2$ in (3.4.8) and this suggest the use of the so-called *Debye* model*, specified by

$$g(\omega) = \frac{9N}{\omega_{\mathrm{D}}^3}\omega^2 \eta(\omega_{\mathrm{D}} - \omega) = \left\{ \begin{array}{ll} \dfrac{9N\omega^2}{\omega_{\mathrm{D}}^3} & \text{for } \omega < \omega_{\mathrm{D}} \\ 0 & \text{for } \omega > \omega_{\mathrm{D}} \end{array} \right\}, \tag{3.5.7}$$

where $\eta(\ )$ is Heaviside's† unit step function and ω_{D} is the *Debye cut-off frequency.* This mode density $g(\omega)$ is properly normalized,

$$\int_0^\infty \mathrm{d}\omega\, g(\omega) = 3N\,, \tag{3.5.8}$$

and catches some bulk features of solids quite well. Any more detailed description of a particular solid requires more precise knowledge of the phonon mode density $g(\omega)$, which knowledge is provided by the methods of solid-state physics.

The cut-off frequency in the Debye model recognizes that a solid cannot support oscillations with wavelengths shorter than the distance between atoms. Let us see how this works out. We modify the photon expression of (3.4.8) and write

$$g(\omega) = \frac{3}{2}\frac{V}{\pi^2 c^3}\omega^2 \quad \text{for } \omega < \omega_{\mathrm{D}}\,, \tag{3.5.9}$$

where c is now the speed of sound and the factor $\frac{3}{2}$ accounts for the polarizations of sound waves (two transverse, one longitudinal) as compared with two polarizations for light (both transverse). The comparison with Debye's expression gives

$$\frac{9N}{\omega_{\mathrm{D}}^3} = \frac{3}{2}\frac{V}{\pi^2 c^3} \tag{3.5.10}$$

or

$$\lambda_{\mathrm{D}} = \frac{2\pi c}{\omega_{\mathrm{D}}} = 2\pi\left(\frac{\frac{V}{N}}{6\pi^2}\right)^{\frac{1}{3}} = \left(\frac{4\pi}{3}\frac{V}{N}\right)^{\frac{1}{3}} \tag{3.5.11}$$

for the Debye wavelength λ_{D}. Since $\dfrac{V}{N}$ = (volume per atom) $\cong$ (distance between atoms)^3, this says that, indeed, the shortest wavelength is of the size of the inter-atom spacing.

---

*Peter Joseph William DEBYE (1884–1966) †Oliver HEAVISIDE (1850–1925)

Now, the average excitation for a phonon mode is given by the same expression as that for a photon mode in (3.4.4),

$$\langle n_j \rangle = \frac{1}{e^{\beta\hbar\omega_j} - 1}, \tag{3.5.12}$$

so that we get

$$\begin{aligned} U = \sum_{j=1}^{3N} \langle n_j \rangle \hbar\omega_j &= \int_0^\infty d\omega\, g(\omega) \frac{\hbar\omega}{e^{\beta\hbar\omega} - 1} \\ &= \frac{9N}{\omega_D^3} \int_0^{\omega_D} d\omega\, \omega^2 \frac{\hbar\omega}{e^{\beta\hbar\omega} - 1} \end{aligned} \tag{3.5.13}$$

for the internal energy. Upon substituting $x = \beta\hbar\omega$ in the integral and introducing the Debye temperature T_D in accordance with

$$\hbar\omega_D = k_B T_D, \tag{3.5.14}$$

we have

$$U = 9Nk_B \frac{T^4}{T_D^3} \int_0^{T_D/T} dx\, \frac{x^3}{e^x - 1}. \tag{3.5.15}$$

For low temperatures, $T \ll T_D$, the upper limit of the integration is so large that we can replace it by ∞ and so get $\frac{\pi^4}{15}$ for this integral, as we did in (3.4.10). Then

$$T \ll T_D: \qquad U = \frac{3\pi^4}{5} N k_B \frac{T^4}{T_D^3}. \tag{3.5.16}$$

For high temperatures, $T \gg T_D$, all x values are small and the replacement

$$\frac{x^3}{e^x - 1} \cong \frac{x^3}{x + \frac{1}{2}x^2 + \frac{1}{6}x^3} \cong x^2 - \frac{1}{2}x^3 + \frac{1}{12}x^4 \tag{3.5.17}$$

is permissible, so that

$$\int_0^{T_D/T} dx\, \frac{x^3}{e^x - 1} \cong \frac{1}{3}\left(\frac{T_D}{T}\right)^3 - \frac{1}{8}\left(\frac{T_D}{T}\right)^4 + \frac{1}{60}\left(\frac{T_D}{T}\right)^5 \tag{3.5.18}$$

and

$$T \gg T_{\mathrm{D}}: \qquad U = 3Nk_{\mathrm{B}}T\left(1 - \frac{3}{8}\frac{T_{\mathrm{D}}}{T} + \frac{1}{20}\left(\frac{T_{\mathrm{D}}}{T}\right)^2\right) \tag{3.5.19}$$

is the high-temperature approximation.

The Debye temperature depends on the volume (through ω_{D}) and, therefore, the heat capacity for constant volume is

$$C_V = \left(\frac{\partial U}{\partial T}\right)_V = \begin{cases} \dfrac{12\pi^4}{5}Nk_{\mathrm{B}}\left(\dfrac{T}{T_{\mathrm{D}}}\right)^3 & \text{for } T \ll T_{\mathrm{D}}\,, \\ 3Nk_{\mathrm{B}}\left(1 - \dfrac{1}{20}\left(\dfrac{T_{\mathrm{D}}}{T}\right)^2\right) & \text{for } T \gg T_{\mathrm{D}}\,, \end{cases} \tag{3.5.20}$$

or, if we turn attention to the specific heat,

$$\frac{C_V}{Nk_{\mathrm{B}}} = \begin{cases} \dfrac{12\pi^4}{5}\left(\dfrac{T}{T_{\mathrm{D}}}\right)^3 & \text{for } T \ll T_{\mathrm{D}}\,, \\ 3 - \dfrac{3}{20}\left(\dfrac{T_{\mathrm{D}}}{T}\right)^2 & \text{for } T \gg T_{\mathrm{D}}\,. \end{cases} \tag{3.5.21}$$

We observe the same $\propto T^3$ behavior for low temperatures as for the photon gas in (3.4.12), but a very different high-temperature form. This is a consequence of the cut-off frequency in the mode density in (3.5.7).

3.6 Bose–Einstein condensation

As the last topic regarding ideal boson gases, let us now take a closer look at the Bose–Einstein condensation that we mentioned briefly after (3.3.5). The grand canonical partition function for the ideal boson gas is stated in (3.3.4),

$$\beta PV = \log Z = -\sum_j \log\left(1 - \mathrm{e}^{-\beta(\varepsilon_j - \mu)}\right) = -\sum_j \log\left(1 - z\,\mathrm{e}^{-\beta\varepsilon_j}\right), \tag{3.6.1}$$

where we recall that the ε_js are the single-particle energies, and

$$z = \mathrm{e}^{\beta\mu} \tag{3.6.2}$$

is the so-called *fugacity*, a common and convenient short-hand for $\mathrm{e}^{\beta\mu}$. We shall regard the summation over j as the integration over phase-space with the single-particle energy given by the kinetic energy,

$$\varepsilon_j \to \frac{\boldsymbol{p}^2}{2m}\,. \tag{3.6.3}$$

For $\boldsymbol{p}=0$, we have the ground-state energy $\varepsilon_0=0$, for which

$$\mathrm{e}^{-\beta\varepsilon_0} = 1 > \mathrm{e}^{-\beta\varepsilon_j} \quad \text{for} \quad j \neq 0\,. \tag{3.6.4}$$

Accordingly, the fugacity is restricted by

$$0 < z < 1 \tag{3.6.5}$$

because the $j=0$, $\boldsymbol{p}=0$ term in (3.6.1) diverges as $z\to 1$.

In order to facilitate the study of the $z\to 1$ limit for the boson gas, we put the $\boldsymbol{p}=0$ term aside when converting the summation over j into a phase-space integral in accordance with

$$\beta PV = \log Z \to -\int \frac{(\mathrm{d}\boldsymbol{r})\,(\mathrm{d}\boldsymbol{p})}{(2\pi\hbar)^3}\log\left(1 - z\,\mathrm{e}^{-\frac{\beta}{2m}\boldsymbol{p}^2}\right) - \underbrace{\log(1-z)}_{\boldsymbol{p}\,=\,0\ \text{term}}\,. \tag{3.6.6}$$

For integrals of this kind, we notice that $\int(\mathrm{d}\boldsymbol{r}) = V$ is the volume, and we substitute

$$y = \frac{\beta}{2m}\boldsymbol{p}^2\,,\quad p = |\boldsymbol{p}| = \sqrt{\frac{2my}{\beta}}\,,\quad \mathrm{d}y = \frac{\beta}{m}\mathrm{d}p\,p$$

$$\frac{(\mathrm{d}\boldsymbol{p})}{(2\pi\hbar)^3} = \frac{4\pi\,\mathrm{d}p\,p^2}{(2\pi\hbar)^3} = \frac{4\pi}{(2\pi\hbar)^3}\frac{m}{\beta}\mathrm{d}y\sqrt{\frac{2my}{\beta}} = \frac{2}{\sqrt{\pi}}\frac{\mathrm{d}y\,\sqrt{y}}{\lambda^3} \tag{3.6.7}$$

for the integration over momentum, where

$$\lambda = \hbar\sqrt{\frac{2\pi\beta}{m}} \tag{3.6.8}$$

is the so-called *thermal wavelength*, roughly equal to the de Broglie* wavelength associated with kinetic energy $k_{\mathrm{B}}T$. Then

$$\begin{aligned}\beta PV = \log Z &= -\frac{2}{\sqrt{\pi}}\frac{V}{\lambda^3}\int_0^\infty \mathrm{d}y\,y^{\frac{1}{2}}\log\left(1 - z\,\mathrm{e}^{-y}\right) - \log(1-z)\\ &= \frac{V}{\lambda^3}g_{\frac{5}{2}}(z) - \log(1-z)\end{aligned} \tag{3.6.9}$$

*Prince Louis-Victor DE BROGLIE (1892–1987)

with

$$g_{\frac{5}{2}}(z) = -\frac{2}{\sqrt{\pi}} \int_0^\infty \mathrm{d}y \, y^{\frac{1}{2}} \log\left(1 - z\,\mathrm{e}^{-y}\right) = \sum_{k=1}^{\infty} \frac{z^k}{k^{\frac{5}{2}}} \,. \tag{3.6.10}$$

This is a function of the fugacity from a family of such functions, the so-called *boson functions*, defined by

$$g_\alpha(z) = \sum_{k=1}^{\infty} \frac{z^k}{k^\alpha} \quad \text{for} \quad 0 \le z < 1 \,, \tag{3.6.11}$$

which are related to each other by the recurrence relations

$$g_{\alpha-1}(z) = z\frac{\partial}{\partial z} g_\alpha(z) \quad \text{and} \quad g_{\alpha+1}(z) = \int_0^z \mathrm{d}z' \, \frac{g_\alpha(z')}{z'} \,. \tag{3.6.12}$$

The limiting value for $z \to 1$ is finite for $\alpha > 1$,

$$\alpha > 1 : \qquad g_\alpha(1) = \sum_{k=1}^{\infty} \frac{1}{k^\alpha} = \zeta(\alpha) \,, \tag{3.6.13}$$

with the Riemann* zeta function $\zeta(\ )$, but $g_\alpha(z) \to \infty$ as $z \to 1$ for $\alpha \le 1$. In particular, we have the values

$$\begin{aligned} g_{\frac{5}{2}}(1) &= \zeta\left(\frac{5}{2}\right) = 1.341487 \,, \\ g_{\frac{3}{2}}(1) &= \zeta\left(\frac{3}{2}\right) = 2.612375 \,, \end{aligned} \tag{3.6.14}$$

given here to six decimals.

We show the equivalence of the integral and the power series in (3.6.10) with the aid of the Taylor† expansion of the logarithm,

$$-\log(1-x) = x + \frac{1}{2}x^2 + \frac{1}{3}x^3 + \cdots = \sum_{k=1}^{\infty} \frac{x^k}{k} \,, \tag{3.6.15}$$

and Euler's‡ integral for the factorial,

$$\nu! = \int_0^\infty \mathrm{d}t \, t^\nu \, \mathrm{e}^{-t} \quad \text{for} \quad \nu > -1 \,. \tag{3.6.16}$$

*Georg Friedrich Bernhard RIEMANN (1826–1866) †Brook TAYLOR (1685–1731)
‡Leonhard EULER (1707–1783)

Here are the steps:

$$
\begin{aligned}
&-\frac{2}{\sqrt{\pi}}\int_0^\infty \mathrm{d}y\, y^{\frac{1}{2}} \log\left(1 - z\,\mathrm{e}^{-y}\right) \\
&= \frac{2}{\sqrt{\pi}}\int_0^\infty \mathrm{d}y\, y^{\frac{1}{2}} \sum_{k=1}^\infty \frac{1}{k}\left(z\,\mathrm{e}^{-y}\right)^k = \sum_{k=1}^\infty \frac{z^k}{k}\frac{2}{\sqrt{\pi}}\int_0^\infty \mathrm{d}y\, y^{\frac{1}{2}}\,\mathrm{e}^{-ky} \\
&\overset{ky=t}{=} \sum_{k=1}^\infty \frac{z^k}{k^{\frac{5}{2}}}\frac{2}{\sqrt{\pi}}\underbrace{\int_0^\infty \mathrm{d}t\, t^{\frac{1}{2}}\,\mathrm{e}^{-t}}_{=(\frac{1}{2})! = \frac{1}{2}\sqrt{\pi}} = \sum_{k=1}^\infty \frac{z^k}{k^{\frac{5}{2}}}\,.
\end{aligned}
\tag{3.6.17}
$$

Now returning to (3.6.9), we note that the $\boldsymbol{p} = 0$ term carries no weight in the integral over $y \propto \boldsymbol{p}^2$ because the factor $y^{\frac{1}{2}}$ suppresses the vicinity of $y = 0$ and, therefore, the $\boldsymbol{p} = 0$ term is not counted twice in (3.6.9) or in

$$
\langle N\rangle = \left(z\frac{\partial}{\partial z}\log\bigl(Z(\beta, V, z)\bigr)\right)_{\beta,V} = \frac{V}{\lambda^3}g_{\frac{3}{2}}(z) + \frac{z}{1-z}\,; \tag{3.6.18}
$$

together (3.6.9) and (3.6.18) make up the pair of (2.9.21). The term

$$
\frac{z}{1-z} = \left.\frac{z}{\mathrm{e}^{\beta\epsilon} - z}\right|_{\epsilon=0} = \langle N_0\rangle \tag{3.6.19}
$$

is the expected number of bosons in the ground state, and

$$
\langle N\rangle - \langle N_0\rangle = \frac{V}{\lambda^3}g_{\frac{3}{2}}(z) \tag{3.6.20}
$$

is the expected number of bosons in the excited states.

Upon recalling that $0 \le g_{\frac{3}{2}}(z) \le \zeta\left(\frac{3}{2}\right)$, we conclude that the maximal number of bosons in the excited states is

$$
\frac{V}{\lambda^3}\zeta\left(\frac{3}{2}\right) = 2.612\frac{V}{\lambda^3} \tag{3.6.21}
$$

for the given volume V and the given temperature $T = \dfrac{2\pi}{mk_{\mathrm{B}}}\left(\dfrac{\hbar}{\lambda}\right)^2$, and if the total number of bosons exceeds this value, the bosons that cannot be accommodated in the excited states, all occupy the ground state. There is then a finite fraction of all bosons in the ground state, the others being distributed over the excited states, and we have the coexistence of the *condensed phase* (bosons assembled in the ground state) and the *gaseous phase* (bosons distributed over the excited states). This is the phenomenon of *Bose–Einstein condensation.* It is a phase transition of the kind that we discussed in Section 1.16 at the example of the van der Waals gas and its range of parameters (volume range for given temperature) for the coexistence region.

Dropping the $\langle\ \rangle$ brackets for expected values, we have this situation:

if $N < \frac{V}{\lambda^3}\zeta\left(\frac{3}{2}\right)$, the fugacity is given by $\lambda^3\frac{N}{V} = g_{\frac{3}{2}}(z)$ and there is no particular occupancy of the ground state, that is: there are $O(1)$ bosons in the ground state, not $O(N)$;

if $N > \frac{V}{\lambda^3}\zeta\left(\frac{3}{2}\right)$, the fugacity is $z = 1$ and we have $N_0 = N - \frac{V}{\lambda^3}\zeta\left(\frac{3}{2}\right)$ bosons in the ground state; in the limit of $T \to 0$, $\beta \to \infty$, $\lambda \to \infty$, *all* bosons are in the ground state.

Note that, when excited states are available for all bosons, there cannot be an excess of bosons in the ground state because that would reduce the entropy, in contradiction to the maximum property of Section 1.3.

For given particle number N and fixed volume V, that is: given specific volume $v = \frac{V}{N}$ or given particle density $\rho = \frac{N}{V} = \frac{1}{v}$, the transition occurs at the critical temperature $T_c = (k_B\beta_c)^{-1}$ that is determined by

$$\lambda^3\Big|_{T=T_c} = \left(\hbar\sqrt{\frac{2\pi\beta_c}{m}}\right)^3 = v\zeta\left(\frac{3}{2}\right), \tag{3.6.22}$$

that is

$$T_c = \frac{2\pi\hbar^2}{mk_B}\left[v\zeta\left(\frac{3}{2}\right)\right]^{-\frac{2}{3}}. \tag{3.6.23}$$

The condensed phase then contains

$$N_0 = N - \frac{Nv}{\lambda^3}\zeta\left(\frac{3}{2}\right) = N - N\left(\frac{T}{T_c}\right)^{\frac{3}{2}} \tag{3.6.24}$$

bosons, so that

$$\frac{N_0}{N} = \begin{cases} 1 - \left(\frac{T}{T_c}\right)^{\frac{3}{2}} & \text{for } T < T_c \\ 0 & \text{for } T > T_c \end{cases} \tag{3.6.25}$$

is the temperature-dependent condensate fraction:

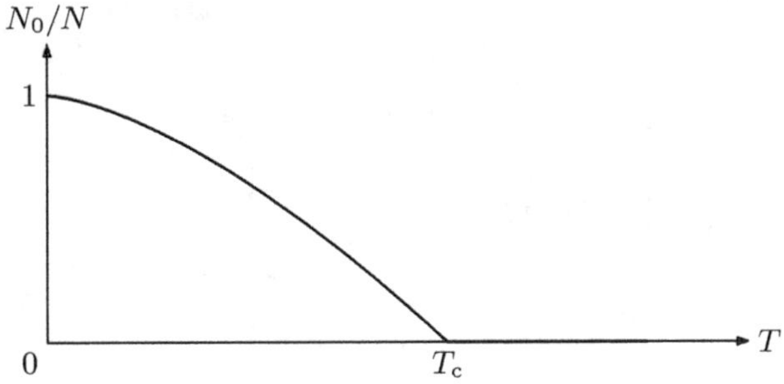

(3.6.26)

We can also think of keeping the temperature fixed while increasing the density. Then Bose–Einstein condensation happens when the density is larger than

$$\rho_c = \frac{1}{\lambda^3}\zeta\left(\frac{3}{2}\right), \tag{3.6.27}$$

that is: when the specific volume is less than its critical value

$$v_c = \frac{\lambda^3}{\zeta(\frac{3}{2})}. \tag{3.6.28}$$

In this way of looking at it, we have

$$\frac{N_0}{N} = \begin{cases} 1 - \dfrac{v}{v_c} & \text{for } v < v_c \\ 0 & \text{for } v > v_c \end{cases} \tag{3.6.29}$$

for the volume-dependent condensate fraction:

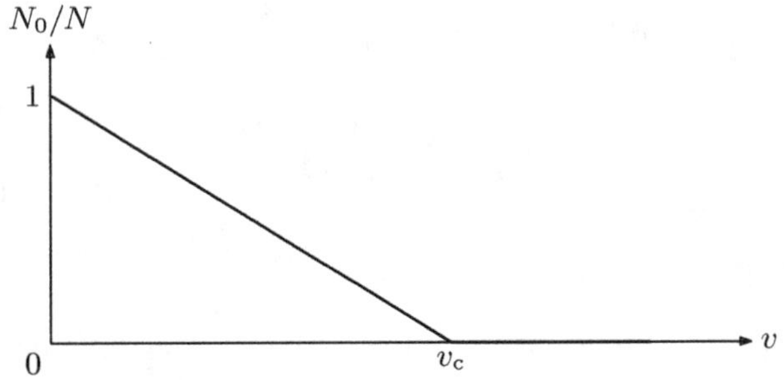

(3.6.30)

There is no account here for the finite size of the bosonic particles (atoms), and corrections are needed for extremely small values of v, the volume per boson.

We get the internal energy $U = \langle E \rangle = \sum_j \langle n_j \rangle \varepsilon_j$ as the β derivative of $-\log Z$,

$$\begin{aligned} U &= -\left(\frac{\partial}{\partial\beta}\log\bigl(Z(\beta,V,z)\bigr)\right)_{V,z} \\ &= -\left(\frac{\partial}{\partial\beta}\left[\frac{V}{\lambda^3}g_{\frac{5}{2}}(z) - \log(1-z)\right]\right)_{V,z} \\ &= \frac{3}{2\beta}\frac{V}{\lambda^3}g_{\frac{5}{2}}(z)\,, \end{aligned} \tag{3.6.31}$$

where we made use of $\lambda \propto \beta^{\frac{1}{2}}$, $\lambda^{-3} \propto \beta^{-\frac{3}{2}}$. The specific internal energy is, therefore,

$$\frac{U}{N} = \begin{cases} \frac{3}{2}k_{\mathrm{B}}T\frac{v}{\lambda^3}g_{\frac{5}{2}}(z) & \text{for } T > T_{\mathrm{c}} \\ \frac{3}{2}k_{\mathrm{B}}T\frac{v}{\lambda^3}g_{\frac{5}{2}}(1) & \text{for } T < T_{\mathrm{c}} \end{cases} \tag{3.6.32}$$

after we supplement the $T > T_{\mathrm{c}}$ expression by the corresponding expression with $z = 1$ for $T < T_{\mathrm{c}}$. For the heat capacity for constant volume, see Exercise 16,

$$C_V = \left(\frac{\partial U}{\partial T}\right)_{V,N}, \tag{3.6.33}$$

we then obtain

$$\begin{aligned} \frac{C_V}{k_{\mathrm{B}}N} &= \begin{cases} \left(\frac{\partial}{\partial T}\left(\frac{3}{2}T\frac{v}{\lambda^3}g_{\frac{5}{2}}(z)\right)\right)_v & \text{for } T > T_{\mathrm{c}} \\ \left(\frac{\partial}{\partial T}\left(\frac{3}{2}T\frac{v}{\lambda^3}g_{\frac{5}{2}}(1)\right)\right)_v & \text{for } T < T_{\mathrm{c}} \end{cases} \\ &= \begin{cases} \frac{15}{4}\frac{v}{\lambda^3}g_{\frac{5}{2}}(z) + \frac{3}{2}T\frac{v}{\lambda^3}\left(\frac{\partial}{\partial T}g_{\frac{5}{2}}(z)\right)_v & \text{for } T > T_{\mathrm{c}} \\ \frac{15}{4}\frac{v}{\lambda^3}g_{\frac{5}{2}}(1) & \text{for } T < T_{\mathrm{c}} \end{cases} \end{aligned} \tag{3.6.34}$$

where $T\dfrac{1}{\lambda^3} \propto T^{\frac{5}{2}}$ is taken into account, and for the remaining differentiation we must regard the fugacity as given by

$$g_{\frac{3}{2}}(z) = \frac{\lambda^3}{v} = \frac{\lambda_{\mathrm{c}}^3}{v}\frac{\lambda^3}{\lambda_{\mathrm{c}}^3} = g_{\frac{3}{2}}(1)\left(\frac{T_{\mathrm{c}}}{T}\right)^{\frac{3}{2}} \propto T^{-\frac{3}{2}} \tag{3.6.35}$$

when $T > T_c$. Therefore,

$$\begin{aligned}\left(\frac{\partial}{\partial T} g_{\frac{3}{2}}(z)\right)_v &= -\frac{3}{2T}\frac{\lambda^3}{v} = -\frac{3}{2T} g_{\frac{3}{2}}(z) \\ &= \left(\frac{\partial z}{\partial T}\right)_v \frac{\partial}{\partial z} g_{\frac{3}{2}} = \left(\frac{\partial z}{\partial T}\right)_v \frac{1}{z} g_{\frac{1}{2}}(z)\,, \end{aligned} \tag{3.6.36}$$

where (3.6.12) entered, or

$$\left(\frac{\partial z}{\partial T}\right)_v = -\frac{3z}{2T}\frac{g_{\frac{3}{2}}(z)}{g_{\frac{1}{2}}(z)}\,, \tag{3.6.37}$$

and

$$\begin{aligned}\left(\frac{\partial}{\partial T} g_{\frac{5}{2}}(z)\right)_v &= \left(\frac{\partial z}{\partial T}\right)_v \frac{\partial}{\partial z} g_{\frac{5}{2}}(z) = -\frac{3}{2T}\frac{g_{\frac{3}{2}}(z)}{g_{\frac{1}{2}}(z)} \underbrace{z\frac{\partial}{\partial z} g_{\frac{5}{2}}(z)}_{= g_{\frac{3}{2}}(z) = \frac{\lambda^3}{v}} \\ &= -\frac{3}{2T}\frac{\lambda^3}{v}\frac{g_{\frac{3}{2}}(z)}{g_{\frac{1}{2}}(z)} \end{aligned} \tag{3.6.38}$$

follows. Accordingly, we have

$$\frac{C_V}{k_B N} = \begin{cases} \dfrac{15}{4}\dfrac{v}{\lambda^3} g_{\frac{5}{2}}(z) - \dfrac{9}{4}\dfrac{g_{\frac{3}{2}}(z)}{g_{\frac{1}{2}}(z)} & \text{for } T > T_c\,, \\ \dfrac{15}{4}\dfrac{v}{\lambda^3} g_{\frac{5}{2}}(1) & \text{for } T < T_c\,, \end{cases} \tag{3.6.39}$$

which is continuous at $T = T_c$ since $\dfrac{1}{g_{\frac{1}{2}}(z)} \to 0$ for $z \to 1$. With

$$\frac{v}{\lambda^3} = \frac{1}{g_{\frac{3}{2}}(1)}\left(\frac{T}{T_c}\right)^{\frac{3}{2}} \tag{3.6.40}$$

for $T < T_c$, this reads

$$\frac{C_V}{k_B N} = \begin{cases} \dfrac{15}{4}\dfrac{g_{\frac{5}{2}}(z)}{g_{\frac{3}{2}}(z)} - \dfrac{9}{4}\dfrac{g_{\frac{3}{2}}(z)}{g_{\frac{1}{2}}(z)} & \text{for } T > T_c\,, \\[2ex] \dfrac{15}{4}\dfrac{g_{\frac{5}{2}}(1)}{g_{\frac{3}{2}}(1)}\left(\dfrac{T}{T_c}\right)^{\frac{3}{2}} & \text{for } T < T_c\,, \end{cases} \tag{3.6.41}$$

where the relation in (3.6.35) determines the fugacity z for $T > T_c$, that is

$$z = g_{\frac{3}{2}}^{-1}(\lambda^3/v)\,. \tag{3.6.42}$$

For very low temperatures, we have $C_V \propto T^{\frac{3}{2}}$; for very large temperatures — when $1 \gg \dfrac{\lambda^3}{v} = g_{\frac{3}{2}}(z) \cong z \cong g_\alpha(z)$ — we have

$$\frac{C_V}{k_{\mathrm{B}}N} = \frac{15}{4} - \frac{9}{4} = \frac{3}{2}, \tag{3.6.43}$$

which is the value of the classical ideal gas in (1.7.8) for $\gamma = \dfrac{5}{3}$ (single component, no internal structure), and the value at $T = T_{\mathrm{c}}$ is

$$\left.\frac{C_V}{k_{\mathrm{B}}N}\right|_{T=T_{\mathrm{c}}} = \frac{15}{4}\frac{g_{\frac{5}{2}}(1)}{g_{\frac{3}{2}}(1)} = \frac{15}{4}\frac{\zeta(\frac{5}{2})}{\zeta(\frac{3}{2})} = \frac{15}{4}\frac{1.342}{2.612} = 1.926\,, \tag{3.6.44}$$

which give us this graph:

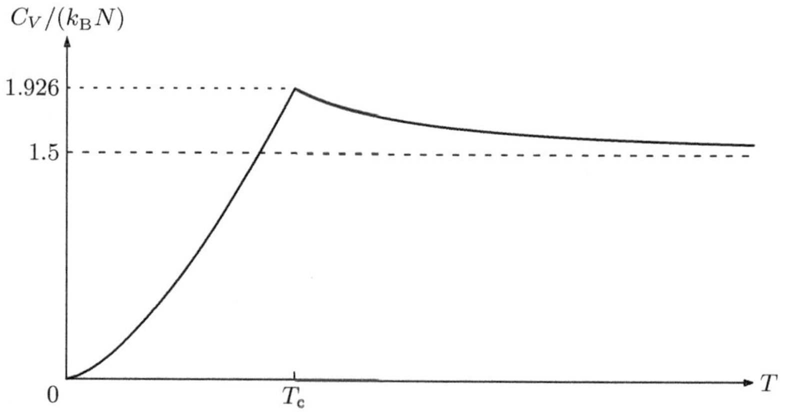

(3.6.45)

A few further aspects of ideal boson gases below $T < T_{\mathrm{c}}$ are the subject matter of Exercises 67–76. We now turn to gases of fermions.

3.7 Fermion gas

If we ignore the somewhat esoteric neutrinos, there are no massless fermions and, therefore, there are no fermion analogs of the photon gas in Section 3.4 or the phonon gas in Section 3.5. For massive fermions — electrons and fermionic atoms are important examples — we have the grand canonical partition function of (3.3.7) and the average occupation numbers of (3.3.8),

$$\beta PV = \log\bigl(Z(\beta,V,z)\bigr) = \sum_j \log\Bigl(1 + z\,\mathrm{e}^{-\beta\varepsilon_j}\Bigr)\,,$$

$$\langle n_j\rangle = \frac{z}{\mathrm{e}^{\beta\varepsilon_j} + z}\,, \tag{3.7.1}$$

where $z = \mathrm{e}^{\beta E_{\mathrm{F}}}$ is the fugacity of (3.6.2). We follow the tradition of denoting the chemical potential by the so-called *Fermi energy* E_{F}, rather than the symbol μ, in the context of the fermion gas.

Here, the fugacity can take on any positive value without restriction. There is no fermionic analog of the bosonic Bose–Einstein condensation. Accordingly, there is no need to isolate the $\boldsymbol{p} = 0$ term in the analogs of (3.6.6) and (3.6.9),

$$\begin{aligned}\beta PV &= \int \frac{(\mathrm{d}\boldsymbol{r})\,(\mathrm{d}\boldsymbol{p})}{(2\pi\hbar)^3} \log\left(1 + z\,\mathrm{e}^{-\frac{\beta}{2m}\boldsymbol{p}^2}\right)\\ \overset{(3.6.7)}&{=} \frac{2}{\sqrt{\pi}}\frac{V}{\lambda^3}\int_0^\infty \mathrm{d}y\, y^{\frac{1}{2}} \log\Bigl(1 + z\,\mathrm{e}^{-y}\Bigr)\\ &= \frac{V}{\lambda^3} f_{\frac{5}{2}}(z)\,,\end{aligned} \tag{3.7.2}$$

where λ is the thermal wavelength of (3.6.8) and $f_{\frac{5}{2}}(\;)$ is the $\alpha = \frac{5}{2}$ member in the family of *fermion functions* defined by

$$f_\alpha(z) = \sum_{k=1}^{\infty} \frac{(-1)^{k+1}}{k^\alpha} z^k\,. \tag{3.7.3}$$

As a consequence of the alternating sign, this series converges for $z = 1$ if $\alpha > 0$ and can be extended by analytical continuation to all $z > 0$, in marked contrast to the series for $g_\alpha(z)$ in (3.6.11). For $|z| < 1$, we have the formal identity

$$f_\alpha(z) = -g_\alpha(-z)\,, \tag{3.7.4}$$

although for both $f_\alpha(\;)$ and $g_\alpha(\;)$ the physical values of the arguments are never negative; see also Exercise 81. The analogs of the recurrence relations in (3.6.12),

$$f_{\alpha-1}(z) = z\frac{\partial}{\partial z} f_\alpha(z) \quad\text{and}\quad f_{\alpha+1}(z) = \int_0^z \mathrm{d}z'\,\frac{f_\alpha(z')}{z'}\,, \tag{3.7.5}$$

have exactly the same form.

We get the average particle number by differentiation

$$\langle N\rangle = \left(z\frac{\partial}{\partial z}\log\bigl(Z(\beta,V,z)\bigr)\right)_{\beta,V}$$
$$= \frac{V}{\lambda^3}z\frac{\partial}{\partial z}f_{\frac{5}{2}}(z) = \frac{V}{\lambda^3}f_{\frac{3}{2}}(z)\,. \tag{3.7.6}$$

With the particle density $\rho = \dfrac{\langle N\rangle}{V}$, this is

$$\rho\lambda^3 = f_{\frac{3}{2}}(z)\,. \tag{3.7.7}$$

In the spirit of (2.9.21), we should read this as determining the fugacity for given density and temperature (in $\lambda \propto T^{-\frac{1}{2}}$), and should then insert $z(\rho,\lambda)$ into (3.7.2) to establish the thermal equation of state, that is:

$$\frac{\beta P}{\rho} = \frac{1}{\rho\lambda^3}f_{\frac{5}{2}}(z)\bigg|_{z=f_{\frac{3}{2}}^{-1}(\rho\lambda^3)} = \frac{f_{\frac{5}{2}}(z)}{f_{\frac{3}{2}}(z)}\bigg|_{z=f_{\frac{3}{2}}^{-1}(\rho\lambda^3)}, \tag{3.7.8}$$

and so obtain the analog of $\dfrac{\beta P}{\rho} = 1$ for the classical ideal gas in (2.9.19).

In view of the discussion of the dilute-gas limit in Section 3.3.3, we expect that the ideal fermion gas behaves similar to the classical ideal gas when the density is low or the temperature high, that is: when $\rho\lambda^3 \ll 1$. Then, also the fugacity is small, $z \ll 1$, and we have

$$\rho\lambda^3 = z + O(z^2) \tag{3.7.9}$$

and

$$\frac{\beta P}{\rho} = \frac{z - z^2/\sqrt{32} + O(z^3)}{z - z^2/\sqrt{8} + O(z^3)} = 1 + \frac{1}{\sqrt{32}}z + O(z^2)\,, \tag{3.7.10}$$

so that

$$z = \rho\lambda^3 + O\Bigl(\bigl(\rho\lambda^3\bigr)^2\Bigr) \tag{3.7.11}$$

and

$$\frac{\beta P}{\rho} = 1 + \frac{1}{\sqrt{32}}\rho\lambda^3 + \cdots\,, \tag{3.7.12}$$

where the ellipsis stands for terms proportional to $\bigl(\rho\lambda^3\bigr)^2$ and higher powers. This exhibits the equation of state for the classical ideal gas together with the leading correction that results from the Fermi–Dirac statistics.

The other extreme is that of very low temperature, when $\rho\lambda^3 \gg 1$ and $z \gg 1$, and we need the large-z approximation of $f_\alpha(z)$. We get it by first observing that, for $\alpha > -1$,

$$
\begin{aligned}
f_\alpha(z) &= \sum_{k=1}^{\infty}(-1)^{k+1}kz^k \underbrace{\frac{1}{\alpha!}\int_0^\infty \mathrm{d}y\, y^\alpha\, \mathrm{e}^{-ky}}_{=k^{-(1+\alpha)}} \\
&= \frac{1}{\alpha!}\int_0^\infty \mathrm{d}y\, y^\alpha \frac{z\,\mathrm{e}^{-y}}{\left(1+z\,\mathrm{e}^{-y}\right)^2}
\end{aligned}
\tag{3.7.13}
$$

as a consequence of Euler's factorial integral in (3.6.16) and the geometric series

$$
\sum_{k=1}^{\infty}(-1)^{k+1}kx^k = -x\frac{\partial}{\partial x}\sum_{k=0}^{\infty}(-1)^k x^k = -x\frac{\partial}{\partial x}\frac{1}{1+x} = \frac{x}{(1+x)^2}\,. \tag{3.7.14}
$$

In the second step, we write $z = \mathrm{e}^{y_0}$ and note that

$$
\frac{\left(1+z\,\mathrm{e}^{-y}\right)^2}{z\,\mathrm{e}^{-y}} = \left(\mathrm{e}^{\frac{1}{2}(y-y_0)} + \mathrm{e}^{\frac{1}{2}(y_0-y)}\right)^2 = 4\cosh\!\left(\frac{y-y_0}{2}\right)^2, \tag{3.7.15}
$$

which takes us to

$$
f_\alpha\big(z = \mathrm{e}^{y_0}\big) = \frac{1}{\alpha!}\int_0^\infty \mathrm{d}y\, \frac{y^\alpha}{4\cosh\!\left(\frac{y-y_0}{2}\right)^2}\,. \tag{3.7.16}
$$

In the third step, we recognize that the factor $\dfrac{1}{\cosh(\cdots)^2}$ is very narrowly peaked at $y = y_0 \gg 1$ and, therefore, it is permissible to replace y^α by y_0^α in the numerator and extend the lower limit of the integration to $-\infty$, and so we arrive at

$$
\begin{aligned}
f_\alpha\big(z = \mathrm{e}^{y_0}\big) &\cong \frac{y_0^\alpha}{\alpha!}\int_{-\infty}^{\infty} \mathrm{d}y\, \frac{1}{4\cosh\!\left(\frac{y-y_0}{2}\right)^2} \\
&= \frac{y_0^\alpha}{\alpha!}\,\frac{1}{2}\tanh\!\left(\frac{y-y_0}{2}\right)\bigg|_{y=-\infty}^{\infty} = \frac{y_0^\alpha}{\alpha!}\,.
\end{aligned}
\tag{3.7.17}
$$

In summary, this establishes

$$
f_\alpha(z) \cong \frac{1}{\alpha!}\big(\log z\big)^\alpha \quad \text{for} \quad z \gg 1\,. \tag{3.7.18}
$$

This derivation of the large-z approximation for $f_\alpha(z)$ is valid for $\alpha > -1$, but the recurrence relations in (3.7.5) lift this restriction.

As an immediate consequence, we find

$$\rho\lambda^3 = \frac{1}{(\frac{3}{2})!}(\log z)^{\frac{3}{2}} = \frac{4}{3\sqrt{\pi}}(\beta E_\mathrm{F})^{\frac{3}{2}} \tag{3.7.19}$$

and

$$P = \frac{\rho}{\beta}\frac{f_{\frac{5}{2}}(z)}{f_{\frac{3}{2}}(z)} = \frac{\rho}{\beta}\frac{(\frac{3}{2})!}{(\frac{5}{2})!}(\log z)^{\frac{5}{2}-\frac{3}{2}} = \frac{2}{5}\rho E_\mathrm{F}\,, \tag{3.7.20}$$

which is the thermal equation of state at low temperature, $k_\mathrm{B}T \ll E_\mathrm{F}$. Since $\lambda \propto \beta^{-\frac{1}{2}}$, there is a finite density of the fermion gas for $T \to 0$,

$$\rho = \frac{4}{3\sqrt{\pi}}\left(\frac{\beta E_\mathrm{F}}{\lambda^2}\right)^{\frac{3}{2}} = \frac{4}{3\sqrt{\pi}}\left(\frac{mE_\mathrm{F}}{2\pi\hbar^2}\right)^{\frac{3}{2}} = \frac{1}{6\pi^2}k_\mathrm{F}^3\,, \tag{3.7.21}$$

where k_F is the *Fermi wave number*, defined by

$$E_\mathrm{F} = \frac{(\hbar k_\mathrm{F})^2}{2m}\,. \tag{3.7.22}$$

We note in passing that the corresponding wavelength $\dfrac{2\pi}{k_\mathrm{F}} = \left(\dfrac{4\pi}{3\rho}\right)^{\frac{1}{3}}$ is strongly reminiscent of the Debye wavelength in (3.5.11).

There is also a finite pressure at $T = 0$, namely

$$P = \frac{2}{5}\rho E_\mathrm{F} = \frac{2}{5}\frac{k_\mathrm{F}^3}{6\pi^2}\frac{(\hbar k_\mathrm{F})^2}{2m} = \frac{1}{30\pi^2}\frac{\hbar^2}{m}k_\mathrm{F}^5\,. \tag{3.7.23}$$

An ideal gas of fermions has finite pressure at zero temperature because only one fermion can have $\boldsymbol{p} = 0$ (or a few if there are internal degrees of freedom), all others have $\boldsymbol{p} \neq 0$.

Regarding the energy density, we note that

$$\begin{aligned} u = \frac{\langle E\rangle}{V} &= -\frac{1}{V}\left(\frac{\partial}{\partial\beta}\log\bigl(Z(\beta,V,z)\bigr)\right)_{V,z} \\ &= -\frac{\partial\lambda^{-3}}{\partial\beta}f_{\frac{5}{2}}(z) = \frac{3}{2\beta\lambda^3}f_{\frac{5}{2}}(z) = \frac{3}{2}P\,, \end{aligned} \tag{3.7.24}$$

reminiscent of, yet different from, the photon-gas relation $u = 3P$ that we found in (3.4.16). Here, the fugacity is determined by (3.7.7), that is:

$z = f_{\frac{3}{2}}^{-1}(\rho\lambda^3)$, the analog of (3.6.42). We get the corresponding expression for the constant-volume heat capacity,

$$\frac{C_V}{k_B N} = \frac{15}{4}\frac{f_{\frac{5}{2}}(z)}{f_{\frac{3}{2}}(z)} - \frac{9}{4}\frac{f_{\frac{3}{2}}(z)}{f_{\frac{1}{2}}(z)} \quad \text{with} \quad z = f_{\frac{3}{2}}^{-1}(\rho\lambda^3) \tag{3.7.25}$$

by the same reasoning that established the $T > T_c$ result in (3.6.41). As we did in (3.7.9)–(3.7.12), we use the small-z approximation

$$f_\alpha(z) = z - 2^{-\alpha}z^2 + \cdots = z\bigl(1 - 2^{-\alpha}z + O(z^2)\bigr) \tag{3.7.26}$$

to find the heat capacity at large temperatures or small densities,

$$\begin{aligned}\rho\lambda^3 \ll 1: \quad \frac{C_V}{k_B N} &= \frac{15}{4}\bigl(1 - 2^{-\frac{5}{2}}z + 2^{-\frac{3}{2}}z\bigr) \\ &\quad - \frac{9}{4}\bigl(1 - 2^{-\frac{3}{2}}z + 2^{-\frac{1}{2}}z\bigr) + O(z^2) \\ &= \frac{3}{2}\Bigl(1 - \frac{z}{\sqrt{128}} + O(z^2)\Bigr) \\ &= \frac{3}{2}\Bigl(1 - \frac{\rho\lambda^3}{\sqrt{128}} + \cdots\Bigr),\end{aligned} \tag{3.7.27}$$

where the ellipsis stands for terms proportional to $\bigl(\rho\lambda^3\bigr)^2$ and higher powers. This is, of course, identical with what we obtain from (3.7.12) when taking $u = \frac{3}{2}P$ into account.

For large densities or low temperatures, such that $\rho\lambda^3 \gg 1$ and $\log z = \beta E_F \gg 1$, the large-z approximation in (3.7.18) is not sufficiently accurate, inasmuch as it gives

$$\frac{C_V}{k_B N} \cong \frac{15}{4}\frac{\frac{3}{2}!}{\frac{5}{2}!}\log z - \frac{9}{4}\frac{\frac{1}{2}!}{\frac{3}{2}!}\log z = 0 \tag{3.7.28}$$

and does not tell us the low-temperature heat capacity correctly. We need to improve on (3.7.18). For this purpose, we substitute $y = y_0 + 2\vartheta$, $\mathrm{d}y = 2\mathrm{d}\vartheta$ in (3.7.16) and so get

$$f_\alpha\bigl(z = \mathrm{e}^{y_0}\bigr) = \frac{1}{\alpha!}\int_{-\frac{1}{2}y_0}^{\infty} \mathrm{d}\vartheta\, \frac{(y_0 + 2\vartheta)^\alpha}{2(\cosh\vartheta)^2}, \tag{3.7.29}$$

where only the vicinity of $\vartheta = 0$ contributes substantially when $y_0 \gg 1$. With this in mind, we put the lower integration limit to $-\infty$ and expand the numerator to second order in ϑ,

$$(y_0 + 2\vartheta)^\alpha \cong y_0^\alpha + 2\alpha\vartheta y_0^{\alpha-1} + 2\alpha(\alpha-1)\vartheta^2 y_0^{\alpha-2}. \tag{3.7.30}$$

With

$$\int_{-\infty}^{\infty} \frac{\mathrm{d}\vartheta}{2(\cosh\vartheta)^2}\,(1,\vartheta,\vartheta^2) = \left(1,0,\frac{\pi^2}{12}\right) \tag{3.7.31}$$

this results in

$$f_\alpha\bigl(z = \mathrm{e}^{y_0}\bigr) \cong \frac{y_0^\alpha}{\alpha!}\left(1 + \frac{\pi^2}{6}\frac{\alpha(\alpha-1)}{y_0^2}\right), \tag{3.7.32}$$

which are the first terms in an expansion developed by Sommerfeld.* We use it on the right-hand side of (3.7.25) and find

$$\begin{aligned}\frac{C_V}{k_\mathrm{B}N} &\cong \frac{3y_0}{2}\left(1 + \frac{\pi^2}{6}\frac{15}{4y_0^2} - \frac{\pi^2}{6}\frac{3}{4y_0^2}\right)\\ &\quad - \frac{3y_0}{2}\left(1 + \frac{\pi^2}{6}\frac{3}{4y_0^2} + \frac{\pi^2}{6}\frac{1}{4y_0^2}\right) = \frac{\pi^2}{2y_0}\end{aligned} \tag{3.7.33}$$

or

$$\frac{C_V}{k_\mathrm{B}N} \cong \frac{\pi^2}{2}\frac{T}{T_\mathrm{F}} \quad \text{for} \quad T \ll T_\mathrm{F} = \frac{E_\mathrm{F}}{k_\mathrm{B}} \tag{3.7.34}$$

with the *Fermi temperature* T_F. For an ultracold ideal gas of fermions, the heat capacity is proportional to the temperature, whereas it is proportional to $T^{\frac{3}{2}}$ for ultracold bosons, as we found in (3.6.41).

3.8 Electron gas in a metal

We can apply these insights about the ideal gas of fermions to the conduction electrons in a metal. It is true that the electrons repel each other because they carry electric charge, but this repulsive interaction is remarkably weak. This is so as a consequence of the Fermi–Dirac statistics which requires the electrons to fill many single-particle states, so that the unoccupied states with lowest single-particle energy will have a kinetic energy of many $k_\mathrm{B}T$ units (numbers to follow shortly). The excitation into these states give rise to the fluctuations important for the thermodynamical properties of this electron gas. In this situation, then, the density of electrons is

*Arnold SOMMERFELD (1868–1951)

so high and so many single-particle states are occupied, that the interaction energy is very small compared with the kinetic energy of the lowest-energy unoccupied states, and we can get a decent description by ignoring the interaction altogether.

According to (3.7.2) and (3.7.6), we have

$$\begin{aligned} P &= \frac{1}{\beta V}\log Z = \frac{2}{\beta\lambda^3} f_{\frac{5}{2}}(z)\,, \\ \text{and}\quad \rho &= \frac{\langle N\rangle}{V} = \frac{2}{\lambda^3} f_{\frac{3}{2}}(z) \end{aligned} \tag{3.8.1}$$

with the fugacity $z = \mathrm{e}^{\beta E_\mathrm{F}}$, where the additional factor of 2 accounts for the spin multiplicity — we have two spin states for each orbital state of an electron. It is expedient to perform an integration by parts in (3.7.13), based on the identity

$$\frac{z\,\mathrm{e}^{-y}}{(1+z\,\mathrm{e}^{-y})^2} = -\frac{\partial}{\partial y}\frac{z}{\mathrm{e}^{y}+z}\,, \tag{3.8.2}$$

and so establish

$$f_\alpha(z) = \frac{1}{(\alpha-1)!}\int\limits_0^\infty \mathrm{d}y\, y^{\alpha-1}\,\frac{z}{\mathrm{e}^{y}+z} \quad \text{for} \quad \alpha>0 \tag{3.8.3}$$

or

$$f_\alpha\bigl(z=\mathrm{e}^{\beta E_\mathrm{F}}\bigr) = \frac{\beta^\alpha}{(\alpha-1)!}\int\limits_0^\infty \mathrm{d}\varepsilon\,\frac{\varepsilon^{\alpha-1}}{\mathrm{e}^{\beta(\varepsilon-E_\mathrm{F})}+1} \tag{3.8.4}$$

upon recalling the meaning of y in (3.6.7) and switching to the single-particle energy $\varepsilon = \dfrac{y}{\beta} = \dfrac{\boldsymbol{p}^2}{2m}$ as the integration variable. Then the statements in (3.8.1) read

$$\begin{aligned} P &= \frac{1}{3\pi^2}\left(\frac{2m}{\hbar^2}\right)^{\frac{3}{2}}\int\limits_0^\infty \mathrm{d}\varepsilon\,\frac{\varepsilon^{\frac{3}{2}}}{\mathrm{e}^{\beta(\varepsilon-E_\mathrm{F})}+1} \\ \text{and}\quad \rho &= \frac{1}{2\pi^2}\left(\frac{2m}{\hbar^2}\right)^{\frac{3}{2}}\int\limits_0^\infty \mathrm{d}\varepsilon\,\frac{\varepsilon^{\frac{1}{2}}}{\mathrm{e}^{\beta(\varepsilon-E_\mathrm{F})}+1}\,, \end{aligned} \tag{3.8.5}$$

where, more specifically, m is now the electron mass.

The so-called *Fermi function* $\left[e^{\beta(\varepsilon - E_F)} + 1\right]^{-1}$ that appears here is the expected occupation number of a state with single-particle energy ε,

$$\langle n_\varepsilon \rangle = \frac{1}{e^{\beta(\varepsilon - E_F)} + 1}, \tag{3.8.6}$$

which we first encountered in (3.3.8). The Fermi function equals 1 for $\beta(\varepsilon - E_F)$ sufficiently large and negative, that is for $E_F - \varepsilon \ll k_B T = \frac{1}{\beta}$, and it equals 0 for $\beta(\varepsilon - E_F)$ sufficiently large and positive, that is for $\varepsilon - E_F \gg k_B T$. Here is the graph of the Fermi function:

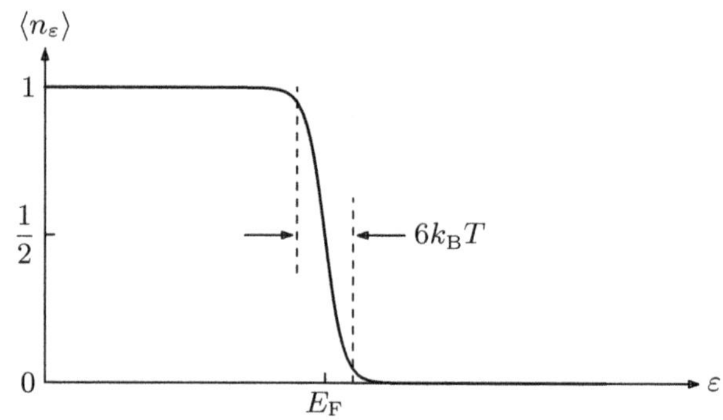

(3.8.7)

The range around E_F, in which the Fermi function drops from 1 to 0 is a few $k_B T$ wide. In particular, we have the step function

$$\underbrace{\frac{1}{e^{\beta(\varepsilon - E_F)} + 1}}_{\beta = \infty,\ k_B T = 0} = \eta(E_F - \varepsilon) = \begin{cases} 1 & \text{for } \varepsilon < E_F \\ 0 & \text{for } \varepsilon > E_F \end{cases} \tag{3.8.8}$$

at temperature zero. Physically this means $k_B T \ll E_F$, or $T \ll T_F$ with the Fermi temperature T_F of (3.7.34).

Let us assume that $T \ll T_F$ and justify this assumption later (in a short while, really). Then

$$\rho = \frac{1}{2\pi^2}\left(\frac{2m}{\hbar^2}\right)^{\frac{3}{2}} \int_0^{E_F} d\varepsilon \sqrt{\varepsilon} = \frac{1}{3\pi^2}\left(\frac{2mE_F}{\hbar^2}\right)^{\frac{3}{2}} = \frac{1}{3\pi^2} k_F^3 \tag{3.8.9}$$

with the Fermi wave number k_F of (3.7.22). This repeats (3.7.21) except for the spin-multiplicity factor of 2. The Fermi energy and the Fermi temperature, in terms of the electron density ρ, are then

$$E_{\mathrm{F}} = \frac{\hbar^2}{2m}(3\pi^2\rho)^{\frac{2}{3}}, \quad T_{\mathrm{F}} = \frac{E_{\mathrm{F}}}{k_{\mathrm{B}}} = \frac{\hbar^2}{2mk_{\mathrm{B}}}(3\pi^2\rho)^{\frac{2}{3}}. \tag{3.8.10}$$

The coefficient $\dfrac{\hbar^2}{2mk_{\mathrm{B}}}(3\pi^2)^{\frac{2}{3}}$ equals $4.23 \times 10^{-15}\,\mathrm{m}^2\,\mathrm{K}$, so that

$$T_{\mathrm{F}}[\mathrm{K}] = 4.23 \times 10^5 \left(\rho[\mathrm{\AA}^{-3}]\right)^{\frac{2}{3}} \tag{3.8.11}$$

for the Fermi temperature in kelvin* with the electron density in electrons per cubic angstrom† ($1\,\mathrm{\AA} = 10^{-10}\,\mathrm{m}$). In a metal, the spacing between atoms is of the order of one to two angstrom and we would typically have one conduction electron per metal atom so that ρ is one electron per a few $\mathrm{\AA}^3$, and the Fermi temperature is several tens of thousands of kelvin, much larger than room temperature. It follows that usually $T \ll T_{\mathrm{F}}$, and the zero-temperature approximation that replaces the Fermi function by a step function is well justified.

As an application, we consider the energy density

$$u = \frac{3}{2}P = \frac{3}{\beta\lambda^3} f_{\frac{5}{2}}\left(z = \mathrm{e}^{\beta E_{\mathrm{F}}}\right) = \frac{1}{2\pi^2}\left(\frac{2m}{\hbar^2}\right)^{\frac{3}{2}} \int_0^\infty \mathrm{d}\varepsilon\, \frac{\varepsilon^{\frac{3}{2}}}{\mathrm{e}^{\beta(\varepsilon - E_{\mathrm{F}})} + 1}, \tag{3.8.12}$$

where (3.7.24) and the expression for the pressure P in (3.8.5) are recalled. The large-z approximation in (3.7.18) yields

$$u = \frac{3}{\beta\lambda^3}\frac{(\beta E_{\mathrm{F}})^{\frac{5}{2}}}{(\frac{5}{2})!} = \frac{\hbar^2 k_{\mathrm{F}}^5}{10\pi^2 m} = \frac{\hbar^2}{10\pi^2 m}(3\pi^2\rho)^{\frac{5}{3}}, \tag{3.8.13}$$

which is the $T \to 0$ limit. The next-to-leading term of (3.7.32) is irrelevant for the energy density u but not for its T derivative, which gives us the heat capacity of (3.7.34),

$$\frac{C_V}{k_{\mathrm{B}}N} = \frac{\pi^2}{2}\frac{T}{T_{\mathrm{F}}} \quad \text{for} \quad T \ll T_{\mathrm{F}}, \tag{3.8.14}$$

where the spin-multiplicity factor of 2 cancels as it doubles both C_V and N. This linear dependence of the heat capacity, and thus of the specific heat, on the temperature is a well established experimental fact about metals.

---

*William THOMSON, 1st Baron Kelvin (1824–1907)
†Jonas Anders ÅNGSTRÖM (1814–1874)

3.9 Pressure and energy density

For the photon gas, we found $u = 3P$ in (3.4.16), whereas it was $u = \frac{3}{2}P$ for the ideal fermion gas in (3.7.24) and also for the ideal boson gas, as we see when combining (3.6.9) and (3.6.31). This difference originates in the different count of single-particle states per energy range. For the photons of Section 3.4, we had

$$\frac{V}{\pi^2 c^3}\omega^2 \mathrm{d}\omega = \frac{V}{\pi^2(\hbar c)^3}(\hbar\omega)^2 \mathrm{d}\hbar\omega \propto \varepsilon^2 \mathrm{d}\varepsilon \tag{3.9.1}$$

in (3.4.8) for the number of states between the single-photon energies $\hbar\omega = \varepsilon$ and $\varepsilon + \mathrm{d}\varepsilon$. For the bosons of Section 3.6 and the fermions of Section 3.7, with their nonrelativistic energy $\varepsilon = \frac{\boldsymbol{p}^2}{2m}$, we had

$$\frac{V}{(2\pi\hbar)^3}4\pi p^2 \mathrm{d}p = \frac{2}{\sqrt{\pi}}\frac{V\beta^{\frac{3}{2}}}{\lambda^3}\varepsilon^{\frac{1}{2}}\mathrm{d}\varepsilon \tag{3.9.2}$$

in (3.6.7).

These are the $\kappa = 3$ and $\kappa = \frac{3}{2}$ cases of the general power law

$$\big(\#\text{ of states between } \varepsilon \text{ and } \varepsilon + \mathrm{d}\varepsilon\big) = \frac{1}{(\kappa-1)!}\frac{V}{V_0}\frac{\varepsilon^{\kappa-1}\mathrm{d}\varepsilon}{E_0^{\kappa}} \tag{3.9.3}$$

with a suitable reference volume V_0 and a reference energy E_0. The logarithm of the grand canonical partition function for the state density (3.9.3) is

$$\begin{aligned} \beta PV = \log Z &= \sum_j \frac{1}{\zeta}\log\left(1 + \zeta z\,\mathrm{e}^{-\beta E_j}\right) \\ &= \frac{V}{V_0 E_0^{\kappa}}\int\limits_0^{\infty}\frac{\mathrm{d}\varepsilon\,\varepsilon^{\kappa-1}}{(\kappa-1)!}\frac{1}{\zeta}\log\left(1 + \zeta z\,\mathrm{e}^{-\beta\varepsilon}\right), \end{aligned} \tag{3.9.4}$$

where $\zeta = 1$ for fermions (FD statistics), $\zeta = -1$ for bosons (BE statistics) and the limit $\zeta \to 0$ applies for classical particles (MB statistics), as in (3.3.14). With the substitution $y = \beta\varepsilon$ this becomes

$$\beta PV = \log\big(Z(\beta, V, z)\big) = \frac{V}{V_0(\beta E_0)^{-\kappa}}\int\limits_0^{\infty}\frac{\mathrm{d}y\,y^{\kappa-1}}{(\kappa-1)!}\frac{1}{\zeta}\log\left(1 + \zeta z\,\mathrm{e}^{-y}\right) \tag{3.9.5}$$

so that

$$P = \frac{1}{\beta V_0(\beta E_0)^{-\kappa}} \begin{cases} g_{\kappa+1}(z) \text{ for } \zeta = 1 \text{ (FD)}, \\ f_{\kappa+1}(z) \text{ for } \zeta = -1 \text{ (BE)}, \\ \quad z \quad\ \text{ for } \zeta \to 0 \text{ (MB)}. \end{cases} \tag{3.9.6}$$

Irrespective of the value of ζ, we find

$$u = \frac{\langle E \rangle}{V} = -\frac{1}{V}\left(\frac{\partial}{\partial \beta}\log Z\right)_{V,z} = \frac{\kappa}{\beta V}\log Z = \kappa P \tag{3.9.7}$$

because the β dependence of $\log Z$ is solely in the factor $\beta^{-\kappa}$.

In summary, whenever there is a power law of the form (3.9.3) in the relevant energy range, the energy density and the pressure will be related to each other by $u = \kappa P$. The relativistic photon gas with $\kappa = 3$ and the nonrelativistic particle gases with $\kappa = \frac{3}{2}$ are the particular cases that we encountered in Sections 3.4, 3.6, and 3.7.

3.10 The Third Law

The heat capacity at low temperatures,

$$\begin{aligned} &C_V \propto T^3 &&\text{for photons in (3.4.12)}, \\ &C_V \propto T^{\frac{3}{2}} &&\text{for bosons in (3.6.41)}, \\ &C_V \propto T &&\text{for fermions in (3.7.34)}, \end{aligned} \tag{3.10.1}$$

vanishes for $T \to 0$ in all cases. This is an illustration of the *Third Law of Thermodynamics*:

Third Law of Thermodynamics

The entropy $S(T, X)$ has a unique specific value and a vanishing log-T gradient in the limit $T \to 0$, that is

(i) $S(T \to 0, X) = S_0$ is proportional to the system size but otherwise independent of X; and (3.10.2)

(ii) $C_X(T \to 0) = \left(T\dfrac{\partial S}{\partial T}\right)_X (T \to 0, X) = 0$ for all X.

Here, X stands for all other thermodynamical variables, such as volume, particle number, or magnetization.

The entropy at $T = 0$ is $S_0 = k_{\mathrm{B}} \log\bigl(\Omega(E_{\mathrm{gs}})\bigr)$ where $\Omega(E_{\mathrm{gs}})$ is the (rather small) number of microstates with ground-state energy E_{gs}. For example, for a gas of noninteracting spin-zero bosons, this is $\Omega(E_{\rm gs})=1$; for a gas of noninteracting spin-$\frac{1}{2}$ fermions, this is $\Omega(E_{\rm gs})=2^N$, the spin multiplicity of the N-fermion ground state; for a gas of atoms with $n_{\rm HF}$ degenerate hyperfine ground states, this is $\Omega(E_{\rm gs})=n_{\rm HF}^N$.

Arguably the most important consequence of the Third Law is that we cannot reach $T=0$. The essence of the argument is this: Consider a cooling process that exploits a sequence of isothermal and isentropic processes,

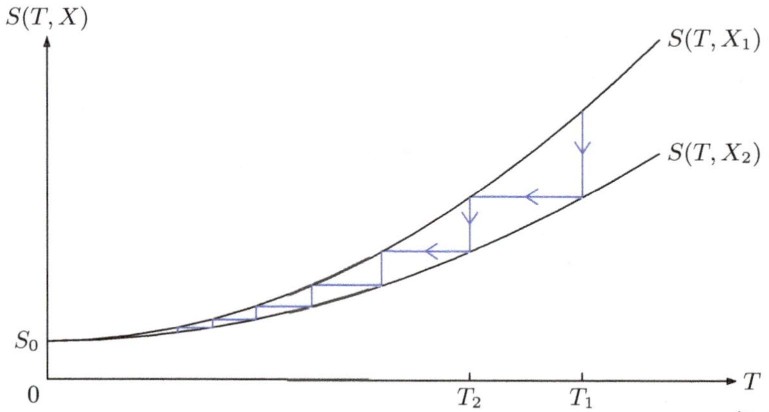

(3.10.3)

where $S(T,X)=S_0+\alpha(X)T^\kappa$ with $\alpha(X)>0$ and $\kappa>0$ for sufficiently low temperatures. In the isothermal step ($\downarrow$), we go from $S(T_1,X_1)$ to $S(T_1,X_2)<S(T_1,X_1)$,

$$S_0+\alpha(X_1)T_1^\kappa \to S_0+\alpha(X_2)T_1^\kappa \tag{3.10.4}$$

with $\alpha(X_1)>\alpha(X_2)>0$, and in the isentropic step ($\leftarrow$), we go from $S(T_1,X_2)$ to $S(T_2,X_1)=S(T_1,X_2)$,

$$T_1\to T_2 \quad \text{with} \quad S_0+\alpha(X_2)T_1^\kappa = S_0+\alpha(X_1)T_2^\kappa\,. \tag{3.10.5}$$

The combination of both steps amounts to

$$X_1\to X_1 \quad \text{and} \quad T_1\to T_2=\left[\frac{\alpha(X_2)}{\alpha(X_1)}\right]^{\frac{1}{\kappa}}<T_1\,, \tag{3.10.6}$$

and if we repeat this k times, we lower the temperature by a factor of $\bigl[\alpha(X_2)/\alpha(X_1)\bigr]^{k/\kappa}$. This can bring us as close to $T=0$ as we wish but not to $T=0$ itself.

Chapter 4

Systems with interaction

4.1 Thomas–Fermi model

As a first example of a nonideal gas, we now turn to the so-called statistical model of the atom, also known as the Thomas*–Fermi (TF) model, in which the electrons are treated as a very cold gas of fermions. We recall from (3.8.13) that the kinetic-energy density of a homogeneous electron gas is

$$\frac{\hbar^2}{10\pi^2 m}\left(3\pi^2\rho\right)^{5/3}, \tag{4.1.1}$$

where ρ is the electron density and m is the electron mass; we disregard the correction $\propto (T/T_{\mathrm{F}})^2$ associated with (3.7.32), which is relevant for the heat capacity in (3.8.14) but not for the kinetic energy as a whole. Now, if we have an inhomogeneous electron gas, we can try to use

$$E_{\mathrm{kin}}[\rho] = \int(\mathrm{d}\boldsymbol{r})\,\frac{\hbar^2}{10\pi^2 m}\left[3\pi^2\rho(\boldsymbol{r})\right]^{5/3} \tag{4.1.2}$$

as an approximation for the kinetic energy. This approximation is expected to be reliable if the density $\rho(\boldsymbol{r})$ changes slowly in space, so that the density is almost homogeneous over finite volumes that contain a good number of electrons, even if overall changes of $\rho(\boldsymbol{r})$ over large distances are substantial.

For the electron-electron interaction, we use the electrostatic interaction energy of a charge density $-e\rho(\boldsymbol{r})$, with the electron charge $-e$, as the basic approximation,

$$E_{\mathrm{int}}[\rho] = \frac{1}{2}\int(\mathrm{d}\boldsymbol{r})(\mathrm{d}\boldsymbol{r}')\,\frac{\left[-e\rho(\boldsymbol{r})\right]\left[-e\rho(\boldsymbol{r}')\right]}{|\boldsymbol{r}-\boldsymbol{r}'|} \tag{4.1.3}$$

*Llewellyn Hilleth THOMAS (1903–1992)

(in cgs units). For an atom with nuclear charge Ze, there is also the electrostatic interaction energy between the nucleus and the electrons,

$$\int(\mathrm{d}\boldsymbol{r})\,\frac{Ze}{r}\left[-e\rho(\boldsymbol{r})\right], \tag{4.1.4}$$

which is not an approximation for a point nucleus with no internal structure. Putting the three pieces together, we get the *TF functional*:

$$\begin{aligned} E_{\mathrm{TF}}[\rho] &= E_{\mathrm{kin}}[\rho] + \int(\mathrm{d}\boldsymbol{r})\,\frac{Ze}{r}\left[-e\rho(\boldsymbol{r})\right] + E_{\mathrm{int}}[\rho] \\ &= \frac{\hbar^2}{10\pi^2 m}\int(\mathrm{d}\boldsymbol{r})\,\left[3\pi^2\rho(\boldsymbol{r})\right]^{5/3} - Ze^2\int(\mathrm{d}\boldsymbol{r})\,\frac{\rho(\boldsymbol{r})}{r} \\ &\quad + \frac{e^2}{2}\int(\mathrm{d}\boldsymbol{r})(\mathrm{d}\boldsymbol{r}')\,\frac{\rho(\boldsymbol{r})\,\rho(\boldsymbol{r}')}{|\boldsymbol{r}-\boldsymbol{r}'|}\,. \end{aligned} \tag{4.1.5}$$

The TF density for a neutral atom is then obtained by minimizing $E_{\mathrm{TF}}[\rho]$ under the constraint that the count of electrons equals the atomic number Z, the number of protons in the nucleus,

$$\int(\mathrm{d}\boldsymbol{r})\,\rho(\boldsymbol{r}) = Z\,. \tag{4.1.6}$$

Accordingly, permissible variations of $\rho(\boldsymbol{r})$ in $E_{\mathrm{TF}}[\rho]$ are those for which

$$\int(\mathrm{d}\boldsymbol{r})\,\delta\rho(\boldsymbol{r}) = 0 \tag{4.1.7}$$

and, therefore, the stationary condition

$$\begin{aligned} 0 &= \delta E_{\mathrm{TF}}[\rho] \\ &= \int(\mathrm{d}\boldsymbol{r})\,\delta\rho(\boldsymbol{r})\left\{\frac{\hbar^2}{2m}\left[3\pi^2\rho(\boldsymbol{r})\right]^{2/3} - \frac{Ze^2}{r} + e^2\int(\mathrm{d}\boldsymbol{r}')\,\frac{\rho(\boldsymbol{r}')}{|\boldsymbol{r}-\boldsymbol{r}'|}\right\} \end{aligned} \tag{4.1.8}$$

implies that $\{\cdots\} = \text{constant}$. Since this constant is related to the integral of $\delta\rho(\boldsymbol{r})$, the change in the count of electrons, we recognize that it is the chemical potential μ, and so arrive at the integral equation obeyed by the actual electron density,

$$\frac{\hbar^2}{2m}\left[3\pi^2\rho(\boldsymbol{r})\right]^{2/3} - \frac{Ze^2}{r} + e^2\int(\mathrm{d}\boldsymbol{r}')\,\frac{\rho(\boldsymbol{r}')}{|\boldsymbol{r}-\boldsymbol{r}'|} = \mu\,. \tag{4.1.9}$$

We introduce

$$V(\boldsymbol{r}) = -\frac{Ze^2}{r} + e^2\int(\mathrm{d}\boldsymbol{r}')\,\frac{\rho(\boldsymbol{r}')}{|\boldsymbol{r}-\boldsymbol{r}'|}\,, \tag{4.1.10}$$

the electrostatic energy of a probe electron at $\boldsymbol{r}$ in the joint electrostatic potentials of the nuclear point charge and the charged electron gas with particle density $\rho(\boldsymbol{r})$, and then have

$$\frac{\hbar^2}{2m}\left[3\pi^2\rho(\boldsymbol{r})\right]^{2/3} = \mu - V(\boldsymbol{r})\,. \tag{4.1.11}$$

Clearly, this can only apply where $V(\boldsymbol{r}) < \mu$, since $\rho(\boldsymbol{r})$ is nonnegative. We have $\rho(\boldsymbol{r}) = 0$ where $V(\boldsymbol{r}) > \mu$. This is obvious from the physical significance of $\rho(\boldsymbol{r})$; we could also be more formal about it and have $\rho(\boldsymbol{r}) \geq 0$ as another constraint, enforced by writing the density as a square, $\rho(\boldsymbol{r}) = \psi(\boldsymbol{r})^2$, so that $\delta\rho = 2\delta\psi\,\psi$ in (4.1.7) and (4.1.8), and it follows that the integral equation (4.1.9) applies where $\rho(\boldsymbol{r}) > 0$ but not where $\rho(\boldsymbol{r}) = 0$. We do not want to indicate this explicitly and so overburden the notation; we just keep it in mind.

Accordingly, we write

$$\begin{aligned}\rho(\boldsymbol{r}) &= \frac{1}{3\pi^2}\left[\frac{2m}{\hbar^2}\bigl(\mu - V(\boldsymbol{r})\bigr)\right]_+^{3/2} \\ &= \begin{cases} \dfrac{1}{3\pi^2}\left[\dfrac{2m}{\hbar^2}\bigl(\mu - V(\boldsymbol{r})\bigr)\right]^{3/2} & \text{where } V(\boldsymbol{r}) < \mu \\ 0 & \text{where } V(\boldsymbol{r}) > \mu \end{cases}\end{aligned} \tag{4.1.12}$$

and so have one way of expressing $\rho(\boldsymbol{r})$ in terms of $V(\boldsymbol{r})$. A second expression is obtained from Gauss's* law of electrostatics, here in the form

$$-\frac{1}{4\pi}\boldsymbol{\nabla}^2\left(V(\boldsymbol{r}) + \frac{Ze^2}{r}\right) = e^2\rho(\boldsymbol{r})\,, \tag{4.1.13}$$

and the two relations together establish

$$-\frac{1}{4\pi}\boldsymbol{\nabla}^2\left(V(\boldsymbol{r}) + \frac{Ze^2}{r}\right) = \frac{e^2}{3\pi^2}\left[\frac{2m}{\hbar^2}\bigl(\mu - V(\boldsymbol{r})\bigr)\right]_+^{3/2}. \tag{4.1.14}$$

This differential equation for $V(\boldsymbol{r})$ is easier to work with than the integral equation for $\rho(\boldsymbol{r})$ in (4.1.9). Of course, the two equations are perfectly equivalent.

For the neutral atom that we have in mind, the Coulomb potential of the nucleus is isotropic and nothing else is there to break the rotational invariance. Therefore, the solutions for $\rho(\boldsymbol{r})$ and $V(\boldsymbol{r})$ will also be isotropic, that is: they depend only on $r = |\boldsymbol{r}|$, the distance from the nucleus, but

*Karl Friedrich GAUSS (1777–1855)

not on the direction of $\boldsymbol{r}$. So we have $V(\boldsymbol{r}) = V(r)$ and $\rho(\boldsymbol{r}) = \rho(r)$, and it is expedient to introduce the *TF function* $F(x)$ in accordance with

$$V(r) = \mu - \frac{Ze^2}{r}F(x) \quad \text{for} \quad x = \frac{r}{b}, \tag{4.1.15}$$

with a reference length b that we will choose shortly. In the vicinity of the nucleus, we have

$$\begin{aligned} V(r) &= -\frac{Ze^2}{r} + e^2 \int (\mathrm{d}\boldsymbol{r}') \, \frac{\rho(r')}{r'} + [\text{terms that vanish for } r \to 0] \\ &= \mu - \frac{Ze^2}{r}\bigl(F(0) + xF'(0) + \cdots\bigr) \end{aligned} \tag{4.1.16}$$

and infer that

$$F(0) = 1 \quad \text{and} \quad \frac{Ze^2}{b}F'(0) = \mu - e^2 \int (\mathrm{d}\boldsymbol{r}') \, \frac{\rho(r')}{r'} \, . \tag{4.1.17}$$

Then, there is no contribution proportional to the Dirac delta function $\delta(\boldsymbol{r}) = -\frac{1}{4\pi}\boldsymbol{\nabla}^2\frac{1}{r}$ in

$$\begin{aligned} -\frac{1}{4\pi}\boldsymbol{\nabla}^2\Bigl(V(\boldsymbol{r}) + \frac{Ze^2}{r}\Bigr) &= -\frac{1}{4\pi}\boldsymbol{\nabla}^2\Bigl(\mu + \frac{Ze^2}{r}\bigl(1 - F(x)\bigr)\Bigr) \\ &= -\frac{1}{4\pi}\frac{1}{r}\Bigl(\frac{\partial}{\partial r}\Bigr)^2\Bigl(r\mu + Ze^2\bigl(1 - F(x)\bigr)\Bigr) \\ &= \frac{1}{4\pi}\frac{Ze^2}{b^3}\frac{1}{x}\Bigl(\frac{\mathrm{d}}{\mathrm{d}x}\Bigr)^2 F(x) = \frac{1}{4\pi}\frac{Ze^2}{b^3}\frac{1}{x}F''(x) \, . \end{aligned} \tag{4.1.18}$$

For the right-hand side in (4.1.14), we have

$$\frac{e^2}{3\pi^2}\left[\frac{2m}{\hbar^2}\frac{Ze^2}{bx}F(x)\right]_+^{3/2} = \frac{e^2}{3\pi^2}\left[\frac{2Z}{a_0 bx}F(x)\right]_+^{3/2} \tag{4.1.19}$$

where

$$a_0 = \frac{\hbar^2}{me^2} = 0.5292 \, \text{Å} \tag{4.1.20}$$

is the first Bohr* radius, the natural length unit of atomic physics. Equating the two expressions gives

$$F''(x) = \frac{4}{3\pi}\frac{b^3}{Z}\left(\frac{2Z}{a_0 b}\right)^{3/2}\frac{\bigl[F(x)\bigr]_+^{3/2}}{x^{1/2}} \tag{4.1.21}$$

*Niels Henrik David BOHR (1885–1962)

or

$$F''(x) = \frac{\left[F(x)\right]_+^{3/2}}{x^{1/2}} = \begin{cases} \dfrac{F(x)^{3/2}}{x^{1/2}} & \text{if } F(x) > 0 \\ 0 & \text{if } F(x) < 0 \end{cases} \tag{4.1.22}$$

if we choose b such that the prefactor on the right-hand side equals unity, that is

$$b = \frac{1}{2}\left(\frac{3\pi}{4}\right)^{2/3} a_0 Z^{-1/3} = \underbrace{0.8853a_0}_{=0.47\,\text{Å}} Z^{-1/3}\,. \tag{4.1.23}$$

We note, in particular, that $b \propto Z^{-1/3}$, which states that the length scale of atoms shrinks by a factor $Z^{-1/3}$ as the atomic number Z increases. In this sense, the heavier atoms are smaller than the lighter ones.

The differential equation for the TF function $F(x)$ in (4.1.22) is known as the *TF equation* — a universal equation, as it is the same for all atoms, irrespective of their atomic number Z. The TF equation is a second-order differential equation with the solution obeying the boundary condition $F(x=0) = F(0) = 1$, as we established in (4.1.17). In the search for a second condition, let us look at the outside of the atom, where $\rho(r) = 0$ and $F(x) < 0$. Let us assume that the transition from $F(x) > 0$ to $F(x) < 0$ happens at distance $r_0 = x_0 b$. Then — this is an application of Newton's* shell theorem if you like — we have

$$\begin{aligned} r > r_0:\ V(r) &= -\frac{Ze^2}{r} + \frac{e^2}{r}\underbrace{\int (\mathrm{d}\boldsymbol{r}')\,\rho(\boldsymbol{r}')}_{=Z} = 0 \\ &= \mu - \frac{Ze^2}{r}F(x) = \mu - \frac{Ze^2}{bx}F(x)\,, \end{aligned} \tag{4.1.24}$$

so that

$$F(x) = \mu\frac{b}{Ze^2}x \quad \text{for} \quad x > x_0\,. \tag{4.1.25}$$

On the other hand, we have $F(x_0) = 0$ and

$$F(x) = F'(x_0)(x - x_0) \quad \text{for} \quad x > x_0 \tag{4.1.26}$$

since $F''(x) = 0$ there. It follows that

$$F'(x_0) = \mu\frac{b}{Ze^2} \quad \text{and} \quad x_0 F'(x_0) = 0\,, \tag{4.1.27}$$

*Sir Isaac NEWTON (1643–1727)

implying

$$F'(x_0) = 0 \quad \text{and} \quad \mu = 0\,. \tag{4.1.28}$$

Now, if $F(x_0) = 0$ and $F'(x_0) = 0$, then also $F''(x_0) = 0$ and all higher derivatives vanish at x_0 as well, so that $F(x) \equiv 0$ everywhere, at odds with $F(0) = 1$. It follows that we cannot have $F(x_0) = 0$ and $F'(x_0) = 0$ at a finite distance x_0, and the actual second boundary condition is $F(x) \to 0$ as $x \to \infty$.

Solutions to the TF equation (4.1.22) with different values of the initial slope $F'(0) < 0$ look like this:

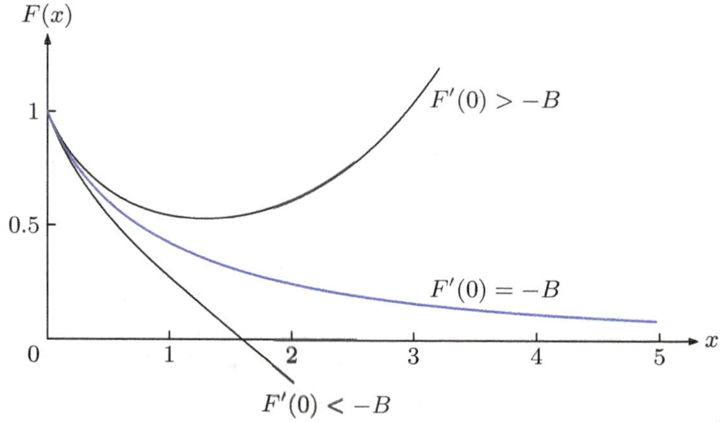

(4.1.29)

The TF function has the initial slope $-F'(0) = B = 1.588$, known as *Baker's* constant.* Solutions with $F'(0) > -B$ grow beyond limits and are not useful for anything. Solutions with $F'(0) < -B$ reach $F(x_0) = 0$ for a finite x_0 and have $F'(x_0) < 0$ there; these solutions are needed for the description of positive ions (number of electrons less than the atomic number Z), as discussed in Exercise 85.

Baker's constant $B = -F'(0)$ has an important physical meaning. We regard the minimum of the TF functional as a function of Z,

$$\underset{\rho(\boldsymbol{r})}{\text{Min}}\,\{E_{\text{TF}}[\rho]\} = E_{\text{TF}}[\rho_{\text{TF}}] = E(Z) \tag{4.1.30}$$

with the isotropic TF electron density $\rho_{\text{TF}}(\boldsymbol{r}) = \rho_{\text{TF}}(r)$ from (4.1.12) with

*Edward Bigelow BAKER (1906–1991)

(4.1.15), that is

$$\rho_{\mathrm{TF}}(r) = \frac{1}{3\pi^2}\left[\frac{2Z}{a_0 r}F\left(\frac{r}{b}\right)\right]^{3/2}, \tag{4.1.31}$$

and ask how $E(Z)$ responds to changes of Z. Since $E_{\mathrm{TF}}[\rho]$ is stationary at $\rho = \rho_{\mathrm{TF}}$, all first-order responses come from the explicit Z dependence of the functional, not from the implicit dependence through the Z dependence of ρ_{TF}. That is:

$$\begin{aligned}
\delta E(Z) &= -\delta Z\, e^2 \int (\mathrm{d}\boldsymbol{r}) \frac{\rho_{\mathrm{TF}}(r)}{r} \\
\overset{(4.1.17)}{} &= \delta Z \left[\frac{Ze^2}{b}\underbrace{F'(0)}_{=-B} - \underbrace{\mu}_{=0}\right] = -\delta Z\, \frac{Ze^2}{b} B \\
&= -\delta Z\, Z^{4/3} \frac{e^2}{a_0} \frac{B}{\frac{1}{2}\left(\frac{3\pi}{4}\right)^{2/3}}
\end{aligned} \tag{4.1.32}$$

where

$$\frac{e^2}{a_0} = 2\,\mathrm{Ry} = 27.2\,\mathrm{eV} \tag{4.1.33}$$

is twice the Rydberg* constant (1 Ry is the ionization energy of hydrogen); this e^2/a_0 is the natural energy unit of atomic physics. Accordingly, we get

$$E(Z) = -\frac{6}{7}\left(\frac{4}{3\pi}\right)^{2/3} B\, Z^{7/3}\frac{e^2}{a_0} = -0.7687\, Z^{7/3}\frac{e^2}{a_0} \tag{4.1.34}$$

for the total energy — the negative of the binding energy of all electrons — of a neutral atom in the TF approximation.

For hydrogen ($Z = 1$, $E = -\frac{1}{2}\frac{e^2}{a_0}$) this is not very good (and why should it be?) but it is quite reasonable for atoms with more electrons. Very good agreement is, however, only achieved after the deficiencies of the TF model are corrected for. There is the failure of the TF approximation for small r, because the density changes rapidly near the nucleus and the assumption of a nearly homogeneous electron gas breaks down. Then, there is the exchange energy in addition to the electrostatic energy of $E_{\mathrm{int}}[\rho]$ in (4.1.3), and there are inhomogeneity corrections to the TF functional for the kinetic energy $E_{\mathrm{kin}}[\rho]$ in (4.1.2). A detailed discussion of these matters

---

*Janne RYDBERG (1854–1919)

is beyond the scope of these lecture notes, but we can at least take note of the outcome:

$$E(Z) = \left(-0.7687\, Z^{7/3} + \frac{1}{2} Z^2 - 0.2699\, Z^{5/3}\right) \frac{e^2}{a_0} \,. \tag{4.1.35}$$

The term $\propto Z^2$ is the so-called Scott[*] correction for the strongly bound electrons; the term $\propto Z^{5/3}$ results from the exchange energy and the inhomogeneity correction and was established by Schwinger.[†] The energy formula (4.1.35) is in error of less than 0.1% for $Z \geq 56$, less than 0.2% for $Z \geq 23$, less than 1% for $Z \geq 5$. We leave it at that.

4.2 One-dimensional Ising model

A model that has been the vehicle of discussions in statistical mechanics and other areas for decades, and still is today, is the so-called *Ising*[‡] *model.* In its one-dimensional version, we have a chain of particles

$$\begin{array}{cccccccccccc} \bigcirc & \bigcirc & \cdots & \bigcirc & \bigcirc & \bigcirc & \bigcirc & \bigcirc & \bigcirc & \cdots & \bigcirc & \bigcirc \\ 1 & 2 & & & & j & j+1 & & & & & N \end{array} \tag{4.2.1}$$

with each particle having two internal states of energy $\pm\frac{1}{2}E_0$, contributing

$$\frac{1}{2}E_0 \sum_j s_j \quad \text{with} \quad s_j = 1 \text{ or } -1 \tag{4.2.2}$$

to the energy of the chain. A physical realization could be a string of magnetic dipoles in an external magnetic field and the value of s_j indicates whether the jth dipole is aligned or anti-aligned with the magnetic field; more about this at the end of this section.

There is also a next-neighbor interaction energy of $-Js_js_{j+1}$ with $J > 0$ for the pair $(j, j+1)$ which gives lower energy $-J$ to aligned pairs, $s_j = s_{j+1} = \pm 1$, and higher energy $+J$ to anti-aligned pairs, $s_j = -s_{j+1} = \pm 1$,

$$-Js_js_{j+1} = \begin{cases} -J & \text{if } s_j = s_{j+1}\ ,\ s_js_{j+1} = 1 \\ +J & \text{if } s_j = -s_{j+1}\ ,\ s_js_{j+1} = -1 \end{cases} \tag{4.2.3}$$

[*]John Moffett Cuthbert SCOTT (1911–1974) [†]Julian SCHWINGER (1918–1994)
[‡]Ernst ISING (1900–1998)

and the resulting total energy of the *Ising chain* is

$$E = \frac{1}{2}E_0 \sum_{j=1}^{N} s_j - J \sum_{j=1}^{N-1} s_j s_{j+1} \,. \tag{4.2.4}$$

One can add an interaction term $-Js_N s_1$ to impose "periodic boundary conditions" by which one identifies s_{N+1} with s_1. In this case, the geometry is that of a circle

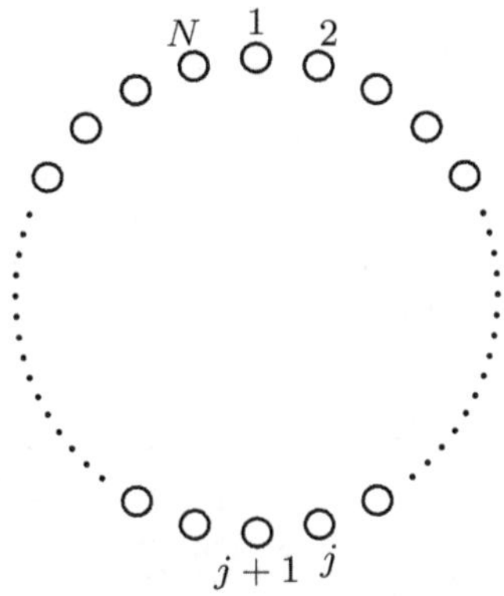

(4.2.5)

and the total energy of this *Ising ring* is

$$\begin{aligned} E &= \frac{1}{2}E_0 \sum_{j=1}^{N} s_j - J \sum_{j=1}^{N-1} s_j s_{j+1} - Js_N s_1 \\ &= \frac{1}{2}E_0 \sum_{j=1}^{N} s_j - J \sum_{j=1}^{N} s_j s_{j+1} \quad \text{with} \quad s_{N+1} \equiv s_1 \,. \end{aligned} \tag{4.2.6}$$

Regarding the thermodynamical properties, the two configurations (chain or ring) are essentially the same, as we shall see, because the extra energy contribution is of relative size $\frac{1}{N}$ and negligible for $N \gg 1$. It is often a matter of convenience whether one works with the chain or the ring geometry, as a particular calculation can be easier for either one of the two.

We wish to evaluate the partition function of the canonical ensemble,

$$Q(\beta, E_0, J, N) = \sum_k \mathrm{e}^{-\beta E_k} \tag{4.2.7}$$

where each configuration k is specified by the list $(s_1, s_2, ..., s_N)$ of s_j values,

$$Q(\beta, E_0, J, N) = \sum_{s_1=\pm 1} \sum_{s_2=\pm 1} \cdots \sum_{s_N=\pm 1} e^{-\frac{1}{2}\beta E_0 \sum_{j=1}^{N} s_j + \beta J \sum_{j=1}^{N(-1)} s_j s_{j+1}}, \tag{4.2.8}$$

where the second sum over j stops at $j = N - 1$ for the chain but includes the term $s_N s_{N+1} \equiv s_N s_1$ for the ring. The $J = 0$ case of the noninteracting Ising model is the model of (2.5.22). We quote the $J = 0$ partition function from (2.5.30),

$$Q(\beta, E_0, J = 0, N) = \left[2 \cosh\left(\tfrac{1}{2}\beta E_0\right)\right]^N, \tag{4.2.9}$$

and find the free energy per particle,

$$\left.\frac{F}{N}\right|_{J=0} = -\frac{1}{\beta N} \log\big(Q(\beta, E_0, 0, N)\big) = -k_B T \log\big(2\cosh\big(\tfrac{1}{2}\beta E_0\big)\big). \tag{4.2.10}$$

Let us now supplement this with the canonical partition function for $E_0 = 0$ and $J \neq 0$, corresponding to the physical model of magnetic particles with no external magnetic field, which is

$$Q(\beta, E_0 = 0, J, N) = \sum_{s_1=\pm 1} \sum_{s_2=\pm 1} \cdots \sum_{s_N=\pm 1} e^{\beta J(s_1 s_2 + s_2 s_3 + \ldots + s_{N-1} s_N)} \tag{4.2.11}$$

for the Ising chain. We observe immediately that changing the signs of all s_js makes no difference to the sum,

$$\sum_{j=1}^{N-1} s_j s_{j+1} \to \sum_{j=1}^{N-1} (-s_j)(-s_{j+1}) = \sum_{j=1}^{N-1} s_j s_{j+1}, \tag{4.2.12}$$

so that we could evaluate the N-fold sum for $s_1 = 1$ only and double the result. Rather than considering then the 2^{N-1} different values of $s_2, s_3, \ldots, s_N$, we can, equivalently, consider the 2^{N-1} different values of $s_1 s_2, s_2 s_3, s_3 s_4, \ldots, s_{N-1} s_N$. Put differently, we specify the values of $s_1, s_2, ..., s_N$ by stating the values of

$$s_1, \; q_1 = s_1 s_2, \; q_2 = s_2 s_3, \ldots, \; q_{N-1} = s_{N-1} s_N \tag{4.2.13}$$

inasmuch as $s_2 = s_1 q_1$, $s_3 = s_1 q_1 q_2$, $s_4 = s_1 q_1 q_2 q_3$, and so forth. Therefore, we get

$$Q(\beta,0,J,N) = \sum_{s_1=\pm1}\sum_{q_1=\pm1}\sum_{q_2=\pm1}\cdots\sum_{q_{N-1}=\pm1} \mathrm{e}^{\beta J(q_1+q_2+\ldots+q_{N-1})}$$
$$= \sum_{s_1=\pm1}\left(\sum_{q=\pm1}\mathrm{e}^{\beta Jq}\right)^{N-1} = 2\left(\mathrm{e}^{\beta J}+\mathrm{e}^{-\beta J}\right)^{N-1}$$
$$= 2\big[2\cosh(\beta J)\big]^{N-1} = \frac{1}{\cosh(\beta J)}\big[2\cosh(\beta J)\big]^{N}. \tag{4.2.14}$$

The resulting free energy per particle is

$$\left.\frac{F}{N}\right|_{E_0=0} = -k_\mathrm{B}T\log\big(2\cosh(\beta J)\big) + \underbrace{\frac{k_\mathrm{B}T}{N}\log\big(\cosh(\beta J)\big)}_{\to 0}$$
$$= -k_\mathrm{B}T\log\big(2\cosh(\beta J)\big), \tag{4.2.15}$$

where $N \gg 1$, as always.

Since $s_N s_1 = q_1 q_2 \cdots q_{N-1}$, this method is not well suited for dealing with the Ising ring, and it is also not convenient when $E_0 \neq 0$. Let us, therefore, rederive $Q(\beta,0,J,N)$ with another method, one that lends itself more easily to generalizations. We regard the four values of

$$\mathrm{e}^{\beta J s_j s_{j+1}} \quad \text{for } s_j=\pm1 \text{ and } s_{j+1}=\pm1 \tag{4.2.16}$$

as the matrix elements of a 2×2 matrix,

$$M = \begin{pmatrix} M_{++} & M_{+-} \\ M_{-+} & M_{--}\end{pmatrix} = \begin{pmatrix} \mathrm{e}^{\beta J} & \mathrm{e}^{-\beta J} \\ \mathrm{e}^{-\beta J} & \mathrm{e}^{\beta J}\end{pmatrix}, \tag{4.2.17}$$

and the rules of matrix multiplication give

$$\sum_{s_2=\pm1}\sum_{s_3=\pm1}\cdots\sum_{s_{N-1}=\pm1} \mathrm{e}^{\beta J(s_1s_2+s_2s_3+\ldots+s_{N-1}s_N)}$$
$$= \sum_{s_2=\pm1}\sum_{s_3=\pm1}\cdots\sum_{s_{N-1}=\pm1} M_{s_1s_2}M_{s_2s_3}\ldots M_{s_{N-1}s_N}$$
$$= \left(M^{N-1}\right)_{s_1s_N}. \tag{4.2.18}$$

Then

$$Q(\beta,0,J,N) = \sum_{s_1=\pm1}\sum_{s_N=\pm1}\left(M^{N-1}\right)_{s_1s_N} = (\,1\;1\,)M^{N-1}\binom{1}{1} \tag{4.2.19}$$

for the Ising chain and

$$Q(\beta,0,J,N)=\sum_{s_1=\pm1}\sum_{s_N=\pm1}\left(M^{N-1}\right)_{s_1s_N}M_{s_Ns_1}=\mathrm{tr}\{M^N\}\qquad(4.2.20)$$

for the Ising ring.

The eigenvalues and eigencolumns of matrix M are exhibited by

$$M\begin{pmatrix}1\\ \pm1\end{pmatrix}=\begin{pmatrix}\mathrm{e}^{\beta J}&\mathrm{e}^{-\beta J}\\ \mathrm{e}^{-\beta J}&\mathrm{e}^{\beta J}\end{pmatrix}\begin{pmatrix}1\\ \pm1\end{pmatrix}=\begin{pmatrix}1\\ \pm1\end{pmatrix}\begin{Bmatrix}2\cosh(\beta J)\\ 2\sinh(\beta J)\end{Bmatrix}.\qquad(4.2.21)$$

Rather immediately, then, we have

$$\begin{aligned}Q(\beta,0,J,N)&=\begin{cases}2\big[2\cosh(\beta J)\big]^{N-1} & \text{for the Ising chain}\\ \big[2\cosh(\beta J)\big]^{N}+\big[2\sinh(\beta J)\big]^{N} & \text{for the Ising ring}\end{cases}\\ &=\big[2\cosh(\beta J)\big]^{N}\times\begin{cases}\cosh(\beta J)^{-1} & \text{(chain)}\\ 1+\tanh(\beta J)^{N} & \text{(ring)}\end{cases}\end{aligned}\qquad(4.2.22)$$

and only the common $[\cdots]^N$ factor is relevant in

$$\left.\frac{F}{N}\right|_{E_0=0}=-k_{\mathrm{B}}T\log\Big(Q(\beta,0,J,N)^{1/N}\Big)=-k_{\mathrm{B}}T\log\big(2\cosh(\beta J)\big)\qquad(4.2.23)$$

since the additional terms are negligibly small for $N\gg1$.

After warming up with the $J=0$ and $E_0=0$ situations, let us now deal with the general case in which $E_0\neq0$ and also $J\neq0$. In preparation for the matrix method, for which symmetric, hermitian matrices are best, we write the energy in (4.2.4) or (4.2.6) in the following way:

$$\begin{aligned}E_k&=\frac{1}{2}E_0\sum_{j=1}^{N}s_j-J\sum_{j=1}^{N-1}s_js_{j+1}+\underbrace{\big[-Js_Ns_1\big]}_{\text{ring only}}\qquad(4.2.24)\\ &=\sum_{j=1}^{N-1}\Big(\frac{1}{4}E_0(s_j+s_{j+1})-Js_js_{j+1}\Big)+\frac{1}{4}E_0\,(s_N+s_1)+\big[-Js_Ns_1\big].\end{aligned}$$

This gives

$$\begin{aligned}\mathrm{e}^{-\beta E_k}&=\left(\prod_{j=1}^{N-1}\mathrm{e}^{-\frac{1}{4}\beta E_0(s_j+s_{j+1})+\beta Js_js_{j+1}}\right)\mathrm{e}^{-\frac{1}{4}\beta E_0(s_N+s_1)+[\beta Js_Ns_1]}\\ &=M_{s_1s_2}M_{s_2s_3}\ldots M_{s_{N-1}s_N}\widetilde{M}_{s_Ns_1}\end{aligned}\qquad(4.2.25)$$

with

$$M = \begin{pmatrix} M_{++} & M_{+-} \\ M_{-+} & M_{--} \end{pmatrix} = \begin{pmatrix} \mathrm{e}^{-\frac{1}{2}\beta E_0 + \beta J} & \mathrm{e}^{-\beta J} \\ \mathrm{e}^{-\beta J} & \mathrm{e}^{\frac{1}{2}\beta E_0 + \beta J} \end{pmatrix} \tag{4.2.26}$$

and

$$\widetilde{M} = M_{J=0} = \begin{pmatrix} \mathrm{e}^{-\frac{1}{2}\beta E_0} & 1 \\ 1 & \mathrm{e}^{\frac{1}{2}\beta E_0} \end{pmatrix} \tag{4.2.27}$$

for the Ising chain, while $\widetilde{M} = M$ for the Ising ring. Accordingly, the canonical partition function is

$$\begin{aligned} Q(\beta, E_0, J, N) &= \sum_{s_1=\pm 1} \sum_{s_2=\pm 1} \cdots \sum_{s_N=\pm 1} \mathrm{e}^{-\beta E_k} \\ &= \sum_{s_1} \sum_{s_2} \cdots \sum_{s_N} M_{s_1 s_2} M_{s_2 s_3} \cdots M_{s_{N-1} s_N} \widetilde{M}_{s_N s_1} \\ &= \mathrm{tr}\Big\{ M^{N-1} \widetilde{M} \Big\}. \end{aligned} \tag{4.2.28}$$

Clearly, we get the expressions in (4.2.19) and (4.2.20) for $E_0 = 0$, and the $J = 0$ result in (4.2.9) is recovered upon noting that

$$M_{J=0} = \widetilde{M} = \begin{pmatrix} \mathrm{e}^{-\frac{1}{4}\beta E_0} \\ \mathrm{e}^{\frac{1}{4}\beta E_0} \end{pmatrix} \begin{pmatrix} \mathrm{e}^{-\frac{1}{4}\beta E_0} & \mathrm{e}^{\frac{1}{4}\beta E_0} \end{pmatrix} \tag{4.2.29}$$

and

$$\begin{aligned} \mathrm{tr}\Big\{ (M_{J=0})^N \Big\} &= \left[\begin{pmatrix} \mathrm{e}^{-\frac{1}{4}\beta E_0} & \mathrm{e}^{\frac{1}{4}\beta E_0} \end{pmatrix} \begin{pmatrix} \mathrm{e}^{-\frac{1}{4}\beta E_0} \\ \mathrm{e}^{\frac{1}{4}\beta E_0} \end{pmatrix} \right]^N \\ &= \left[2\cosh\left(\tfrac{1}{2}\beta E_0\right) \right]^N. \end{aligned} \tag{4.2.30}$$

We handle the powers of M in

$$Q(\beta, E_0, J, N) = \begin{cases} \mathrm{tr}\big\{ M^{N-1} M_{J=0} \big\} & \text{for the Ising chain} \\ \mathrm{tr}\big\{ M^N \big\} & \text{for the Ising ring} \end{cases} \tag{4.2.31}$$

by first writing this 2x2 matrix in terms of its eigenvalues $\lambda_\pm$ and the projectors associated with them,

$$M = \lambda_+ \underbrace{\frac{M - \lambda_- \mathbf{1}_2}{\lambda_+ - \lambda_-}}_{\text{selects } \lambda_+} + \lambda_- \underbrace{\frac{M - \lambda_+ \mathbf{1}_2}{\lambda_- - \lambda_+}}_{\text{selects } \lambda_-}, \tag{4.2.32}$$

where $\mathbf{1}_2 = \begin{pmatrix} 1 & 0 \\ 0 & 1 \end{pmatrix}$ is the 2×2 unit matrix. We determine the eigenvalues from

$$\lambda_{\pm} = \frac{1}{2}\mathrm{tr}\{M\} \pm \frac{1}{2}\sqrt{\mathrm{tr}\{M\}^2 - 4\det\{M\}} \tag{4.2.33}$$

with

$$\begin{aligned} \lambda_+ + \lambda_- &= \mathrm{tr}\{M\} = 2\,\mathrm{e}^{\beta J}\cosh\bigl(\tfrac{1}{2}\beta E_0\bigr)\,, \\ \lambda_+\lambda_- &= \det\{M\} = 2\sinh(2\beta J)\,, \end{aligned} \tag{4.2.34}$$

which gives

$$\begin{aligned} \lambda_{\pm} &= \lambda_{\pm}(\beta E_0, \beta J) \\ &= \mathrm{e}^{\beta J}\cosh\bigl(\tfrac{1}{2}\beta E_0\bigr) \pm \sqrt{\mathrm{e}^{2\beta J}\cosh\bigl(\tfrac{1}{2}\beta E_0\bigr)^2 - 2\sinh(2\beta J)} \\ &= \mathrm{e}^{\beta J}\cosh\bigl(\tfrac{1}{2}\beta E_0\bigr) \pm \sqrt{\mathrm{e}^{2\beta J}\sinh\bigl(\tfrac{1}{2}\beta E_0\bigr)^2 + \mathrm{e}^{-2\beta J}}\,. \end{aligned} \tag{4.2.35}$$

Accordingly, the Nth power of the 2×2 matrix M is

$$M^N = \lambda_+^N \frac{M - \lambda_-\mathbf{1}_2}{\lambda_+ - \lambda_-} + \lambda_-^N \frac{M - \lambda_+\mathbf{1}_2}{\lambda_- - \lambda_+}\,, \tag{4.2.36}$$

and the resulting canonical partition functions are

$$Q(\beta, E_0, J, N) = \mathrm{tr}\bigl\{M^N\bigr\} = \lambda_+^N + \lambda_-^N = \lambda_+^N\left[1 + \left(\frac{\lambda_-}{\lambda_+}\right)^N\right] \tag{4.2.37}$$

for the Ising ring, and

$$\begin{aligned} Q(\beta, E_0, J, N) &= \mathrm{tr}\bigl\{M^{N-1}M_{J=0}\bigr\} \\ &= \lambda_+^{N-1}\frac{\mathrm{tr}\{MM_{J=0} - \lambda_- M_{J=0}\}}{\lambda_+ - \lambda_-} \\ &\quad + \lambda_-^{N-1}\frac{\mathrm{tr}\{MM_{J=0} - \lambda_+ M_{J=0}\}}{\lambda_- - \lambda_+} \\ &= \frac{\lambda_+^N}{\lambda_+ - \lambda_-}\left[\frac{1}{\lambda_+}\left(1 - \left(\frac{\lambda_-}{\lambda_+}\right)^{N-1}\right)\mathrm{tr}\{MM_{J=0}\}\right. \\ &\qquad \left. - \left(\frac{\lambda_-}{\lambda_+} - \left(\frac{\lambda_-}{\lambda_+}\right)^{N-1}\right)\mathrm{tr}\{M_{J=0}\}\right] \end{aligned} \tag{4.2.38}$$

for the Ising chain, where

$$\mathrm{tr}\{M_{J=0}\} = 2\cosh\bigl(\tfrac{1}{2}\beta E_0\bigr) \tag{4.2.39}$$

and

$$\operatorname{tr}\{MM_{J=0}\} = 2\,\mathrm{e}^{\beta J}\cosh(\beta E_0) + 2\,\mathrm{e}^{-\beta J}\,. \tag{4.2.40}$$

In fact, much of this detail is not relevant, inasmuch as $\dfrac{\lambda_-}{\lambda_+} < 1$ and, therefore,

$$-\beta F = \log Q = N\big[\log\lambda_+ + \{\text{terms that are negligible for } N \gg 1\}\big] \tag{4.2.41}$$

so that

$$\frac{F}{N} = -k_{\mathrm{B}}T\log\big(\lambda_+(\beta E_0, \beta J)\big) \tag{4.2.42}$$

for the free energy per particle. The special cases with $J = 0$ and $E_0 = 0$ in (4.2.10) and (4.2.23) are recovered, of course, since

$$\lambda_+\Big|_{\beta J\,=\,0} = 2\cosh\big(\tfrac{1}{2}\beta E_0\big) \quad\text{and}\quad \lambda_+\Big|_{\beta E_0\,=\,0} = 2\cosh(\beta J) \tag{4.2.43}$$

are the respective particular cases of (4.2.35).

As an application, let us consider the situation of N magnetic dipoles $\boldsymbol{\mu}$ in an external magnetic field $\boldsymbol{B}$. A single dipole has the energy

$$-\boldsymbol{\mu}\cdot\boldsymbol{B} = \begin{cases} -|\boldsymbol{\mu}||\boldsymbol{B}| & \text{if the dipole is aligned with the field,}\\ +|\boldsymbol{\mu}||\boldsymbol{B}| & \text{if it is anti-aligned.}\end{cases} \tag{4.2.44}$$

We associate $s = +1$ with the aligned dipole and $s = -1$ with the anti-aligned dipole. Then this magnetic energy is

$$\frac{1}{2}E_0 s \quad\text{with}\quad E_0 = -2|\boldsymbol{\mu}||\boldsymbol{B}|\,. \tag{4.2.45}$$

The magnetization of the whole Ising chain is

$$\boldsymbol{M} = |\boldsymbol{\mu}|\boldsymbol{e}\left\langle\sum_{j=1}^{N} s_j\right\rangle \tag{4.2.46}$$

where $\boldsymbol{e} = \boldsymbol{B}/|\boldsymbol{B}|$ is the unit vector for the direction of the magnetic field and

$$\begin{aligned}\left\langle\sum_{j=1}^{N} s_j\right\rangle &= \frac{1}{Q}\sum_k \mathrm{e}^{-\beta E_k}\sum_{j=1}^{N} s_j = -2\left(\frac{\partial}{\partial(\beta E_0)}\log Q\right)_{\beta J,N}\\ &= -2N\left(\frac{\partial}{\partial(\beta E_0)}\log\lambda_+\right)_{\beta J}.\end{aligned} \tag{4.2.47}$$

Rather than differentiating the expression for λ_+ in (4.2.35), we exploit the relations in (4.2.34) to establish first

$$\begin{aligned} \frac{\partial\lambda_+}{\partial(\beta E_0)} + \frac{\partial\lambda_-}{\partial(\beta E_0)} &= \mathrm{e}^{\beta J}\sinh\bigl(\tfrac{1}{2}\beta E_0\bigr)\,, \\ \lambda_-\frac{\partial\lambda_+}{\partial(\beta E_0)} + \lambda_+\frac{\partial\lambda_-}{\partial(\beta E_0)} &= 0 \end{aligned} \tag{4.2.48}$$

and then

$$\frac{1}{\lambda_+}\frac{\partial\lambda_+}{\partial(\beta E_0)} = \frac{\partial\log\lambda_+}{\partial(\beta E_0)} = \frac{\mathrm{e}^{\beta J}\sinh\bigl(\frac{1}{2}\beta E_0\bigr)}{\lambda_+ - \lambda_-} \tag{4.2.49}$$

so that

$$\begin{aligned} \frac{\boldsymbol{M}}{N} &= -2|\boldsymbol{\mu}|\boldsymbol{e}\frac{\mathrm{e}^{\beta J}\sinh\bigl(\frac{1}{2}\beta E_0\bigr)}{\lambda_+ - \lambda_-} \\ &= -|\boldsymbol{\mu}|\boldsymbol{e}\frac{\sinh\bigl(\frac{1}{2}\beta E_0\bigr)}{\sqrt{\sinh\bigl(\frac{1}{2}\beta E_0\bigr)^2 + \mathrm{e}^{-4\beta J}}} \\ &= \frac{\sinh\bigl(\beta|\boldsymbol{\mu}||\boldsymbol{B}|\bigr)\,|\boldsymbol{\mu}|\boldsymbol{e}}{\sqrt{\sinh\bigl(\beta|\boldsymbol{\mu}||\boldsymbol{B}|\bigr)^2 + \mathrm{e}^{-4\beta J}}}\,, \end{aligned} \tag{4.2.50}$$

where $E_0 = -2|\boldsymbol{\mu}||\boldsymbol{B}|$ from (4.2.45) is inserted in the last step.

The main observation is Ising's result of the 1920s, namely that $\boldsymbol{M} \to 0$ when $\boldsymbol{B} \to 0$: There is no residual magnetization of the Ising chain for vanishing magnetic field — the one-dimensional Ising model does not exhibit spontaneous magnetization. This is different in two and three dimensions, but we need to prepare some tools first before we can deal with those much more complicated situations.

It is, in fact, possible to evaluate the canonical partition function for the two-dimensional Ising model with the aid of matrix multiplication techniques analogous, but not very similar, to the one used above for the one-dimensional Ising model. That was accomplished by Onsager* in the 1940s for the $E_0 = 0$ case, and supplemented by Yang's† solution for $E_0 \neq 0$ some twenty years later. These impressive works were true *tours de force*, and they established that there is a phase transition in the two-dimensional Ising model that gives spontaneous magnetization below a critical temperature. We shall not reproduce Onsager's argument here, but shall rather discuss methods for obtaining approximate solutions — methods that can be applied to other situations as well.

---

*Lars ONSAGER (1903–1976) †Chen-Ning (Frank) YANG (b. 1922)

4.3 Renormalization group method (Ising chain)

The first approximation method is the *renormalization group method*, which was pioneered and developed by Kadanoff* and Wilson.† We use the one-dimensional Ising model for a first illustration of the renormalization group reasoning. Let us thus reconsider the canonical partition function for $E_0 = 0$, which we now regard as a function of $K = \beta J$, the interaction strength, and N, the number of particles,

$$\begin{aligned} Q(K,N) &= \sum_{\text{all } s_j} \mathrm{e}^{K(s_1 s_2 + s_2 s_3 + s_3 s_4 + \cdots)} \\ &= \sum_{\substack{s_j \\ (j\text{ odd})}} \sum_{s_2} \mathrm{e}^{K(s_1 s_2 + s_2 s_3)} \sum_{s_4} \mathrm{e}^{K(s_3 s_4 + s_4 s_5)} \cdots , \end{aligned} \tag{4.3.1}$$

where we have rearranged the summations such that we can immediately evaluate the sums over $s_2, s_4, s_6, \ldots$ — all s_js with even j, while leaving the sums over $s_1, s_3, s_5, \ldots$, those with odd j for later. This gives

$$\begin{aligned} Q(K,N) = \sum_{\substack{s_j \\ (j\text{ odd})}} \Bigl[& 2\cosh\bigl(K(s_1+s_3)\bigr)\, 2\cosh\bigl(K(s_3+s_5)\bigr) \\ & \times 2\cosh\bigl(K(s_5+s_7)\bigr)\, 2\cosh\bigl(K(s_7+s_9)\bigr) \cdots \Bigr] , \end{aligned} \tag{4.3.2}$$

which we wish to rewrite such that it resembles the original expression for $Q(K,N)$, but now with half as many s_js to sum over. Upon writing

$$2\cosh\bigl(K(s+s')\bigr) = g(K)\,\mathrm{e}^{K'ss'} \tag{4.3.3}$$

with $g(K)$ and K' to be determined, we then have

$$\begin{aligned} Q(K,N) &= g(K)^{\frac{1}{2}N} \sum_{\substack{s_j \\ (j\text{ odd})}} \mathrm{e}^{K'(s_1 s_3 + s_3 s_5 + s_5 s_7 + \cdots)} \\ &= g(K)^{\frac{1}{2}N} Q\bigl(K', \tfrac{1}{2}N\bigr) , \end{aligned} \tag{4.3.4}$$

which relates $Q(K,N)$ to the partition function for half as many particles and a modified interaction strength K'. This is really possible because we get two equations for K' and $g(K)$ by putting $ss' = \pm 1$,

$$\begin{aligned} ss' = 1\,,\ s+s' = \pm 2 :&\quad 2\cosh(2K) = g(K)\,\mathrm{e}^{K'} , \\ ss' = -1\,,\ s+s' = 0 :&\quad 2 = g(K)\,\mathrm{e}^{-K'} . \end{aligned} \tag{4.3.5}$$

*Leo Philip KADANOFF (1937–2015) †Kenneth Geddes WILSON (1936–2013)

Accordingly, we have

$$g(K) = 2\sqrt{\cosh(2K)} \tag{4.3.6}$$

and

$$e^{2K'} = \cosh(2K) \quad \text{or} \quad \tanh(K') = \tanh(K)^2 . \tag{4.3.7}$$

We note that

$$K' < K \tag{4.3.8}$$

since $\tanh(K)^2 < \tanh(K)$ and $\tanh(\ )$ is monotonically growing.

The step from $Q(K,N)$ to $Q\bigl(K', \frac{1}{2}N\bigr)$ reduces the particle number as well as the interaction strength. We can repeat this transformation and reduce K further until it is so small that $Q(K,N) \cong Q(0,N) = 2^N$ is a permissible approximation. Or we can reverse the process and read

$$Q\bigl(K', \tfrac{1}{2}N\bigr) = g(K)^{-\frac{1}{2}N} Q(K,N) \tag{4.3.9}$$

as relating $Q\bigl(K', \frac{1}{2}N\bigr)$ to the partition function for twice as many particles and a larger interaction strength.

To be more systematic about this, we note that $-\beta F = \log Q$ is extensive, so that $\log\bigl(Q(K,N)\bigr) = Nq(K)$ with the intensive quantity $q(K)$. For $q(K)$, we have

$$Nq(K) = \frac{N}{2}\log\bigl(g(K)\bigr) + \frac{N}{2}q(K') \tag{4.3.10}$$

or

$$\begin{aligned} q(K) &= \frac{1}{2}\log\bigl(g(K)\bigr) + \frac{1}{2}q(K') \\ &= \frac{1}{2}q(K') + \frac{1}{2}K' + \frac{1}{2}\log 2 , \end{aligned} \tag{4.3.11}$$

where $g(K) = 2\,e^{K'}$ is used. For an iteration that converts $K_{\text{old}} \equiv K'$ into $K_{\text{new}} \equiv K$, this says

$$q(K_{\text{new}}) = \frac{1}{2}q(K_{\text{old}}) + \frac{1}{2}K_{\text{old}} + \frac{1}{2}\log 2 \tag{4.3.12}$$

with $e^{2K_{\text{old}}} = \cosh(2K_{\text{new}})$ or

$$K_{\text{new}} = \tanh^{-1}\left(\sqrt{\tanh(K_{\text{old}})}\right) . \tag{4.3.13}$$

The iteration should start with a small value of $K_{\text{old}} = K'$, such as $K' = 10^{-2}$ and $q(K_{\text{old}}) = \log 2$ as corresponds to $Q = 2^N = e^{Nq}$.

We note that the map $K_{\text{old}} \to K_{\text{new}}$ has two fixed points: $K = 0$ and $K = \infty$ whereas $K_{\text{new}} > K_{\text{old}}$ for all intermediate values. The iteration will, therefore, take us from the initial K value, $0 < K \ll 1$ and $q(K) = \log 2$, to ever larger K values, eventually reaching $K \gg 1$ and $q(K) = K$. The limiting values of $q(K \to 0) = \log 2$ and $q(K \to \infty) = K$ are available from the $q(K_{\text{old}}) \to q(K_{\text{new}})$ map in (4.3.12). This plot reports the outcome of such an iteration:

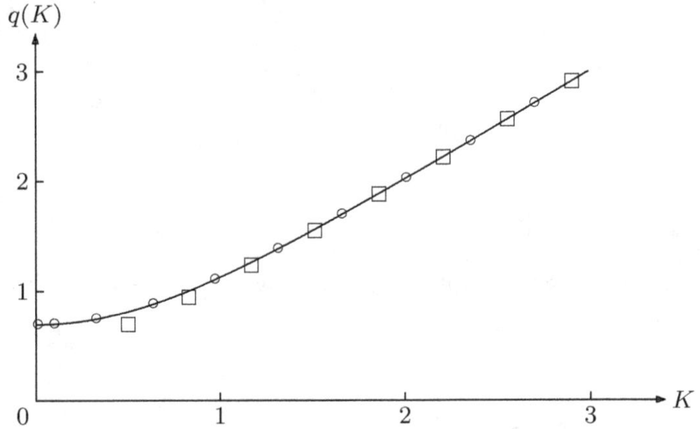

(4.3.14)

The solid line shows the exact result $q(K) = \log\bigl(2\cosh(K)\bigr)$ and the sequences of squares or circles are the successive values obtained from the iteration starting with $q = \log 2$ and $K = 0.5$ (squares $\square$) or $q = \log 2$ and $K = 0.01$ (circles $\circ$). In the $\square$ case, we see that the error in q is decreasing; this is also so in the $\circ$ case, but there the error is so small to begin with that the decrease is not recognizable in the plot.

The error reduction can be understood quite easily. The map $K_{\text{old}} \to K_{\text{new}}$ in (4.3.13) is exact and there is no essential numerical error. The map

$$q_{\text{old}} \to q_{\text{new}} = \frac{1}{2}q_{\text{old}} + \frac{1}{2}K_{\text{old}} + \frac{1}{2}\log 2 \qquad (4.3.15)$$

is such that the error $\epsilon_{\text{old}} = q(K_{\text{old}}) - q_{\text{old}}$ is turned into

$$\epsilon_{\text{new}} = q(K_{\text{new}}) - q_{\text{new}} = \frac{1}{2}\epsilon_{\text{old}}\,, \qquad (4.3.16)$$

so that after n iteration steps we only have 2^{-n} times the initial error.

The reverse situation of the error being doubled in each step is the case for the reverse iteration from larger K values to smaller ones. We must, therefore, always start from $0 < K \ll 1$ and use the iteration scheme of (4.3.13), in which K grows.

Before moving on to the two-dimensional Ising model, let us briefly summarize. First, we had the Ising chain (or ring) with next-neighbor interaction:

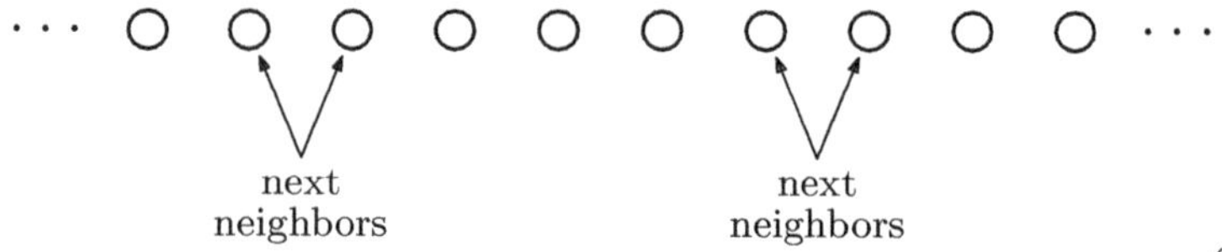

(4.3.17)

and a canonical partition function

$$Q(K, N) = \sum_{\text{all sites}} \mathrm{e}^{K \sum_{(\mathrm{nn})} ss'} \quad \text{with} \quad K = \beta J \tag{4.3.18}$$

where $\sum_{(\mathrm{nn})}$ stands for the sum over next-neighbor pairs of chain sites. Second, we explicitly summed over every second site:

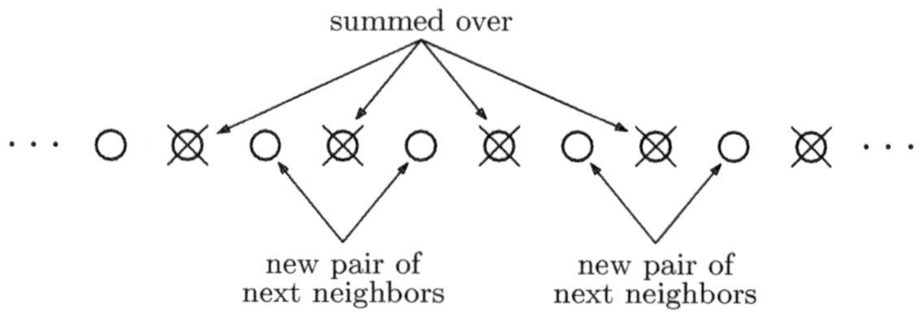

(4.3.19)

and the remaining chain with half as many sites has new pairs of next neighbors. Third, we observed that the sum over the new pairs of next neighbors can be written such that it looks exactly like the original sum except for a different value of the interaction strength K and an overall factor,

$$Q(K, N) = g(K)^{\frac{1}{2}N} Q\big(K', \tfrac{1}{2}N\big) \tag{4.3.20}$$

with $\tanh(K') = \tanh(K)^2$ and $g(K) = 2\sqrt{\cosh(2K)} = 2\,\mathrm{e}^{K'}$. Fourth, remembering that $-\beta F = \log Q$ is extensive whereas $K = \beta J$ is intensive, we wrote $Q(K, N) = \mathrm{e}^{Nq(K)}$ and concluded that

$$q(K) = \frac{1}{2}q(K') + \frac{1}{2}\log\big(g(K)\big) = \frac{1}{2}q(K') + \frac{1}{2}K' + \frac{1}{2}\log 2\,. \tag{4.3.21}$$

Fifth, starting from $K' \ll 1$, $q(K') \cong \log 2$, we found other values of $q(K)$ by iteration.

The crucial step is the summation over a finite fraction of the sites (here: half of them) followed by making the result look like a scaled version (new K value) of the original problem. Such a scaling is known as a *renormalization* and since the map $K, q(K) \to K', q(K')$ can be concatenated, which is a group property, one calls the equations of the map the "renormalization group equations." Wilson turned renormalization group techniques into a powerful tool for studying phase transitions in the 1970s, an effort that was eventually rewarded by the 1982 Nobel Prize in Physics. An important precursor to Wilson's work were ideas developed by Kadanoff in the 1960s.

4.4 Renormalization group method (Ising square lattice)

We use the two-dimensional Ising model for an illustration of the more general situation. Rather than the one-dimensional chain, we now have a regular square array of sites:

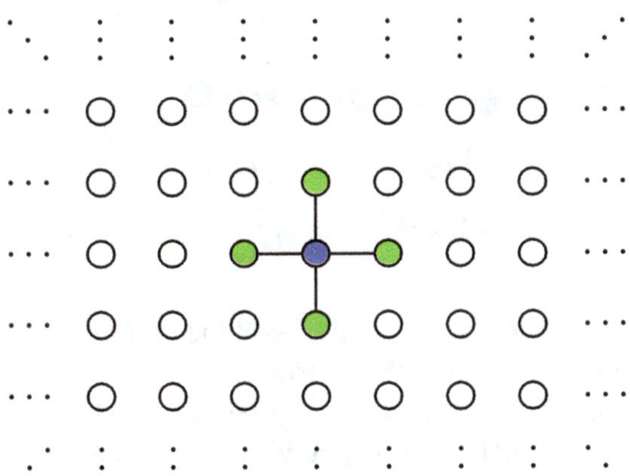

(4.4.1)

As indicated for a selected site (blue filled circle), which could be any site, each site has four next-neighbor sites (green filled circles). The interaction energy is a sum over next-neighbor pairs,

$$\frac{1}{2}\sum_{j,j'} J_{jj'} s_j s_{j'} = J \sum_{(\text{nn})} s s' \quad \text{with} \quad J_{jj'} = \begin{cases} J & \text{for next-neighbor sites,} \\ 0 & \text{else,} \end{cases} \tag{4.4.2}$$

where we note the factor $\frac{1}{2}$ that prevents double counting of the pairs in the double sum over sites j, j', whereas there is no double counting in the (nn) summation over next-neighbor pairs.

The canonical partition function has the same general structure as before,

$$Q(K,N) = \sum_{\text{all sites}} \mathrm{e}^{K\sum_{(\mathrm{nn})} ss'} \quad \text{with} \quad K = \beta J\,. \tag{4.4.3}$$

Again, we can sum over every other site and obtain a smaller array of the same geometry:

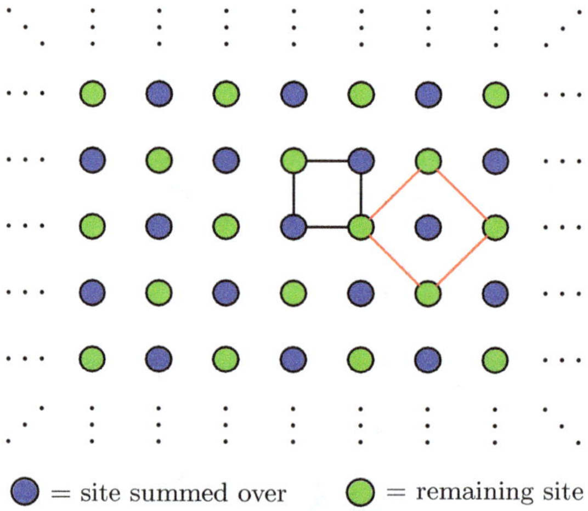

(4.4.4)

The physical square lattice is made up of all blue and green sites; the black square with two blue and two green corners shows one lattice cell. After summing over all blue sites, the lattice of green sites is left over, and this is also a square lattice with the red square identifying one of the cells. Thus, the array of the remaining sites has the same geometry as the original array, with the picture rotated by 45 degrees.

Let us focus on one blue site with its four nearest neighbors:

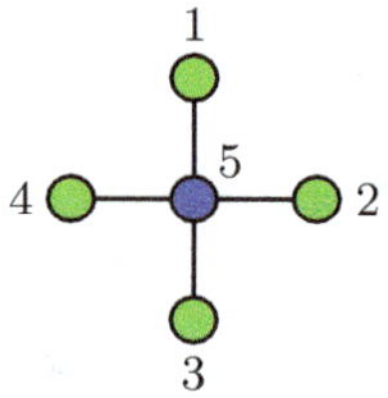

(4.4.5)

In the canonical partition function $Q(K,N)$ we have the factor

$$\sum_{s_5=\pm 1} \mathrm{e}^{K(s_1 s_5 + s_2 s_5 + s_3 s_5 + s_4 s_5)} = 2\cosh\bigl((s_1+s_2+s_3+s_4)K\bigr) \quad (4.4.6)$$

associated with the new square of sites 1234 that has site 5 at its center. There is one such factor for each new square, so that

$$Q(K,N) = \sum_{\substack{\text{remaining}\\ \text{sites}}} \prod_{\substack{\text{all new}\\ \text{squares}}} 2\cosh\left(K \sum_{\substack{\text{one}\\ \text{square}}} s\right) \quad (4.4.7)$$

with half as many remaining sites $\left(\frac{1}{2}N\right)$ as there were original sites (N). Upon denoting by s_1, s_2, s_3, s_4 the s values at the corners of a new square, as in (4.4.5), and trying to repeat what was so successful for the one-dimensional Ising model, we make the ansatz

$$2\cosh\bigl((s_1+s_2+s_3+s_4)K\bigr) = g(K)\,\mathrm{e}^{\frac{1}{2}K'(s_1 s_2 + s_2 s_3 + s_3 s_4 + s_4 s_1)} \quad (4.4.8)$$

which recognizes that (1,2), (2,3), (3,4), and (4,1) are now the sites of next-neighbor pairs. We have $\frac{1}{2}K'$ here, rather than K' as in (4.3.3), because each of these next-neighbor pairs belongs to two squares of the red kind in (4.4.4), and we must not count their contribution to the energy twice. The ansatz (4.4.8), however, does not work, because we have three possible values for the left-hand side,

$$\left.\begin{array}{r}(s_1,s_2,s_3,s_4) = (1,1,1,1)\\ \text{or } (-1,-1,-1,-1)\end{array}\right\} : \ 2\cosh(4K) = g(K)\,\mathrm{e}^{2K'}\,,$$

$$\left.\begin{array}{r}(s_1,s_2,s_3,s_4) = (1,1,1,-1)\\ \text{or } (-1,-1,-1,1)\\ \text{and cyclic permutations}\end{array}\right\} : \ 2\cosh(2K) = g(K)\,,$$

$$\left.\begin{array}{r}(s_1,s_2,s_3,s_4) = (1,1,-1,-1)\\ \text{and cyclic permutations}\end{array}\right\} : \ 2 = g(K)\,,$$

$$\left.\begin{array}{r}(s_1,s_2,s_3,s_4) = (1,-1,1,-1)\\ \text{and cyclic permutations}\end{array}\right\} : \ 2 = g(K)\,\mathrm{e}^{-2K'}\,, \quad (4.4.9)$$

and these four equations contradict each other (unless $K = K' = 0$).

Given that these are four different assignments of ± 1 to the four corners of the square (we keep the invariance under overall sign changes and under cyclic permutation intact), a better ansatz should have three K's rather than just one. This suggests to try

$$\begin{aligned} 2\cosh\bigl((s_1+s_2+s_3+s_4)K\bigr) &= g(K)\,\mathrm{e}^{\frac{1}{2}K'(s_1s_2+s_2s_3+s_3s_4+s_4s_1)} \\ &\quad\times \mathrm{e}^{K''(s_1s_3+s_2s_4)} \\ &\quad\times \mathrm{e}^{K'''s_1s_2s_3s_4}\,, \end{aligned} \tag{4.4.10}$$

where the exponent proportional to K' is the sum over (new) next neighbors, that proportional to K'' is the sum over next-next neighbors, and that proportional to K''' is for the square as a whole. Note that a term proportional to $s_1s_2s_3+s_2s_3s_4+s_3s_4s_1+s_4s_1s_3$ is all right as far as cyclic permutations are concerned but changes sign when all signs are reversed; such a term should not be there.

For the four different assignments of ± 1 to s_1, s_2, s_3, and s_4 in (4.4.9) we now have the four equations

$$\begin{aligned} 2\cosh(4K) &= g(K)\,\mathrm{e}^{2K'+2K''+K'''}\,, \\ 2\cosh(2K) &= g(K)\,\mathrm{e}^{-K'''}\,, \\ 2 &= g(K)\,\mathrm{e}^{-2K''+K'''}\,, \\ 2 &= g(K)\,\mathrm{e}^{-2K'+2K''+K'''}\,, \end{aligned} \tag{4.4.11}$$

for the four unknowns K', K'', K''' and $g(K)$. The solutions are

$$\begin{aligned} \mathrm{e}^{4K'} &= \mathrm{e}^{8K''} = \cosh(4K)\,, \\ \mathrm{e}^{-2K'''} &= \mathrm{e}^{-K'}\cosh(2K)\,, \end{aligned} \tag{4.4.12}$$

and

$$g(K) = 2\cosh(2K)\,\mathrm{e}^{K'''} \tag{4.4.13}$$

or, more explicitly,

$$\begin{aligned} K' &= \frac{1}{4}\log\bigl(\cosh(4K)\bigr)\,, \\ K'' &= \frac{1}{8}\log\bigl(\cosh(4K)\bigr)\,, \\ K''' &= \frac{1}{8}\log\bigl(\cosh(4K)\bigr) - \frac{1}{2}\log\bigl(\cosh(2K)\bigr)\,, \end{aligned} \tag{4.4.14}$$

and

$$\begin{aligned} g(K) &= 2\cosh(2K)^{\frac{1}{2}}\cosh(4K)^{\frac{1}{8}} \\ &= 2\,\mathrm{e}^{K'}\cosh(2K')^{\frac{1}{4}}\,. \end{aligned} \tag{4.4.15}$$

Now, for each new square, we have a factor

$$g(K)\,\underbrace{\mathrm{e}^{\frac{1}{2}K'(s_1s_2+s_2s_3+s_3s_4+s_4s_1)}}_{\text{next neighbors}}\;\underbrace{\mathrm{e}^{K''(s_1s_3+s_2s_4)}}_{\text{next−next neighbors}}\;\underbrace{\mathrm{e}^{K'''s_1s_2s_3s_4}}_{\text{square}} \tag{4.4.16}$$

where the pairs (s_1, s_3) and (s_2, s_4) belong only to the square under consideration, whereas each of the pairs (s_1, s_2), (s_2, s_3), (s_3, s_4), and (s_4, s_1) also belongs to the respective adjacent square — the adjacent red square in (4.4.4), that is. Therefore, they all contribute twice to the product of the factors for all new squares, and we get

$$Q(K,N) = g(K)^{\frac{1}{2}N} \sum_{\substack{\text{remaining}\\ \text{sites}}} \mathrm{e}^{K'\sum_{(\mathrm{nn})} ss' + K''\sum_{(\mathrm{nnn})} ss' + K'''\sum_{(\mathrm{sq})} s_1s_2s_3s_4} \tag{4.4.17}$$

with sums over next neighbors (nn), over next-next neighbors (nnn), and new squares (sq). In effect, then, we reduced the number of sites from N to $\frac{1}{2}N$, but the reduced system has a more complicated interaction with (nnn) and (sq) terms in addition to (nn) terms. It is clear that matters will become even more complicated if we should sum over every other of the remaining sites. Rather than the single rescaling of the one-dimensional Ising model, we now have the situation as summarized in

$$Q(\underset{(\mathrm{nn})}{K},\underset{(\mathrm{nnn})}{0},\underset{(\mathrm{sq})}{0},\underbrace{0,0,0,\ldots}_{\text{(even more complicated)}};N) = g(K)^{\frac{1}{2}N} Q\big(K',K'',K''',0,0,0,\ldots;\tfrac{1}{2}N\big) \tag{4.4.18}$$

and another summation will introduce more nonzero interaction strengths for the "even more complicated" contributions. What is needed is a simplifying physical approximation.

Putting $K'' = 0$ and $K''' = 0$ is too brutal; it brings us back to the renormalization-group equations of the one-dimensional Ising model, with $\cosh(4K) = \mathrm{e}^{4K'}$ replacing $\cosh(2K) = \mathrm{e}^{2K'}$ but, other than this factor of two, there is not much difference. A better attempt neglects the four-site (sq) sums and combines the two pair sums into one,

$$K' \sum_{(\mathrm{nn})} ss' + K'' \sum_{(\mathrm{nnn})} ss' \to \widetilde{K} \sum_{(\mathrm{nn})} ss' \tag{4.4.19}$$

with an effective interaction strength $\widetilde{K}$ to be determined in terms of K' and K''. This replacement is suggested by the observation that the (nn) sum and the (nnn) sum favor $s = s'$ over $s \neq s'$, so both terms tend to enforce aligned values at nearby sites.

To get a value for $\widetilde{K}$, let us consider the extreme situation where all s values are the same, and $ss' = 1$ for all pairs. In the two-dimensional square lattice with $\frac{1}{2}N$ sites, each site has four next neighbors and also four next-next neighbors, so there are N next-neighbor pairs and also N next-next neighbor pairs. Then, for $ss' = 1$ everywhere,

$$K' \sum_{(\mathrm{nn})} ss' + K'' \sum_{(\mathrm{nnn})} ss' = NK' + NK'' \tag{4.4.20}$$

and

$$\widetilde{K} \sum_{(\mathrm{nn})} ss' = N\widetilde{K} \tag{4.4.21}$$

which invites

$$\widetilde{K} = K' + K'' = 3K'' = \frac{3}{2}K' = \frac{3}{8}\log\bigl(\cosh(4K)\bigr)\,. \tag{4.4.22}$$

With this simplification, then, we have

$$\begin{aligned} Q(K,N) &= g(K)^{\frac{1}{2}N} Q\bigl(\widetilde{K}, \tfrac{1}{2}N\bigr) \\ \text{with}\quad g(K) &= 2\,\mathrm{e}^{\frac{2}{3}\widetilde{K}} \cosh\bigl(\tfrac{4}{3}\widetilde{K}\bigr)^{\frac{1}{4}}\,, \end{aligned} \tag{4.4.23}$$

and $Q(K,N) = \mathrm{e}^{Nq(K)}$ gives

$$q(K) = \frac{1}{2}q(\widetilde{K}) + \frac{1}{2}\log\bigl(g(K)\bigr) \tag{4.4.24}$$

with

$$\cosh(4K) = \mathrm{e}^{\frac{8}{3}\widetilde{K}} \quad \text{or} \quad \tanh(2K)^2 = \tanh\bigl(\tfrac{4}{3}\widetilde{K}\bigr)\,. \tag{4.4.25}$$

Accordingly,

$$\begin{aligned} q(K) &= \frac{1}{2}q(\widetilde{K}) + \frac{1}{3}\widetilde{K} + \frac{1}{8}\log\Bigl(\cosh\bigl(\tfrac{4}{3}\widetilde{K}\bigr)\Bigr) + \frac{1}{2}\log 2\,, \\ K &= \frac{1}{2}\tanh^{-1}\Bigl(\tanh\bigl(\tfrac{4}{3}\widetilde{K}\bigr)^{\frac{1}{2}}\Bigr) \end{aligned} \tag{4.4.26}$$

are the recurrence relations for the approximate two-dimensional Ising model.

The seemingly minor modification of the $\widetilde{K} \to K$ map is, in fact, quite substantial. For $K, \widetilde{K}$ small, we have approximately

$$(2K)^2 = \frac{4}{3}\widetilde{K} \quad \text{or} \quad K = \sqrt{\frac{1}{3}\widetilde{K}} > \widetilde{K}\,, \tag{4.4.27}$$

and for $K, \widetilde{K}$ large, we have approximately

$$\frac{1}{2}\,\mathrm{e}^{4K} = \mathrm{e}^{\frac{8}{3}\widetilde{K}} \quad \text{or} \quad K = \frac{2}{3}\widetilde{K} + \frac{1}{4}\log 2 < \widetilde{K}\,. \tag{4.4.28}$$

That is, the iteration moves away from the fixed point $K = \widetilde{K} = 0$ toward larger values, and also away from the fixed point at $K = \widetilde{K} = \infty$ toward smaller values. It follows that there is a third fixed point with $K = \widetilde{K} = K_0 \neq 0, \infty$ with K_0 solving

$$\cosh(4K_0) = \mathrm{e}^{\frac{8}{3}K_0}\,, \tag{4.4.29}$$

which gives $K_0 = 0.507$. For $\widetilde{K} = K_0 + \epsilon$, we have $K = K_0 + \frac{2}{3}\epsilon\coth(4K_0) = K_0 + 0.69\epsilon$, showing that the sequence of K values converges to K_0. The picture is this:

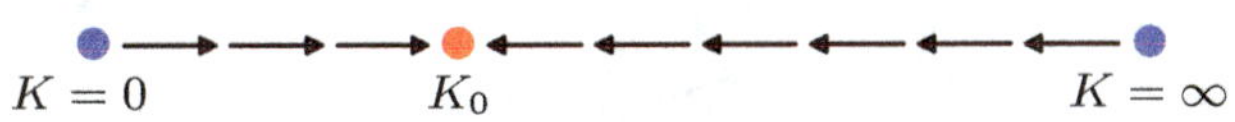

(4.4.30)

which is quite different from what we had for the one-dimensional Ising model, where consecutive K values continue to increase without limit.

Clearly, then, the physics of the two-dimensional Ising model is much richer than that of the one-dimensional model. Since for sufficiently high temperature $T = \dfrac{1}{k_{\mathrm{B}}\beta} = \dfrac{J}{k_{\mathrm{B}}K}$ matters are always similar to those of an ideal gas, we have a gas phase for $K < K_0$ and expect another phase for $K > K_0$ or $T < J/(k_{\mathrm{B}}K_0) \cong 2J/k_{\mathrm{B}}$. In other words, there is a phase transition at $K = K_0$.

4.5 Order-disorder phase transition

To acquire some understanding of this phase transition — and this is the main purpose of renormalization-group techniques — we look at the heat capacity

$$\begin{aligned} C = T\frac{\partial S}{\partial T} = -T\frac{\partial^2 F}{\partial T^2} &= T\Big(\frac{\partial}{\partial T}\Big)^2 \big(k_{\mathrm{B}}TNq(K)\big) \\ &= k_{\mathrm{B}}NT\Big(\frac{\partial}{\partial T}\Big)^2 Tq(K) \end{aligned} \tag{4.5.1}$$

or, with $T\left(\frac{\partial}{\partial T}\right)^2 T = \beta^2\left(\frac{\partial}{\partial \beta}\right)^2$ and $\beta = \frac{K}{J}$,

$$\frac{C}{k_{\mathrm{B}}N} = \beta^2\frac{\partial^2 q}{\partial \beta^2} = K^2\frac{\partial^2 q}{\partial K^2}\,. \tag{4.5.2}$$

Now, a graph of $q(K)$ as obtained from the recurrence relation:

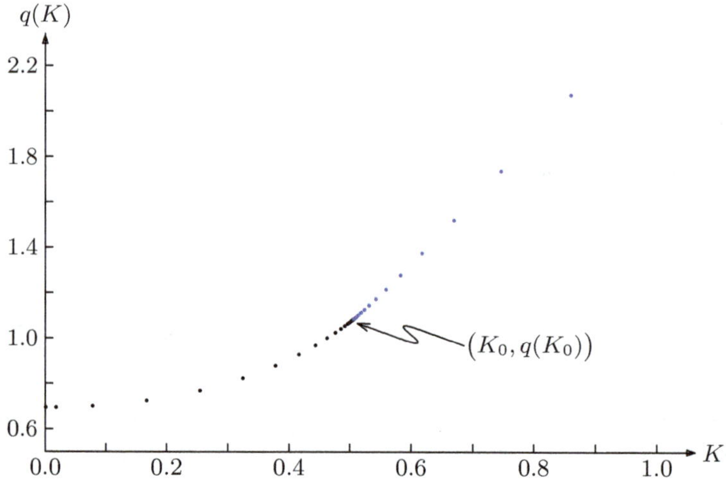

(4.5.3)

shows a smooth function $K \to q(K)$ with no discontinuity or cusp at the point $\big(K_0, q(K_0)\big) = (0.5070, 1.0844)$ in the graph. The sequence of black dots in the plot is produced by starting the iteration with $K = 10^{-3}$ and $q(K) = \log(2) + K^2$; for the sequence of blue dots we start the iteration with $K = 10^2$ and $q(K) = 3K - \frac{3}{4}\log(2)$.

There is, however, the phenomenology of paramagnetic to ferromagnetic phase transitions: Above a critical temperature T_{c} the magnetic dipoles in a metal (iron, for example) are randomly oriented and the solid shows no net magnetization, but at lower temperature the dipoles align, at least partially, and a spontaneous magnetization occurs. The direction of the magnetization vector $\boldsymbol{M}$ is not determined, it is chosen by whatever uncontrolled fluctuations, whereas its size is proportional to a small power ($0.3 \cdots 0.4$ by experience) of $|T - T_{\mathrm{c}}|$, and the specific heat exhibits a singularity proportional to $|T - T_{\mathrm{c}}|^{-\alpha}$ with a small positive value for the critical exponent α. Another example of such an "order-disorder transition" is the well-known case of β-brass, where we have equal amount of copper and zinc but different crystal arrangements above and below 460° C. Below this temperature, each copper atom has as many neighboring copper atoms as it has neighboring zinc atoms; at higher temperatures, the copper and zinc atoms are randomly distributed over the lattice sites.

Such order-disorder phase transitions have no latent heat, so that $\frac{\partial F}{\partial T}$ is continuous at the transition temperature — just as $\frac{\partial q}{\partial K}$ is continuous in (4.5.3). But the second derivative diverges at the transition temperature, consistent with $C \propto |T - T_c|^{-\alpha}$.

In the two-dimensional Ising model, we expect a random distribution of $s = +1$ and $s = -1$ values for $K < K_0$ but there could be a local excess of $s = +1$ values here and matching excess of $s = -1$ values there, as would correspond to a build-up of local order for $K > K_0$. If that is really the case, then we should have

$$q(K) = (\text{smooth function}) + A|K - K_0|^{2-\alpha} \tag{4.5.4}$$

near $K = K_0$ with some constant A and a small positive value for α. The recurrence relations (4.4.26) then require

$$A\,|K - K_0|^{2-\alpha} = \frac{1}{2}A\,\big|\widetilde{K} - K_0\big|^{2-\alpha} \tag{4.5.5}$$

for K and $\widetilde{K}$ very close to K_0, where we safely ignore all smooth contributions to $q(K)$ — those with continuous derivatives. Since

$$\mathrm{d}K\, 4\sinh(4K) = \mathrm{d}\widetilde{K}\,\frac{8}{3}\,\mathrm{e}^{\frac{8}{3}\widetilde{K}} = \mathrm{d}\widetilde{K}\,\frac{8}{3}\cosh(4K) \tag{4.5.6}$$

we have

$$\mathrm{d}\widetilde{K} = \mathrm{d}K\,\frac{3}{2}\tanh(4K) \tag{4.5.7}$$

and

$$\widetilde{K} = K_0 + (K - K_0)\frac{3}{2}\tanh(4K_0) \tag{4.5.8}$$

in the vicinity of $K = \widetilde{K} = K_0$. Accordingly,

$$A\,|K - K_0|^{2-\alpha} = \frac{1}{2}A\left|(K - K_0)\frac{3}{2}\tanh(4K_0)\right|^{2-\alpha}, \tag{4.5.9}$$

which requires

$$\left[\frac{3}{2}\tanh(4K_0)\right]^{2-\alpha} = 2 \tag{4.5.10}$$

or

$$\alpha = 2 - \frac{\log 2}{\log\left(\frac{3}{2}\tanh(4K_0)\right)} = 0.131\,; \tag{4.5.11}$$

the numerical value is for $K_0 = 0.507$, the solution of (4.4.29).

In summary, then, the renormalization-group treatment of the two-dimensional Ising model, with its rather brutal approximation of putting $K''' = 0$ and combining the K' and K'' terms into the single next-neighbor term of strength $\widetilde{K}$, feeds the expectation of an order-disorder transition at the critical temperature

$$T_c \cong \frac{J}{k_B K_0} \tag{4.5.12}$$

with a critical exponent $\alpha \cong 0.131$.

4.6 Onsager's solution

How does this compare with Onsager's solution of the two-dimensional Ising model,

$$q(K) = \log\bigl(2\cosh(K)\bigr) + \frac{1}{\pi}\int_0^{\frac{1}{2}\pi} \mathrm{d}\varphi\, \log\biggl(\frac{1}{2}\Bigl(1 + \sqrt{1-(\gamma\cos\varphi)^2}\Bigr)\biggr) \tag{4.6.1}$$

with $\gamma = \dfrac{2\sinh(2K)}{\cosh(2K)^2} \leq 1$? We note that $\gamma = 1$ only when $\sinh(2K) = 1$, and the argument of the logarithm is never less than $\frac{1}{2}$. Therefore, $q(K)$ is not singular at $K = \frac{1}{2}\sinh^{-1}(1) = 0.4407$. Singularities develop, however, in the derivatives of $q(K)$,

$$\frac{\partial q(K)}{\partial K} = \tanh(K) - \gamma\frac{\partial\gamma}{\partial K}\int_0^{\frac{1}{2}\pi} \frac{\mathrm{d}\varphi}{\pi}\, \frac{(\cos\varphi)^2}{\Bigl(1+\sqrt{1-(\gamma\cos\varphi)^2}\Bigr)\sqrt{1-(\gamma\cos\varphi)^2}}\,. \tag{4.6.2}$$

If we put $\gamma = 1$, then $\sqrt{1-(\gamma\cos\varphi)^2} = \sin\varphi$ and the integral over φ diverges,

$$\int_0^{\frac{1}{2}\pi} \frac{\mathrm{d}\varphi}{\pi}\,\frac{1-\sin\varphi}{\sin\varphi} = \int_0^{\frac{1}{2}\pi}\frac{\mathrm{d}\varphi}{\pi}\left(\frac{1}{\sin\varphi} - 1\right) = \infty\,. \tag{4.6.3}$$

This is, however, multiplied by $\dfrac{\partial\gamma}{\partial K} = 0$ when $\gamma = 1$, so that we get an expression of the "$0 \times \infty$" kind, and a more detailed look at it is needed for establishing its actual value.

At the critical K value, that is: $K_\mathrm{c} = \frac{1}{2}\sinh^{-1}(1)$, we have $\gamma(K_\mathrm{c}) = 1$, $\dfrac{\partial\gamma}{\partial K}(K_\mathrm{c}) = 0$, and $\dfrac{\partial^2\gamma}{\partial K^2}(K_\mathrm{c}) = -8$. Therefore, we can write $\gamma = 1 - \epsilon^2$ for $K \cong K_\mathrm{c}$ with

$$\begin{aligned}\epsilon^2 &= 1 - \gamma(K) = \gamma(K_\mathrm{c}) - \gamma\big(K_\mathrm{c} + (K - K_\mathrm{c})\big) \\ &= -\frac{1}{2}(K - K_\mathrm{c})^2 \frac{\partial^2\gamma}{\partial K^2}(K_\mathrm{c}) = 4(K - K_\mathrm{c})^2 \ll 1\,,\end{aligned} \tag{4.6.4}$$

and $0 < \varphi \ll \frac{1}{2}\pi$, since the singularity of the φ integration occurs at $\varphi = 0$. Then,

$$1 - (\gamma\cos\varphi)^2 = 1 - (1 - 2\epsilon^2)(1 - \varphi^2) = 2\epsilon^2 + \varphi^2 \tag{4.6.5}$$

are the relevant terms in $\sqrt{1 - (\gamma\cos\varphi)^2}$, while we have $(\cos\varphi)^2 \to 1$ and $1 + \sqrt{1 - (\gamma\cos\varphi)^2} \to 1$ in the integral in (4.6.2). We isolate the potential singularity by picking out the contribution from the φ range $0 \cdots \overline{\varphi}$ with a K-independent small positive $\overline{\varphi}$. This gives

$$\begin{aligned}\left[\frac{\partial}{\partial K} q(K)\right]_{\substack{\text{singular}\\ \text{part}}} &= \left(-\gamma\frac{\partial\gamma}{\partial K}\right)_{K \cong K_\mathrm{c}} \int_0^{\overline{\varphi}} \frac{\mathrm{d}\varphi}{\pi} \frac{1}{\sqrt{2\epsilon^2 + \varphi^2}} \\ &= 8(K - K_\mathrm{c})\frac{1}{\pi}\sinh^{-1}\left(\frac{\overline{\varphi}}{\sqrt{2}\,\epsilon}\right)\end{aligned} \tag{4.6.6}$$

or with ϵ from (4.6.4) and $\sinh^{-1}(x) \cong \log(2x)$ for $x \gg 1$,

$$\left[\frac{\partial}{\partial K} q(K)\right]_{\substack{\text{singular}\\ \text{part}}} = -\frac{8}{\pi}(K - K_\mathrm{c})\log\big(|K - K_\mathrm{c}|\big)\,, \tag{4.6.7}$$

which vanishes at $K = K_\mathrm{c}$, so that we have "$0 \times \infty = 0$" in the context of (4.6.2). It follows that $\dfrac{\partial}{\partial K} q(K)$ is continuous but its derivative, the heat capacity

$$\frac{C}{Nk_\mathrm{B}} = K^2 \frac{\partial^2 q(K)}{\partial K^2} = -\frac{8}{\pi} K_\mathrm{c}^2 \log\big(|K - K_\mathrm{c}|\big) \quad \text{for} \quad K \cong K_\mathrm{c}\,, \tag{4.6.8}$$

has a logarithmic singularity at the critical temperature

$$T_c = \frac{1}{k_B \beta_c} = \frac{J}{k_B K_c} = (0.4407)^{-1} \frac{J}{k_B} = 2.269 \frac{J}{k_B} . \tag{4.6.9}$$

We observe that this logarithmic nature of the singularity is not predicted correctly by the renormalization group approximation in Section 4.5, which gave a singularity of the $|K - K_c|^{-\alpha}$ kind. Our simple approximate treatment is qualitatively right — an order-disorder transition near $K = 0.5$ — but fails to correctly identify the logarithmic singularity of the heat capacity.

4.7 Three-dimensional Ising model. Spontaneous magnetization

There is no Onsager-type full solution for the three-dimensional Ising model. One knows from numerical studies that it also exhibits an order-disorder phase transition with

$$C \propto |T - T_c|^{-\alpha} \quad \text{with} \quad \alpha \cong \frac{1}{8} \tag{4.7.1}$$

at the critical temperature. Regarding the magnetization $\boldsymbol{M} \propto \left\langle \sum_j s_j \right\rangle$ of (4.2.46), we have

$$\boldsymbol{M} \propto |T_c - T|^{\underline{\beta}} \quad \text{for} \quad T < T_c \tag{4.7.2}$$

in both the two-dimensional and the three-dimensional Ising model with Onsager's value $\underline{\beta} = \frac{1}{8}$ and $\underline{\beta} \cong 0.313$, respectively.

How do we understand the occurrence of this spontaneous magnetization? The expressions in (4.2.46) and (4.2.47) are equally valid in two and three dimensions,

$$\boldsymbol{M} = |\boldsymbol{\mu}| \boldsymbol{e} \left\langle \sum_j s_j \right\rangle = -2|\boldsymbol{\mu}| \boldsymbol{e} \left(\frac{\partial}{\partial(\beta E_0)} \log Q \right)_{\beta J} \tag{4.7.3}$$

with

$$Q = \sum_k \mathrm{e}^{-\beta E_k} = \sum_{\text{all sites}} \mathrm{e}^{-\frac{1}{2}\beta E_0 \sum_j s_j + \beta J \sum_{(\text{nn})} s s'} , \tag{4.7.4}$$

where $E_0 = -2|\boldsymbol{\mu}||\boldsymbol{B}|$. The expected value

$$\left\langle \sum_j s_j \right\rangle = -2\left(\frac{\partial}{\partial(\beta E_0)} \log Q\right)_{\beta J} = \frac{1}{Q}\sum_k \mathrm{e}^{-\beta E_k}\left(\sum_j s_j\right)_k \tag{4.7.5}$$

is the crucial quantity. If there is no magnetic field, $E_0 = 0$, we have

$$\left\langle \sum_j s_j \right\rangle = \frac{1}{Q} \sum_{\text{all sites}} \mathrm{e}^{\beta J\left(\sum_{(\mathrm{nn})} ss'\right)_k}\left(\sum_j s_j\right)_k . \tag{4.7.6}$$

For each configuration $k \,\widehat{=}\, (s_1, s_2, s_3, \dots)$, this contains the reversed configuration $(-s_1, -s_2, -s_3, \dots)$ with the same Boltzmann weight $\mathrm{e}^{\beta J \sum_{(\mathrm{nn})} ss'}$. Then, should we not get zero for the magnetization $\boldsymbol{M}$ when $\boldsymbol{B} = 0$?

To answer this question, let us consider a weak magnetic field, with its strength $\propto \dfrac{1}{N}$ or only slightly more than that, and we constrain the summation in the partition function to configurations with a fixed value of the magnetization,

$$\widetilde{Q}(M) = \sum_k \mathrm{e}^{-\beta E_k}\, \delta_{\boldsymbol{M}_k, M\boldsymbol{e}} , \tag{4.7.7}$$

where the Kronecker delta selects those configurations for which $\boldsymbol{M}_k = M\boldsymbol{e}$, that is: for which $\left\langle \sum_j s_j \right\rangle$ has a pre-chosen value. The corresponding free energy

$$\widetilde{F}(M) = -k_{\mathrm{B}} T \log\bigl(\widetilde{Q}(M)\bigr) \tag{4.7.8}$$

determines the work $\widetilde{F}(M') - \widetilde{F}(M)$ that is required to reversibly change the magnetization from M to M'. Typically, $\widetilde{F}(M)$ will have a graph like this:

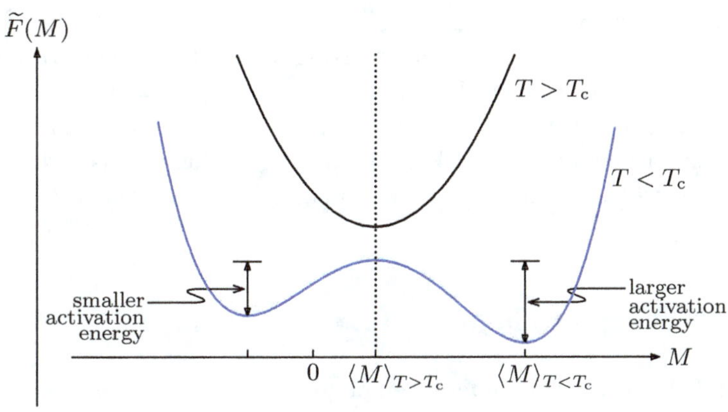

(4.7.9)

with a single minimum for $T > T_c$ and two local minima for $T < T_c$. To cross the barrier between the minima requires an activation energy that is slightly different for the two minima when $\boldsymbol{B} \neq 0$ but equal when the graph is symmetric for $\boldsymbol{B} = 0$:

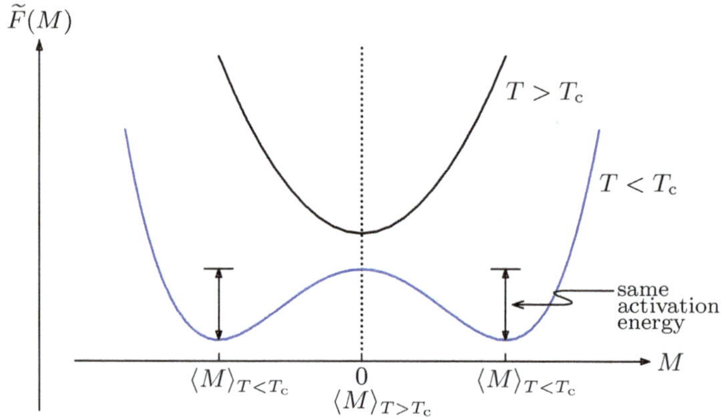

(4.7.10)

In a three-dimensional setting, this activation energy is a surface tension energy proportional to $N^{2/3}$; it is a line tension energy proportional to $N^{1/2}$ for a two-dimensional system, and is, therefore, very large. The Boltzmann factor $\mathrm{e}^{-\beta(\text{activation energy})}$ is vanishingly small, so that fluctuations that would take the system from one local minimum in M to another are so highly unlikely that the local minima correspond to stable configurations of the system. Does this remind you of the familiar phenomenon of hysteresis in iron? It should.

The correct understanding of the $(s_1, s_2, s_3, \dots) \to (-s_1, -s_2, -s_3, \dots)$ argument above is, therefore, that these are equivalent configurations with $\langle M\rangle \to -\langle M\rangle$, while $M = 0$ is not a stable configuration. Fluctuations will immediately drive the system toward one of the local minima.

For this order-disorder transition in the Ising model, the magnetization is the so-called *order parameter*. We have disorder for $\langle M\rangle = 0$ and (partial) order for $\langle M\rangle \neq 0$. An observation of a nonzero magnetization in the absence of an external magnetic field, or of a magnetization that is not simply proportional to the strength of a weak magnetic field, indicates order in the system.

4.8 Mean-field approximation

One way to go about order parameters employs the so-called *mean field*. In the context of the Ising model, we state the energy in terms of the external magnetic field $\boldsymbol{B}$ and the pair interaction strength J,

$$E = \underbrace{-|\boldsymbol{\mu}||\boldsymbol{B}|}_{=\frac{1}{2}E_0}\sum_j s_j - J\sum_{(\mathrm{nn})} ss'$$
$$= \frac{1}{2}E_0\sum_j s_j - \frac{1}{2}\sum_{j,j'} J_{jj'}s_js_{j'}\,, \tag{4.8.1}$$

where we introduce

$$J_{jj'} = \begin{cases} J & \text{for next-neighbor sites} \\ 0 & \text{else} \end{cases} \tag{4.8.2}$$

for notational convenience, and note the factor $\frac{1}{2}$ that multiplies the pair term to avoid double counting of the pairs. The response of E to a change of s_j,

$$\frac{\partial}{\partial s_j}E = \frac{1}{2}E_0 - \sum_{j'} J_{jj'}s_{j'} \equiv \frac{1}{2}\Big(E_0^{(\mathrm{eff})}\Big)_j\,, \tag{4.8.3}$$

identifies the effective magnetic field $|\boldsymbol{B}|_j = -\frac{1}{2|\boldsymbol{\mu}|}\Big(E_0^{(\mathrm{eff})}\Big)_j$ at the jth site, the sum of the external field $\boldsymbol{B}$ and the contribution of all next-neighbor sites. The expected value of $|\boldsymbol{B}|_j$ is the mean field,

$$\Big\langle |\boldsymbol{B}|_j \Big\rangle = -\frac{1}{2|\boldsymbol{\mu}|}\Big\langle \Big(E_0^{(\mathrm{eff})}\Big)_j \Big\rangle$$
$$\text{with}\quad \Big\langle \Big(E_0^{(\mathrm{eff})}\Big)_j \Big\rangle = E_0 - 2\sum_{j'} J_{jj'}\langle s_{j'}\rangle\,. \tag{4.8.4}$$

We now take into account that $\langle s_j\rangle = \langle s_1\rangle \equiv \langle s\rangle$ has the same value at all sites, and that each site has the same number $N_{(\mathrm{nn})}$ of next neighbors,

$$E_0^{(\mathrm{eff})} \equiv \Big\langle \Big(E_0^{(\mathrm{eff})}\Big)_j \Big\rangle = E_0 - 2JN_{(\mathrm{nn})}\langle s\rangle\,, \tag{4.8.5}$$

with the same effective value of E_0 at all sites, $E_0^{(\mathrm{eff})} = -2|\boldsymbol{\mu}|\big|\boldsymbol{B}^{(\mathrm{eff})}\big|$. It is, then, *as if* each site is exposed to an external field $\boldsymbol{B}^{(\mathrm{eff})}$, whose strength depends on $\langle s\rangle$. We determine $\langle s\rangle$ as the expected value with the Boltzmann factor $\mathrm{e}^{-\frac{1}{2}\beta E_0^{(\mathrm{eff})}s}$,

$$\langle s\rangle = \frac{\sum\limits_{s=\pm 1} s\, \mathrm{e}^{-\frac{1}{2}\beta E_0^{(\mathrm{eff})} s}}{\sum\limits_{s=\pm 1} \mathrm{e}^{-\frac{1}{2}\beta E_0^{(\mathrm{eff})} s}} = -\tanh\left(\tfrac{1}{2}\beta E_0^{(\mathrm{eff})}\right)$$
$$= \tanh\left(\beta J N_{(\mathrm{nn})}\langle s\rangle - \tfrac{1}{2}\beta E_0\right). \qquad (4.8.6)$$

The self-consistent value of $\langle s\rangle$ is found as the solution of this equation. We look at it for $E_0 = 0$, no external field,

$$\langle s\rangle = \tanh\left(\beta J N_{(\mathrm{nn})}\langle s\rangle\right). \qquad (4.8.7)$$

Since the left-hand side (LHS) and the right-hand side (RHS) are both odd in $\langle s\rangle$, it is sufficient to consider $\langle s\rangle > 0$, and then the graph is this:

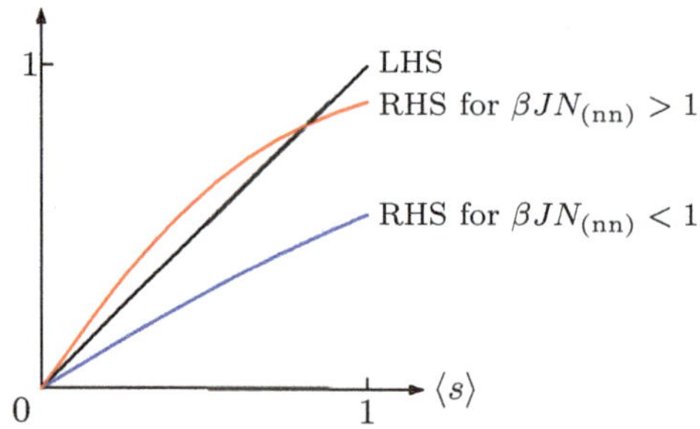

(4.8.8)

There is a solution $\langle s\rangle \neq 0$ (spontaneous magnetization) when $\beta J N_{(\mathrm{nn})} > 1$, but no such solution when $\beta J N_{(\mathrm{nn})} < 1$. Accordingly, the critical temperature $T_\mathrm{c} = \dfrac{1}{k_\mathrm{B}\beta_\mathrm{c}}$ is

$$T_\mathrm{c} = \frac{J N_{(\mathrm{nn})}}{k_\mathrm{B}} \qquad (4.8.9)$$

with

$$N_{(\mathrm{nn})} = \begin{cases} 2 & \text{for the one-dimensional chain,} \\ 4 & \text{for the two-dimensional square lattice,} \\ 6 & \text{for the three-dimensional cubic lattice,} \end{cases} \qquad (4.8.10)$$

which is completely off for the one-dimensional Ising model, where no spontaneous magnetization occurs, and not very good quantitatively for the two- and three-dimensional Ising models:

$$\frac{k_\mathrm{B}T_\mathrm{c}}{J} = \begin{cases} 2 & \text{rather than 0 in 1D,} \\ 4 & \text{rather than } (0.44)^{-1} = 2.27 \text{ in 2D,} \\ 6 & \text{rather than } \cong 4 \text{ in 3D.} \end{cases} \qquad (4.8.11)$$

The critical temperature is over-estimated in each case.

Just below the critical temperature, the relation

$$\beta J N_{(nn)} = \frac{T_c}{T}\beta_c J N_{(nn)} = \frac{T_c}{T} = \frac{1}{2\langle s\rangle}\log\frac{1+\langle s\rangle}{1-\langle s\rangle}$$
$$= 1 + \frac{1}{3}\langle s\rangle^2 + \frac{1}{5}\langle s\rangle^4 + \cdots \qquad (4.8.12)$$

tells us that

$$\langle s\rangle = \sqrt{\frac{3}{T_c}(T_c - T)} \propto (T_c - T)^{\frac{1}{2}}\,, \qquad (4.8.13)$$

that is: the critical exponent equals $\frac{1}{2}$. This value is generic for mean-field theories: whatever the correct value, the mean-field estimate is $\frac{1}{2}$.

For very low temperatures $0 < T \ll T_c$, we have $\langle s\rangle \lesssim 1$ and

$$\frac{T_c}{T} = \frac{1}{2}\log\frac{2}{1-\langle s\rangle} \quad \text{or} \quad \langle s\rangle = 1 - 2\,\mathrm{e}^{-2\frac{T_c}{T}}\,. \qquad (4.8.14)$$

This T dependence of $\langle s\rangle$, both near and far below the critical temperature, is clearly visible in this plot:

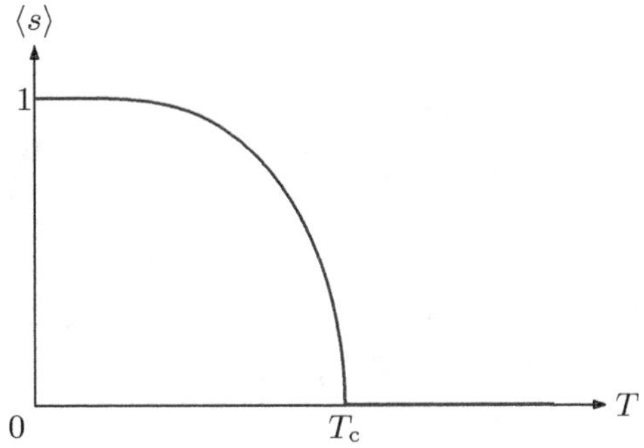

(4.8.15)

It is, of course, possible to refine the mean-field treatment. The simplest extension is obtained by singling out a pair of next-neighbor sites, rather than just one site, and approximating the net effect of all other sites by an effective field. The two-body problem thus defined can be handled, thereby getting an approximate description that deals correctly with single-site and pair-site fluctuation while neglecting more complicated contributions. The discarding of all long-range correlations will, however, still yield wrong critical exponents. The resulting picture is always similar to what we see in (4.8.15).

The mean-field approximation can also be viewed from the perspective of thermodynamical perturbation theory. This is our next topic.

Chapter 5

Perturbation theory

5.1 Large and small contributions

We break up the energy E_k of the kth configuration into a "large and easy" part $E_k^{(0)}$ and a "small" correction $E_k^{(1)}$,

$$E_k = E_k^{(0)} + E_k^{(1)} . \tag{5.1.1}$$

As always, the quality of the approximate treatment by perturbation theory depends on how clever we are in identifying the large and easy $E_k^{(0)}$ and the small $E_k^{(1)}$. For $E_k^{(0)}$, we then have the canonical partition function

$$Q^{(0)}(\beta, X) = \sum_k \mathrm{e}^{-\beta E_k^{(0)}} , \tag{5.1.2}$$

where X stands for all other variables (volume, particle number, ...), but we shall leave the X dependence implicit and write simply $Q^{(0)}(\beta)$, and likewise $Q(\beta)$ for the full partition function,

$$\begin{aligned} Q(\beta) &= \sum_k \mathrm{e}^{-\beta E_k} = \sum_k \mathrm{e}^{-\beta E_k^{(0)} - \beta E_k^{(1)}} \\ &= Q^{(0)}(\beta) \sum_k \frac{\mathrm{e}^{-\beta E_k^{(0)}}}{Q^{(0)}(\beta)} \mathrm{e}^{-\beta E_k^{(1)}} . \end{aligned} \tag{5.1.3}$$

The factor $\mathrm{e}^{-\beta E_k^{(0)}}/Q^{(0)}(\beta)$ is a weight, with which the expected value of $\mathrm{e}^{-\beta E_k^{(1)}}$ is calculated. More generally, we write

$$\langle A_k \rangle^{(0)} = \sum_k \frac{\mathrm{e}^{-\beta E_k^{(0)}}}{Q^{(0)}(\beta)} A_k \tag{5.1.4}$$

for the "(0) mean," or expected value, of a quantity whose value depends on the configuration specified by subscript k. Accordingly, we have

$$Q(\beta) = Q^{(0)}(\beta)\Bigl\langle \mathrm{e}^{-\beta E_k^{(1)}} \Bigr\rangle^{(0)} . \tag{5.1.5}$$

5.2 Gibbs–Bogolyubov inequality

In the next step, we exploit that

$$\mathrm{e}^x \geq 1 + x \quad \text{for all real } x \text{ values,} \tag{5.2.1}$$

which is a well-known inequality that can be demonstrated by a variety of methods, perhaps simplest by noting that $x \mapsto \mathrm{e}^x - 1 - x$ vanishes at $x = 0$ and has a vanishing derivative there while the second derivative is positive for all x. Hence,

$$\begin{aligned}
\Bigl\langle \mathrm{e}^{-\beta E_k^{(1)}} \Bigr\rangle^{(0)} &= \mathrm{e}^{-\beta\left\langle E_k^{(1)}\right\rangle^{(0)}} \Bigl\langle \mathrm{e}^{-\beta\left(E_k^{(1)} - \left\langle E_k^{(1)}\right\rangle^{(0)}\right)} \Bigr\rangle^{(0)} \\
&\geq \mathrm{e}^{-\beta\left\langle E_k^{(1)}\right\rangle^{(0)}} \Bigl\langle \Bigl[1 - \beta\Bigl(E_k^{(1)} - \Bigl\langle E_k^{(1)}\Bigr\rangle^{(0)}\Bigr)\Bigr] \Bigr\rangle^{(0)} \\
&= \mathrm{e}^{-\beta\left\langle E_k^{(1)}\right\rangle^{(0)}} \Bigl[1 - \beta\Bigl\langle E_k^{(1)}\Bigr\rangle^{(0)} + \beta\Bigl\langle E_k^{(1)}\Bigr\rangle^{(0)}\Bigr] \\
&= \mathrm{e}^{-\beta\left\langle E_k^{(1)}\right\rangle^{(0)}} ,
\end{aligned} \tag{5.2.2}$$

and it follows that

$$Q(\beta) \geq Q^{(0)}(\beta)\, \mathrm{e}^{-\beta\left\langle E_k^{(1)}\right\rangle^{(0)}} , \tag{5.2.3}$$

which is known as the *Gibbs–Bogolyubov* inequality*, also called the Gibbs–Bogolyubov–Feynman† bound by some authors.

In view of $-\beta F(\beta) = \log\bigl(Q(\beta)\bigr)$ and $-\beta F^{(0)}(\beta) = \log\bigl(Q^{(0)}(\beta)\bigr)$, the lower bound on the canonical partition function is an upper bound on the free energy,

$$F(\beta) \leq F^{(0)}(\beta) + \Bigl\langle E_k^{(1)} \Bigr\rangle^{(0)} . \tag{5.2.4}$$

This, then, is just another version of the Gibbs–Bogolyubov inequality, completely equivalent to the inequality in (5.2.3).

*Nikolay Nikolayevich BOGOLYUBOV (1909–1992)
†Richard Phillips FEYNMAN (1918–1988)

5.2.1 *Example: Ising model (mean-field approximation)*

For an application to the mean-field treatment of the Ising model, we write

$$\begin{aligned} E_k &= \frac{1}{2}E_0\sum_j s_j - \frac{1}{2}\sum_{j,j'} J_{jj'}s_js_{j'} \\ &= \underbrace{\frac{1}{2}(E_0 - e_0)\sum_j s_j}_{=E_k^{(0)}} + \underbrace{\frac{1}{2}e_0\sum_j s_j - \frac{1}{2}\sum_{j,j'} J_{jj'}s_js_{j'}}_{=E_k^{(1)}} \\ &= E_k^{(0)} + E_k^{(1)} \end{aligned} \tag{5.2.5}$$

with $J_{jj'}$ as in (4.8.1) and an adjustable energy parameter e_0 that we will choose as best as we can. In the mean-field treatment in Section 4.8, we found

$$e_0 = 2JN_{(\mathrm{nn})}\langle s\rangle \tag{5.2.6}$$

in (4.8.5), and now we choose e_0 such that it maximizes the lower bound of the Gibbs–Bogolyubov inequality in (5.2.3). We have

$$Q^{(0)}(\beta) = \sum_k \mathrm{e}^{-\frac{1}{2}\beta(E_0 - e_0)\sum_j s_j} = \left[2\cosh\left(\tfrac{1}{2}\beta(E_0 - e_0)\right)\right]^N \tag{5.2.7}$$

and

$$\begin{aligned} \langle s_j\rangle^{(0)} = \frac{1}{N}\left\langle\sum_{j'=1}^{N} s_{j'}\right\rangle^{(0)} &= \frac{2}{N\beta}\frac{\partial}{\partial e_0}\log\left(Q^{(0)}(\beta)\right) \\ &= -\tanh\left(\tfrac{1}{2}\beta(E_0 - e_0)\right) \equiv \langle s\rangle^{(0)}, \end{aligned} \tag{5.2.8}$$

the same value for all js, as well as $\langle s_js_{j'}\rangle^{(0)} = \langle s_j\rangle^{(0)}\langle s_{j'}\rangle^{(0)} = \left(\langle s\rangle^{(0)}\right)^2$ for $j \neq j'$. Accordingly,

$$\left\langle E_k^{(1)}\right\rangle^{(0)} = \frac{1}{2}e_0N\langle s\rangle^{(0)} - \frac{1}{2}JNN_{(\mathrm{nn})}\left(\langle s\rangle^{(0)}\right)^2, \tag{5.2.9}$$

and the Gibbs–Bogolyubov inequality says

$$\begin{aligned} \log\left(Q(\beta)\right) &\geq \log\left(Q^{(0)}(\beta)\right) - \beta\left\langle E_k^{(1)}\right\rangle^{(0)} \\ &= N\log\Big(2\cosh\left(\tfrac{1}{2}\beta(E_0 - e_0)\right)\Big) \\ &\quad - \frac{1}{2}\beta e_0N\langle s\rangle^{(0)} + \frac{1}{2}\beta JNN_{(\mathrm{nn})}\left(\langle s\rangle^{(0)}\right)^2. \end{aligned} \tag{5.2.10}$$

Our best choice for e_0 is the one for which the right-hand side gives the most useful lower bound, that is: the one for which the right-hand side is largest. Differentiation with respect to e_0 establishes

$$\begin{aligned}&\frac{1}{2}\beta N\langle s\rangle^{(0)} - \frac{1}{2}\beta N\langle s\rangle^{(0)} - \frac{1}{2}\beta N e_0\frac{\partial}{\partial e_0}\langle s\rangle^{(0)}\\&+\beta JNN_{(nn)}\langle s\rangle^{(0)}\frac{\partial}{\partial e_0}\langle s\rangle^{(0)}\\=&-\frac{1}{2}\beta N\Big(e_0 - 2JN_{(\mathrm{nn})}\langle s\rangle^{(0)}\Big)\frac{\partial}{\partial e_0}\langle s\rangle^{(0)} \underset{\text{(want)}}{=} 0\,,\end{aligned} \tag{5.2.11}$$

and this tells us that the best choice is such that $e_0 = 2JN_{(\mathrm{nn})}\langle s\rangle^{(0)}$, which is exactly the mean-field statement in (5.2.6). Accordingly, the mean-field approximation in Section 4.8 is optimal in this sense of giving the largest lower bound in the Gibbs–Bogolyubov inequality.

5.2.2 *Quantum aspects*

So far, we did not explicitly elaborate on quantum-mechanical aspects, except in a general way, for example when assigning one quantum state to phase-space volume $(2\pi\hbar)^{3N}$ in (2.2.1) and (2.3.7), when invoking Bose–Einstein or Fermi–Dirac statistics for indistinguishable particles in Section 3.3, or in the discrete sum in Exercise 61. Let us now take a closer look in the context of the Gibbs–Bogolyubov inequality.

In a quantum-mechanical description, we have the Hamilton operator H for the energy, and its eigenvalues E_k are summed over in the canonical partition function,

$$Q(\beta) = \sum_k \mathrm{e}^{-\beta E_k} \quad \text{with} \quad H|k\rangle = |k\rangle E_k\,, \tag{5.2.12}$$

where $|k\rangle$ is the kth eigenket of H and the energy E_k is its eigenvalue. Recalling that $f(H)|k\rangle = |k\rangle f(E_k)$ for any function $f(H)$ of the Hamilton operator and, as usual, normalizing the eigenkets such that

$$\langle k|k'\rangle = \delta_{kk'}\,, \tag{5.2.13}$$

we thus have

$$\begin{aligned} Q(\beta) &= \sum_k \langle k|\,\mathrm{e}^{-\beta H}|k\rangle \\ &= \mathrm{tr}\left\{\mathrm{e}^{-\beta H}\sum_k |k\rangle\langle k|\right\} = \mathrm{tr}\left\{\mathrm{e}^{-\beta H}\right\}. \end{aligned} \tag{5.2.14}$$

We are relying on three ingredients here: (i) the basic definition of the trace by its linearity,

$$\mathrm{tr}\{A+B\} = \mathrm{tr}\{A\} + \mathrm{tr}\{B\}, \tag{5.2.15}$$

and the value it assigns to a ket-bra,

$$\mathrm{tr}\{|a\rangle\langle b|\} = \langle b|a\rangle\,; \tag{5.2.16}$$

(ii) the cyclic property of the trace,

$$\mathrm{tr}\{AB\} = \mathrm{tr}\{BA\}\,; \tag{5.2.17}$$

and (iii) the completeness of the eigenstates of the Hamilton operator,

$$\sum_k |k\rangle\langle k| = 1\,. \tag{5.2.18}$$

Whereas ingredients (i) and (ii) are of a mathematical nature, (iii) is a physical statement, namely that a measurement of the energy has an outcome — always. The compact final expression

$$Q(\beta) = \mathrm{tr}\left\{\mathrm{e}^{-\beta H}\right\} \tag{5.2.19}$$

makes no reference to the eigenstates and eigenvalues of H and we can evaluate this trace by any method of our choosing and are not bound to always summing over the eigenvalues of H as in the expression we started with in (5.2.12).

The break-up $E_k = E_k^{(0)} + E_k^{(1)}$ in (5.1.1) corresponds to

$$H = H_0 + H_1 \tag{5.2.20}$$

in the quantum setting, but there is no immediate analog to

$$\mathrm{e}^{-\beta E_k} = \mathrm{e}^{-\beta E_k^{(0)}}\,\mathrm{e}^{-\beta E_k^{(1)}} \tag{5.2.21}$$

since, usually, H_0 and H_1 do not commute. Indeed, when $H_0H_1 = H_1H_0$, there is no need for perturbation theory, as the eigenstates of H_0 can be chosen such that they are eigenstates of $H = H_0 + H_1$; therefore, $H_0 H_1 \neq H_1 H_0$ is the generic situation. We must exercise more care, and more skill, when factorizing

$$\mathrm{e}^{-\beta H} = \mathrm{e}^{-\beta H_0 - \beta H_1} = \mathrm{e}^{-\beta H_0}\Bigl(\mathrm{e}^{\beta H_0}\,\mathrm{e}^{-\beta H}\Bigr), \tag{5.2.22}$$

where we need an expression for $\mathrm{e}^{\beta H_0}\,\mathrm{e}^{-\beta H} = \mathrm{e}^{\beta H_0}\,\mathrm{e}^{-\beta H_0 - \beta H_1}$ that accounts for H_1 perturbatively.

Mindful of the eventual evaluation of a trace, let us consider

$$f(\epsilon) = \mathrm{tr}\Bigl\{\mathrm{e}^{A + \epsilon B}\Bigr\} = f(0) + \epsilon f'(0) + \frac{\epsilon^2}{2} f''(\overline{\epsilon}) \tag{5.2.23}$$

for two hermitian* operators A and B, where $0 \leq \overline{\epsilon} \leq \epsilon$ according to Taylor's† theorem. We have

$$f(0) = \mathrm{tr}\Bigl\{\mathrm{e}^{A}\Bigr\} \tag{5.2.24}$$

and find the derivative with respect to ϵ by an application of

$$\delta\,\mathrm{e}^{X} = \int_0^1 \mathrm{d}\alpha\; \mathrm{e}^{\alpha X}\,\delta X\; \mathrm{e}^{(1-\alpha)X}, \tag{5.2.25}$$

the response of an exponential function of an operator to a variation, which is the basic relation of all perturbation theory in quantum mechanics. Here, we use it in the form

$$\frac{\partial}{\partial\epsilon}\,\mathrm{e}^{A + \epsilon B} = \int_0^1 \mathrm{d}\alpha\; \mathrm{e}^{\alpha(A + \epsilon B)}\, B\, \mathrm{e}^{(1-\alpha)(A + \epsilon B)} \tag{5.2.26}$$

with the consequence that

$$f'(0) = \mathrm{tr}\left\{\int_0^1 \mathrm{d}\alpha\; \mathrm{e}^{\alpha A}\, B\, \mathrm{e}^{(1-\alpha)A}\right\} = \mathrm{tr}\Bigl\{B\,\mathrm{e}^{A}\Bigr\} \tag{5.2.27}$$

after another use of the cyclic property of the trace. The $\epsilon \neq 0$ version

$$f'(\epsilon) = \mathrm{tr}\Bigl\{B\,\mathrm{e}^{A + \epsilon B}\Bigr\} \tag{5.2.28}$$

*Charles HERMITE (1822–1901) †Brook TAYLOR (1685–1731)

can be differentiated another time to yield

$$f''(\bar{\epsilon}) = \int_0^1 \mathrm{d}\alpha \; \mathrm{tr}\Big\{B\,\mathrm{e}^{\alpha(A+\bar{\epsilon}B)}B\,\mathrm{e}^{(1-\alpha)(A+\bar{\epsilon}B)}\Big\}$$

$$= \int_0^1 \mathrm{d}\alpha \; \mathrm{tr}\{Y_\alpha^\dagger Y_\alpha\} \geq 0$$

$$\text{with} \quad Y_\alpha = \mathrm{e}^{\frac{1}{2}\alpha(A+\bar{\epsilon}B)}B\,\mathrm{e}^{\frac{1}{2}(1-\alpha)(A+\bar{\epsilon}B)}\,, \tag{5.2.29}$$

since $\mathrm{tr}\{Y_\alpha^\dagger Y_\alpha\} \geq 0$ is the trace of a nonnegative operator.

Upon collecting the pieces we have

$$\begin{aligned}\mathrm{tr}\Big\{\mathrm{e}^{A+\epsilon B}\Big\} &= \mathrm{tr}\Big\{\mathrm{e}^{A}\Big\} + \epsilon\,\mathrm{tr}\Big\{B\,\mathrm{e}^{A}\Big\} + \frac{\epsilon^2}{2}\underbrace{f''(\bar{\epsilon})}_{\geq 0} \\ &\geq \mathrm{tr}\Big\{\mathrm{e}^{A}\Big\} + \epsilon\,\mathrm{tr}\Big\{B\,\mathrm{e}^{A}\Big\}\end{aligned} \tag{5.2.30}$$

and, for $\epsilon = 1$,

$$\mathrm{tr}\Big\{\mathrm{e}^{A+B}\Big\} \geq \mathrm{tr}\Big\{\mathrm{e}^{A}\Big\} + \mathrm{tr}\Big\{B\,\mathrm{e}^{A}\Big\}. \tag{5.2.31}$$

We apply this to

$$A = -\beta H_0 - \beta\langle H_1\rangle_0\,, \quad B = -\beta\big(H_1 - \langle H_1\rangle_0\big) \tag{5.2.32}$$

where

$$\langle(\cdots)\rangle_0 = \frac{\mathrm{tr}\{(\cdots)\,\mathrm{e}^{-\beta H_0}\}}{\mathrm{tr}\{\mathrm{e}^{-\beta H_0}\}} = \frac{\mathrm{tr}\{(\cdots)\,\mathrm{e}^{-\beta H_0}\}}{Q_0(\beta)} \tag{5.2.33}$$

is the analog of the expression for $\langle A_k\rangle^{(0)}$ in (5.1.4). Then

$$\begin{aligned}A + B &= -\beta(H_0 + H_1) = -\beta H\,, \\ \mathrm{tr}\Big\{\mathrm{e}^{A}\Big\} &= \mathrm{tr}\Big\{\mathrm{e}^{-\beta H_0}\Big\}\,\mathrm{e}^{-\beta\langle H_1\rangle_0}\,,\end{aligned} \tag{5.2.34}$$

and

$$\begin{aligned}\mathrm{tr}\Big\{B\,\mathrm{e}^{A}\Big\} &= \mathrm{tr}\Big\{-\beta\big(H_1 - \langle H_1\rangle_0\big)\,\mathrm{e}^{-\beta H_0 - \beta\langle H_1\rangle_0}\Big\} \\ &= -\beta\,\mathrm{e}^{-\beta\langle H_1\rangle_0}Q_0(\beta)\underbrace{\Big\langle\big(H_1 - \langle H_1\rangle_0\big)\Big\rangle_0}_{=0} = 0\,.\end{aligned} \tag{5.2.35}$$

Therefore,

$$\operatorname{tr}\left\{e^{-\beta H}\right\} \geq \operatorname{tr}\left\{e^{-\beta H_0}\right\} e^{-\beta\langle H_1\rangle_0} \tag{5.2.36}$$

or

$$Q(\beta) \geq Q_0(\beta)\, e^{-\beta\langle H_1\rangle_0}\,, \tag{5.2.37}$$

exactly the Gibbs–Bogolyubov inequality in (5.2.3) except for the closely related, if somewhat different, meaning of $\langle H_1\rangle_0$ here and $\left\langle E_k^{(1)}\right\rangle^{(0)}$ there.

5.3 Peierls inequality

While we are at it, let us also look at the so-called *Peierls* inequality*,

$$\operatorname{tr}\left\{e^{-\beta H}\right\} \geq \sum_\gamma e^{-\beta\langle\gamma|H|\gamma\rangle}\,, \tag{5.3.1}$$

where the kets $|\gamma\rangle$ are orthonormal,

$$\langle\gamma|\gamma'\rangle = \delta_{\gamma\gamma'}\,, \tag{5.3.2}$$

but need not be complete. We can, however, consider a complete set of pairwise orthogonal kets $|\gamma\rangle$ because

$$\sum_{\text{all }\gamma\text{s}} e^{-\beta\langle\gamma|H|\gamma\rangle} \geq \sum_{\text{some }\gamma\text{s}} e^{-\beta\langle\gamma|H|\gamma\rangle} \tag{5.3.3}$$

is clearly true since all summands are positive. For a complete set, then, we have

$$\begin{aligned}
\operatorname{tr}\left\{e^{-\beta H}\right\} &= \sum_\gamma \langle\gamma|\, e^{-\beta H}|\gamma\rangle \\
&= \sum_\gamma \langle\gamma|\, e^{-\beta(H - \langle\gamma|H|\gamma\rangle)}|\gamma\rangle\, e^{-\beta\langle\gamma|H|\gamma\rangle} \\
&\geq \sum_\gamma \underbrace{\langle\gamma|\left[1-\beta(H-\langle\gamma|H|\gamma\rangle)\right]|\gamma\rangle}_{=1}\, e^{-\beta\langle\gamma|H|\gamma\rangle} \\
&= \sum_\gamma e^{-\beta\langle\gamma|H|\gamma\rangle}\,,
\end{aligned} \tag{5.3.4}$$

indeed. In fact, the Peierls inequality holds term by term in the sum over γ, essentially as a consequence of $e^x > 1 + x$, once more.

*Sir Rudolf Ernst PEIERLS (1907–1995)

Now, if we choose the eigenstates of H_0 for the set of kets $|\gamma\rangle$, that is

$$H_0|\gamma\rangle = |\gamma\rangle E_{0\gamma}\,, \tag{5.3.5}$$

then $\langle\gamma|H|\gamma\rangle = E_{0\gamma} + \langle\gamma|H_1|\gamma\rangle$ for $H = H_0 + H_1$, and

$$\begin{aligned} Q(\beta) = \mathrm{tr}\Big\{\mathrm{e}^{-\beta H}\Big\} &\geq \sum_\gamma \mathrm{e}^{-\beta E_{0\gamma}}\,\mathrm{e}^{-\beta\langle\gamma|H_1|\gamma\rangle} \\ &= \Big(\sum_{\gamma'} \mathrm{e}^{-\beta E_{0\gamma'}}\Big)\sum_\gamma w_\gamma\,\mathrm{e}^{-\beta\langle\gamma|H_1|\gamma\rangle} \\ &= \underbrace{\mathrm{tr}\Big\{\mathrm{e}^{-\beta H_0}\Big\}}_{=Q_0(\beta)}\sum_\gamma w_\gamma\,\mathrm{e}^{-\beta\langle\gamma|H_1|\gamma\rangle} \end{aligned} \tag{5.3.6}$$

with weights

$$w_\gamma = \frac{\mathrm{e}^{-\beta E_{0\gamma}}}{\sum_{\gamma'}\mathrm{e}^{-\beta E_{0\gamma'}}} = \frac{\mathrm{e}^{-\beta E_{0\gamma}}}{\mathrm{tr}\big\{\mathrm{e}^{-\beta H_0}\big\}} = \frac{\mathrm{e}^{-\beta E_{0\gamma}}}{Q_0(\beta)}\,. \tag{5.3.7}$$

Another use of $\mathrm{e}^x \geq 1 + x$ establishes

$$\sum_\gamma w_\gamma\,\mathrm{e}^{-\beta\big(\langle\gamma|H_1|\gamma\rangle - \sum_{\gamma'} w_{\gamma'}\langle\gamma'|H_1|\gamma'\rangle\big)} \geq 1\,, \tag{5.3.8}$$

and so we have

$$\sum_\gamma w_\gamma\,\mathrm{e}^{-\beta\langle\gamma|H_1|\gamma\rangle} \geq \mathrm{e}^{-\beta\sum_\gamma w_\gamma\langle\gamma|H_1|\gamma\rangle} = \mathrm{e}^{-\beta\langle H_1\rangle_0} \tag{5.3.9}$$

after recognizing that

$$\begin{aligned} \sum_\gamma w_\gamma\langle\gamma|H_1|\gamma\rangle &= \frac{1}{\mathrm{tr}\big\{\mathrm{e}^{-\beta H_0}\big\}}\sum_\gamma \mathrm{e}^{-\beta E_{0\gamma}}\langle\gamma|H_1|\gamma\rangle \\ &= \frac{1}{\mathrm{tr}\big\{\mathrm{e}^{-\beta H_0}\big\}}\sum_\gamma\langle\gamma|\,\mathrm{e}^{-\beta H_0}H_1|\gamma\rangle \\ &= \frac{\mathrm{tr}\big\{\mathrm{e}^{-\beta H_0}H_1\big\}}{\mathrm{tr}\big\{\mathrm{e}^{-\beta H_0}\big\}}\,. \end{aligned} \tag{5.3.10}$$

In the final step, we combine (5.3.6) and (5.3.9) to arrive at the Gibbs–Bogolyubov inequality in (5.2.37).

We can, therefore, regard the Gibbs–Bogolyubov inequality as an implication of the Peierls inequality, which is stronger and, in view of the free choice of orthonormal kets $|\gamma\rangle$, more flexible.

5.4 Minimum principle for the free energy

There is yet another way of establishing the Gibbs–Bogolyubov inequality. It begins with first noting that the free energy $F(\beta)$ obeys a minimum principle,

$$\begin{aligned} F(\beta) &= -\frac{1}{\beta}\log\bigl(Q(\beta)\bigr) = -\frac{1}{\beta}\log\Bigl(\mathrm{tr}\Bigl\{\mathrm{e}^{-\beta H}\Bigr\}\Bigr) \\ &= \underset{\rho}{\mathrm{Min}}\left\{\mathrm{tr}\{H\rho\} + \frac{1}{\beta}\mathrm{tr}\{\rho\log\rho\}\right\}, \end{aligned} \tag{5.4.1}$$

where all nonnegative, unit-trace *statistical operators* ρ participate in the competition,

$$\rho \geq 0\,, \quad \mathrm{tr}\{\rho\} = 1\,, \tag{5.4.2}$$

so that the permissible variations of ρ are constrained by

$$\mathrm{tr}\{\delta\rho\} = 0\,. \tag{5.4.3}$$

On the other hand, we have

$$\mathrm{tr}\{H\delta\rho\} + \frac{1}{\beta}\mathrm{tr}\{\delta\rho\log\rho + \delta\rho\} = 0 \tag{5.4.4}$$

for the extremal ρ and all permissible $\delta\rho$, so that the minimum is obtained for

$$\rho = (\text{constant})\,\mathrm{e}^{-\beta H} = \frac{\mathrm{e}^{-\beta H}}{\mathrm{tr}\bigl\{\mathrm{e}^{-\beta H}\bigr\}} = \mathrm{e}^{-\beta(H\,-\,F(\beta))} \equiv \rho_\beta\,, \tag{5.4.5}$$

where the value of the multiplicative constant is determined by the normalization to unit trace. This ρ_β is usually referred to as the *thermal state* associated with the Hamilton operator H at temperature $T = (k_{\mathrm{B}}\beta)^{-1}$. Since

$$\begin{aligned} &\left(\mathrm{tr}\{H\rho\} + \frac{1}{\beta}\mathrm{tr}\{\rho\log\rho\}\right)\Bigg|_{\rho=\rho_\beta} \\ \log\rho_\beta = -\beta H + \beta F(\beta) \;\longrightarrow\; &= \mathrm{tr}\{H\rho_\beta\} + \frac{1}{\beta}\mathrm{tr}\bigl\{\rho_\beta\bigl[-\beta H + \beta F(\beta)\bigr]\bigr\} \\ &= F(\beta)\,, \end{aligned} \tag{5.4.6}$$

it is true that we get the free energy as the stationary value of $\mathrm{tr}\{H\rho\} + \frac{1}{\beta}\mathrm{tr}\{\rho\log\rho\}$.

To confirm that the extremum is a minimum, we compare the value for an arbitrary ρ with that of ρ_β,

$$\begin{aligned}&\left(\operatorname{tr}\{H\rho\}+\frac{1}{\beta}\operatorname{tr}\{\rho\log\rho\}\right)-\left(\operatorname{tr}\{H\rho_\beta\}+\frac{1}{\beta}\operatorname{tr}\{\rho_\beta\log\rho_\beta\}\right)\\ &=\left(\operatorname{tr}\big\{H\,(\rho-\rho_\beta)\big\}+\frac{1}{\beta}\operatorname{tr}\{\rho\log\rho-\rho_\beta\log\rho_\beta\}\right)\equiv f(\epsilon=1)\end{aligned} \tag{5.4.7}$$

with

$$\begin{aligned}f(\epsilon)&=\operatorname{tr}\big\{H\epsilon\,(\rho-\rho_\beta)\big\}\\ &\quad+\frac{1}{\beta}\operatorname{tr}\big\{\big(\rho_\beta+\epsilon(\rho-\rho_\beta)\big)\log\big(\rho_\beta+\epsilon(\rho-\rho_\beta)\big)-\rho_\beta\log\rho_\beta\big\}\,,\end{aligned} \tag{5.4.8}$$

where

$$\rho_\beta+\epsilon\,(\rho-\rho_\beta)=(1-\epsilon)\,\rho_\beta+\epsilon\rho \quad \text{for} \quad 0\le\epsilon\le 1 \tag{5.4.9}$$

is the convex sum of two statistical operators and, therefore, another statistical operator.

We have $f(0)=0$ by definition and $f'(0)=0$ as a consequence of the stationary property at $\rho=\rho_\beta$, so that it is enough to show that $f''(\epsilon)\ge 0$ for $0\le\epsilon\le 1$. Since

$$f'(\epsilon)=\operatorname{tr}\big\{H(\rho-\rho_\beta)\big\}+\frac{1}{\beta}\operatorname{tr}\big\{(\rho-\rho_\beta)\log\big(\rho_\beta+\epsilon(\rho-\rho_\beta)\big)\big\} \tag{5.4.10}$$

we need to find

$$\begin{aligned}f''(\epsilon)&=\frac{1}{\beta}\operatorname{tr}\left\{(\rho-\rho_\beta)\frac{\partial}{\partial\epsilon}\log\big(\rho_\beta+\epsilon(\rho-\rho_\beta)\big)\right\}\\ &=\frac{1}{\beta}\operatorname{tr}\left\{\frac{\partial A_\epsilon}{\partial\epsilon}\frac{\partial}{\partial\epsilon}\log A_\epsilon\right\}\\ \text{with}\quad A_\epsilon&=\rho_\beta+\epsilon(\rho-\rho_\beta)\,.\end{aligned} \tag{5.4.11}$$

For this purpose, we make use of

$$\begin{aligned}\log A&=\int_1^A\frac{\mathrm{d}a}{a}=\int_1^A\mathrm{d}a\int_0^\infty\frac{\mathrm{d}\alpha}{(\alpha+a)^2}=\int_0^\infty\mathrm{d}\alpha\int_1^A\frac{\mathrm{d}a}{(\alpha+a)^2}\\ &=\int_0^\infty\mathrm{d}\alpha\left(\frac{1}{\alpha+1}-\frac{1}{\alpha+A}\right)\quad\text{for}\quad A>0\end{aligned} \tag{5.4.12}$$

and

$$\delta\frac{1}{B} = -\frac{1}{B}\delta B\frac{1}{B} \quad \text{for all invertible } B\,, \tag{5.4.13}$$

which we combine into

$$\delta \log A = \int_0^\infty \mathrm{d}\alpha\, \frac{1}{\alpha + A}\delta A \frac{1}{\alpha + A}\,. \tag{5.4.14}$$

So we have

$$\begin{aligned} f''(\epsilon) &= \frac{1}{\beta}\mathrm{tr}\left\{ \frac{\partial A_\epsilon}{\partial \epsilon} \int_0^\infty \mathrm{d}\alpha\, \frac{1}{\alpha + A_\epsilon}\frac{\partial A_\epsilon}{\partial \epsilon}\frac{1}{\alpha + A_\epsilon}\right\} \\ &= \frac{1}{\beta}\int_0^\infty \mathrm{d}\alpha\,\mathrm{tr}\left\{\left[(\alpha + A_\epsilon)^{-\frac{1}{2}}\frac{\partial A_\epsilon}{\partial \epsilon}(\alpha + A_\epsilon)^{-\frac{1}{2}}\right]^2\right\} \geq 0\,, \end{aligned} \tag{5.4.15}$$

where the positivity follows from the fact that the operator $[\cdots]$ is hermitian. Indeed, the minimum property of $F(\beta)$ in (5.4.1) is confirmed.

We now take

$$\rho_{0\beta} = \frac{\mathrm{e}^{-\beta H_0}}{\mathrm{tr}\{\mathrm{e}^{-\beta H_0}\}} = \frac{\mathrm{e}^{-\beta H_0}}{Q_0(\beta)} \tag{5.4.16}$$

for the trial ρ, and so find

$$F(\beta) \leq \mathrm{tr}\{H\rho_{0\beta}\} + \frac{1}{\beta}\mathrm{tr}\{\rho_{0\beta}\log\rho_{0\beta}\} = \langle H\rangle_0 + \frac{1}{\beta}\langle\log\rho_{0\beta}\rangle_0 \tag{5.4.17}$$

after noting that the "0" expected value of (5.2.33) is the expectation value in the state that has $\rho_{0\beta}$ as its statistical operator,

$$\big\langle(\cdots)\big\rangle_0 = \mathrm{tr}\{(\cdots)\rho_{0\beta}\}\,. \tag{5.4.18}$$

Now,

$$\langle H\rangle_0 = \langle H_0\rangle_0 + \langle H_1\rangle_0 \tag{5.4.19}$$

and

$$\langle\log\rho_{0\beta}\rangle_0 = -\beta\langle H_0\rangle_0 - \log\big(Q_0(\beta)\big) = -\beta\langle H_0\rangle_0 + \beta F_0(\beta) \tag{5.4.20}$$

and, therefore,

$$F(\beta) \leq F_0(\beta) + \langle H_1\rangle_0\,, \tag{5.4.21}$$

the free-energy version of the Gibbs–Bogolyubov inequality in (5.2.37).

5.5 Ising chain with next-next-neighbors interaction

As an example of perturbation theory based on the Gibbs–Bogolyubov inequality of (5.2.4),

$$F(\beta) \leq F_0(\beta) + \left\langle E_k^{(1)} \right\rangle^{(0)}, \tag{5.5.1}$$

we consider the one-dimensional Ising model with no on-site energy but with an additional weak next-next-neighbor interaction,

$$\begin{aligned} E_k &= -J \sum_{(\text{nn})} ss' - J' \sum_{(\text{nnn})} ss' \\ &= -J \sum_j s_j s_{j+1} - J' \sum_j s_{j-1} s_{j+1}, \end{aligned} \tag{5.5.2}$$

with $J' \ll J$, as symbolized here:

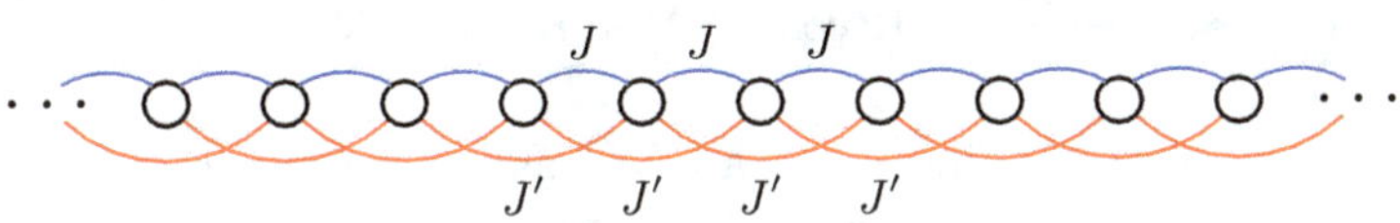

(5.5.3)

With

$$E_k^{(0)} = -J \sum_j s_j s_{j+1} \quad \text{and} \quad E_k^{(1)} = -J' \sum_j s_{j-1} s_{j+1} \tag{5.5.4}$$

we have

$$F_0(\beta) = -\frac{N}{\beta} \log\bigl(2\cosh(\beta J)\bigr) = -\frac{N}{\beta} \log\bigl(2\cosh(K)\bigr), \tag{5.5.5}$$

where $K = \beta J$ as before, and we need

$$\left\langle E_k^{(1)} \right\rangle^{(0)} = -J' \left\langle \sum_j s_{j-1} s_{j+1} \right\rangle^{(0)} = -J' N \bigl\langle s_{j-1} s_{j+1} \bigr\rangle^{(0)}, \tag{5.5.6}$$

where the second version realizes that $\langle s_{j-1} s_{j+1} \rangle^{(0)}$ has the same value for all next-next-neighbor pairs and picks out any one of them.

For the next-neighbor pairs, the partition function

$$Q^{(0)}(K) = \sum_{s_1, s_2, \ldots} \mathrm{e}^{K \sum_j s_j s_{j+1}} = \bigl[2\cosh(K)\bigr]^N \tag{5.5.7}$$

gives

$$\left\langle \sum_j s_j s_{j+1} \right\rangle^{(0)} = N\langle s_j s_{j+1}\rangle^{(0)} = \frac{1}{Q^{(0)}(K)} \frac{\partial}{\partial K} Q^{(0)}(K)$$
$$= N \tanh(K) \,. \tag{5.5.8}$$

Since $Q^{(0)}$ contains one factor $2\cosh(K)$ for each next-neighbor pair, it is suggestive that the pairs are uncorrelated. If they are, we have

$$\begin{aligned}\langle s_{j-1}s_{j+1}\rangle^{(0)} &= \langle (s_{j-1}s_j)(s_j s_{j+1})\rangle^{(0)} = \langle s_{j-1}s_j\rangle^{(0)}\langle s_j s_{j+1}\rangle^{(0)}\\ &= \tanh(K)^2 \,,\end{aligned} \tag{5.5.9}$$

where $(s_j)^2 = 1$ is used. Then, we get

$$\begin{aligned}F(\beta) &= -\frac{N}{\beta}\log\bigl(2\cosh(K)\bigr) - NJ'\tanh(K)^2\\ &= -\frac{N}{\beta}\bigl[\log\bigl(2\cosh(K)\bigr) + K'\tanh(K)^2\bigr]\end{aligned} \tag{5.5.10}$$

with $K' = \beta J' \ll K$ as the first-order approximation for the free energy.

As suggestive as it is, the assumption that next-neighbor pairs in the Ising model are uncorrelated requires a justification. We provide it by regarding the Ising chain (or ring) as a special case of a more general situation, in which the link between the pair $j, j+1$ has strength J_j, different for all links, and the standard Ising model is obtained when putting $J_j = J$ for all j:

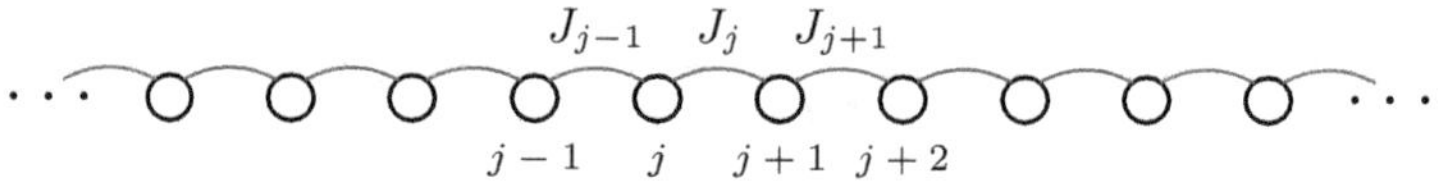

(5.5.11)

so that

$$E_k = -\sum_j J_j s_j s_{j+1} \tag{5.5.12}$$

is now the next-neighbor interaction energy in the kth configuration. The 2×2 matrix for the jth link is then

$$M_j = \begin{pmatrix} e^{\beta J_j} & e^{-\beta J_j} \\ e^{-\beta J_j} & e^{\beta J_j} \end{pmatrix} \tag{5.5.13}$$

and for a chain of N links, we have

$$Q = (1\;1)M_1M_2M_3\cdots M_N\begin{pmatrix}1\\1\end{pmatrix} \tag{5.5.14}$$

instead of $Q = (1\;1)M^N\begin{pmatrix}1\\1\end{pmatrix}$ as in (4.2.19) for a chain with identical links ($N-1$ links there, N links here, but that is of no consequence). Now

$$M_j\begin{pmatrix}1\\1\end{pmatrix} = \begin{pmatrix}1\\1\end{pmatrix}2\cosh(\beta J_j) \tag{5.5.15}$$

so that

$$\begin{aligned} Q &= 2\big[2\cosh(\beta J_1)\big]\,\big[2\cosh(\beta J_2)\big]\cdots\big[2\cosh(\beta J_N)\big] \\ &= 2\prod_{j=1}^{N}\big[2\cosh(\beta J_j)\big]\,. \end{aligned} \tag{5.5.16}$$

For the jth link, then,

$$\big\langle s_js_{j+1}\big\rangle = \frac{\partial}{\partial(\beta J_j)}\log Q = \tanh(\beta J_j)\,, \tag{5.5.17}$$

and for any two different links

$$\begin{aligned} \big\langle s_{j_1}s_{j_1+1}\;s_{j_2}s_{j_2+1}\big\rangle &= \frac{1}{Q}\frac{\partial}{\partial(\beta J_{j_1})}\frac{\partial}{\partial(\beta J_{j_2})}Q \\ &= \tanh(\beta J_{j_1})\tanh(\beta J_{j_2}) \\ &= \big\langle s_{j_1}s_{j_1+1}\big\rangle\big\langle s_{j_2}s_{j_2+1}\big\rangle\,. \end{aligned} \tag{5.5.18}$$

Different pairs are indeed uncorrelated. For $j_1 = j-1$, $j_2 = j$, this is

$$\big\langle s_{j-1}\underbrace{s_js_j}_{=1}s_{j+1}\big\rangle = \big\langle s_{j-1}s_{j+1}\big\rangle = \tanh(\beta J_{j-1})\tanh(\beta J_j)\,. \tag{5.5.19}$$

Upon putting $J_j = J$ for all j and so returning to the standard Ising model, we indeed get the next-next-neighbor expected value of (5.5.9),

$$\big\langle s_{j-1}s_{j+1}\big\rangle = \tanh(\beta J)^2 = \tanh(K)^2\,. \tag{5.5.20}$$

We can improve on the first-order approximation for $F(\beta)$ by using mean-field ideas. For this purpose, we introduce a parameter ϵ whose value we shall choose later and write

$$E_k = \underbrace{-\left(J+\frac{\epsilon}{\beta}\right)\sum_j s_j s_{j+1}}_{=E_k^{(0)}} + \underbrace{\frac{\epsilon}{\beta}\sum_j s_j s_{j+1} - J'\sum_j s_{j-1}s_{j+1}}_{=E_k^{(1)}}$$
$$= E_k^{(0)} + E_k^{(1)} . \tag{5.5.21}$$

For this split of E_k into $E_k^{(0)}$ and $E_k^{(1)}$ we have

$$-\beta E_k^{(0)} = (K+\epsilon)\sum_j s_j s_{j+1} \tag{5.5.22}$$

and

$$-\beta E_k^{(1)} = -\epsilon\sum_j s_j s_{j+1} + K'\sum_j s_{j-1}s_{j+1} , \tag{5.5.23}$$

so that we get

$$Q^{(0)} = \left[2\cosh\left(K+\epsilon\right)\right]^N \tag{5.5.24}$$

and

$$-\beta\left\langle E_k^{(1)}\right\rangle^{(0)} = -\epsilon N\tanh(K+\epsilon) + K'N\tanh(K+\epsilon)^2 , \tag{5.5.25}$$

as follows immediately when replacing K by $K+\epsilon$ in the earlier expressions.

Taken together, these establish the upper bound

$$F(\beta) \le -\frac{N}{\beta}\left[\log\bigl(2\cosh(K+\epsilon)\bigr) - \epsilon\tanh(K+\epsilon) + K'\tanh(K+\epsilon)^2\right] \tag{5.5.26}$$

for whatever value we choose for ϵ. The best choice is that for which the right-hand side is smallest, or the $[\cdots]$ term largest. We differentiate with respect to ϵ,

$$\frac{\partial}{\partial\epsilon}[\cdots] = \tanh(K+\epsilon) - \tanh(K+\epsilon) - \frac{\epsilon}{\cosh(K+\epsilon)^2} + \frac{2K'\tanh(K+\epsilon)}{\cosh(K+\epsilon)^2}$$
$$= -\frac{1}{\cosh(K+\epsilon)^2}\left[\epsilon - 2K'\tanh(K+\epsilon)\right] , \tag{5.5.27}$$

and conclude that the optimal ϵ value solves

$$\epsilon = 2K'\tanh(K+\epsilon) . \tag{5.5.28}$$

The graph

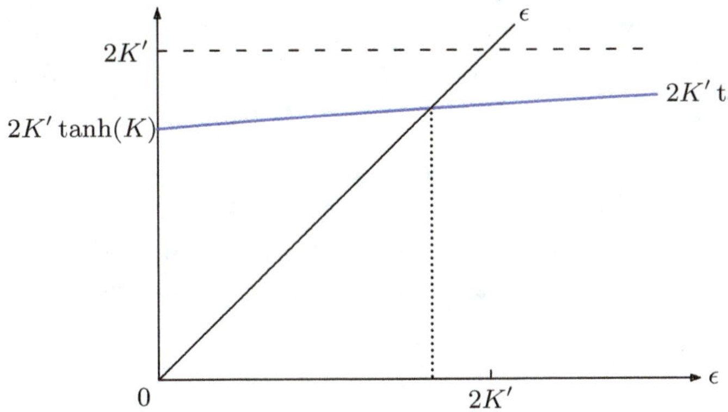

(5.5.29)

shows that the equation has a solution and that $\epsilon \cong 2K'$ is an acceptable sub-optimal value for ϵ if $\tanh(K) \lesssim 1$. For this choice, the approximate free energy is

$$F(\beta) = -\frac{N}{\beta}\Big[\log\big(2\cosh(K+2K')\big) - 2K'\tanh(K+2K') \\ + K'\tanh(K+2K')^2\Big]\,, \tag{5.5.30}$$

which improves quite a bit on the $\epsilon = 0$ approximation in (5.5.10).

5.6 Ising model and two-state quantum system

Before we leave the Ising model behind, it is worth showing a connection with a quantum system, which shall serve as an illustration of such analogies between models in classical statistical mechanics and quantum-mechanical systems.

A two-state system in quantum mechanics is described by a Hamilton operator that can be represented by a 2×2 matrix

$$H = \begin{pmatrix} E & -W \\ -W & -E \end{pmatrix} \quad \text{with real } E \text{ and } W. \tag{5.6.1}$$

More general hermitian 2×2 matrices, with a nonzero trace and complex off-diagonal entries, can be brought into this form by subtracting half the trace times the unit matrix and performing a simple unitary transformation that removes the complex phase of the off-diagonal entries. In short, this matrix with real E and W covers all cases.

With the familiar Pauli* matrices $\sigma_x = \begin{pmatrix} 0 & 1 \\ 1 & 0 \end{pmatrix}$ and $\sigma_z = \begin{pmatrix} 1 & 0 \\ 0 & -1 \end{pmatrix}$, we write

$$H = -W\sigma_x + E\sigma_z \tag{5.6.2}$$

and have

$$Q = \mathrm{tr}\left\{ \mathrm{e}^{-\beta H} \right\} = \mathrm{tr}\left\{ \mathrm{e}^{\beta W\sigma_x - \beta E\sigma_z} \right\} \tag{5.6.3}$$

for the partition function of this simple two-state system. Since the eigenvalues of H are $\pm\sqrt{E^2 + W^2}$, we can, of course, evaluate the trace immediately and get

$$Q = \mathrm{e}^{\beta\sqrt{E^2 + W^2}} + \mathrm{e}^{-\beta\sqrt{E^2 + W^2}} = 2\cosh\left(\beta\sqrt{E^2 + W^2}\right). \tag{5.6.4}$$

But this is beside the point. Rather, let us cope with the problem that σ_x and σ_z do not commute and that, therefore, the exponential function $\mathrm{e}^{-\beta H}$ cannot be factored into $\mathrm{e}^{\beta W\sigma_x}$ times $\mathrm{e}^{-\beta E\sigma_z}$, by the following method, which has many applications in a large variety of contexts.

First, we note that

$$\mathrm{e}^{-\beta H} = \left(\mathrm{e}^{-\frac{\beta}{N}H}\right)^N = \left(\mathrm{e}^{-\epsilon H}\right)^{\frac{\beta}{\epsilon}}, \tag{5.6.5}$$

where the $\epsilon = \dfrac{\beta}{N}$ is a mock value for β that we can make as small as we need it; in a sense this introduces an arbitrary large temperature at which we expect to have simpler matters. In return, we then have a chain of $\mathrm{e}^{-\epsilon H}$ matrices, as if there were N two-state systems, each at temperature NT, rather than one system at temperature T. Second, in the light of Exercise 113, we employ the *split-operator approximation*,

$$\begin{aligned} \mathrm{e}^{-\epsilon H} &= \mathrm{e}^{\epsilon W\sigma_x - \epsilon E\sigma_z} \\ &\cong \mathrm{e}^{-\frac{1}{2}\epsilon E\sigma_z}\,\mathrm{e}^{\epsilon W\sigma_x}\,\mathrm{e}^{-\frac{1}{2}\epsilon E\sigma_z}, \end{aligned} \tag{5.6.6}$$

which introduces an error of order $\epsilon^3 \propto N^{-3}$ in each term, and an overall error of order $N\epsilon^3 \propto N^{-2}$. Accordingly, with sufficiently large N, the overall error will be negligibly small, and perhaps the limit $N \to \infty$ can be executed at a later stage.

---

*Wolfgang PAULI (1900–1958)

Third, since ϵ is tiny, we can approximate the σ_x part, proceeding from

$$\left(\mathrm{e}^{\epsilon W\sigma_x}\right)_{ss'} = \cosh(\epsilon W)\mathbf{1}_2 + \sinh(\epsilon W)\sigma_x = \mathbf{1}_2 + \epsilon W\sigma_x + \mathcal{O}(\epsilon^2)\,, \tag{5.6.7}$$

where $\mathbf{1}_2 = \sigma_x^2 = \sigma_z^2$ is the 2×2 unit matrix. Next we look at matrix elements, thereby labeling the rows and columns by $s, s' = \pm 1$,

$$\begin{aligned}\left(\mathrm{e}^{\epsilon W\sigma_x}\right)_{ss'} &= \delta_{ss'} + \epsilon W(1-\delta_{ss'}) + \mathcal{O}(\epsilon^2) \qquad (5.6.8)\\ &= \mathcal{O}(\epsilon^2) + \sqrt{\epsilon W}\begin{cases}\sqrt{\epsilon W}^{\,-1} & \text{if } s = s' = \pm 1\ ,\ ss' = 1\\ \sqrt{\epsilon W} & \text{if } s = -s' = \pm 1\ ,\ ss' = -1\end{cases}\end{aligned}$$

or

$$\left(\mathrm{e}^{\epsilon W\sigma_x}\right)_{ss'} = \sqrt{\epsilon W}\,\mathrm{e}^{-ss'\log\sqrt{\epsilon W}} + \mathcal{O}(\epsilon^2)\,. \tag{5.6.9}$$

Then, the approximation

$$\begin{aligned}\mathrm{e}^{-\epsilon H} &\cong \begin{pmatrix}\mathrm{e}^{-\frac{1}{2}\epsilon E} & 0\\ 0 & \mathrm{e}^{\frac{1}{2}\epsilon E}\end{pmatrix}\sqrt{\epsilon W}\begin{pmatrix}\mathrm{e}^{-\log\sqrt{\epsilon W}} & \mathrm{e}^{\log\sqrt{\epsilon W}}\\ \mathrm{e}^{\log\sqrt{\epsilon W}} & \mathrm{e}^{-\log\sqrt{\epsilon W}}\end{pmatrix}\begin{pmatrix}\mathrm{e}^{-\frac{1}{2}\epsilon E} & 0\\ 0 & \mathrm{e}^{\frac{1}{2}\epsilon E}\end{pmatrix}\\ &= \sqrt{\epsilon W}\begin{pmatrix}\mathrm{e}^{-\epsilon E - \log\sqrt{\epsilon W}} & \mathrm{e}^{\log\sqrt{\epsilon W}}\\ \mathrm{e}^{\log\sqrt{\epsilon W}} & \mathrm{e}^{\epsilon E - \log\sqrt{\epsilon W}}\end{pmatrix}\end{aligned} \tag{5.6.10}$$

has an error of order ϵ^2, yielding an overall error of order $N\epsilon^2 \propto \frac{1}{N}$, which we can make as small as we need it.

The resulting partition function is

$$Q = \mathrm{tr}\left\{\left[\sqrt{\epsilon W}\begin{pmatrix}\mathrm{e}^{-\epsilon E - \log\sqrt{\epsilon W}} & \mathrm{e}^{\log\sqrt{\epsilon W}}\\ \mathrm{e}^{\log\sqrt{\epsilon W}} & \mathrm{e}^{\epsilon E - \log\sqrt{\epsilon W}}\end{pmatrix}\right]^N\right\}, \tag{5.6.11}$$

which is essentially the partition function of the Ising ring, inasmuch as we get the 2×2 matrix M in (4.2.26) if we identify

$$\epsilon E \equiv \frac{1}{2}\beta E_0\,, \quad \log\sqrt{\epsilon W} \equiv -\beta J \tag{5.6.12}$$

and the sole remaining difference is the factor

$$(\epsilon W)^{\frac{1}{2}N} \equiv \mathrm{e}^{-N\beta J}\,. \tag{5.6.13}$$

It is, in fact, important in the limit $N \to \infty$, as it compensates for the factor $\mathrm{e}^{\beta J}$ in $\lambda_\pm$ in (4.2.35).

Chapter 6

Real gases

6.1 Interacting atoms

For a real gas with an interaction energy $V\big(|\boldsymbol{r}_j - \boldsymbol{r}_k|\big) = V(r_{jk})$ for the jth atom at $\boldsymbol{r}_j$ and the kth atom at $\boldsymbol{r}_k$, the total energy for N atoms is

$$E = \sum_{j=1}^{N} \frac{1}{2m}\boldsymbol{p}_j^2 + \frac{1}{2}\sum_{\substack{j,k=1 \\ (j\neq k)}}^{N} V(r_{jk}) = \sum_{j=1}^{N} \frac{1}{2m}\boldsymbol{p}_j^2 + \sum_{j<k}^{N} V(r_{jk}), \tag{6.1.1}$$

where the factor of $\frac{1}{2}$, or the restriction $j < k$, avoids the double-counting of the (jk) pair. We assume that the forces between atoms are along the line of sight, which is natural (angular momentum conservation), but the general formalism would also work for potential energies that depend on the vector $\boldsymbol{r}_j - \boldsymbol{r}_k$ rather than only its length $r_{jk} = |\boldsymbol{r}_j - \boldsymbol{r}_k|$.

With the Maxwell–Boltzmann weight $\mathrm{e}^{-\beta E}$ and classical counting by a phase-space integral, the canonical partition function is

$$\begin{aligned} Q(\beta, V, N) &= \frac{1}{N!\,(2\pi\hbar)^{3N}} \int (\mathrm{d}\boldsymbol{r}_1)(\mathrm{d}\boldsymbol{p}_1)(\mathrm{d}\boldsymbol{r}_2)(\mathrm{d}\boldsymbol{p}_2)\cdots(\mathrm{d}\boldsymbol{r}_N)(\mathrm{d}\boldsymbol{p}_N)\,\mathrm{e}^{-\beta E} \\ &= \frac{1}{N!\,\lambda^{3N}} \int (\mathrm{d}\boldsymbol{r}_1)(\mathrm{d}\boldsymbol{r}_2)\cdots(\mathrm{d}\boldsymbol{r}_N)\,\mathrm{e}^{-\beta\sum_{j<k}^{N} V(r_{jk})}\,, \end{aligned} \tag{6.1.2}$$

where all momentum integrals are evaluated, each contributing a factor

$$\frac{1}{(2\pi\hbar)^3} \int (\mathrm{d}\boldsymbol{p})\ \mathrm{e}^{-\frac{\beta}{2m}\boldsymbol{p}^2} = \frac{1}{\lambda^3} \tag{6.1.3}$$

with the familiar thermal wavelength $\lambda = \hbar\sqrt{\dfrac{2\pi\beta}{m}}$ of (3.6.8).

For each pair, there is a factor $e^{-\beta V(r_{jk})}$ with the potential energy for this pair. Now, the forces between neutral atoms are strongly repulsive at small distances and weakly attractive at large distances, often well-modeled by the Lennard-Jones* potential:

$$V(r) = V_0 \left[\left(\frac{r_0}{r} \right)^{12} - 2 \left(\frac{r_0}{r} \right)^{6} \right], \tag{6.1.4}$$

which has its minimum $V(r_0) = -V_0$ at distance r_0, is quite small for r substantially larger than r_0 — for example, $V(2r_0) \cong 0.03V_0$ — and very large for r less than r_0 — for example $V(\frac{1}{2}r_0) \cong 4000V_0$. These matters are shown in the graph of $V(r)$:

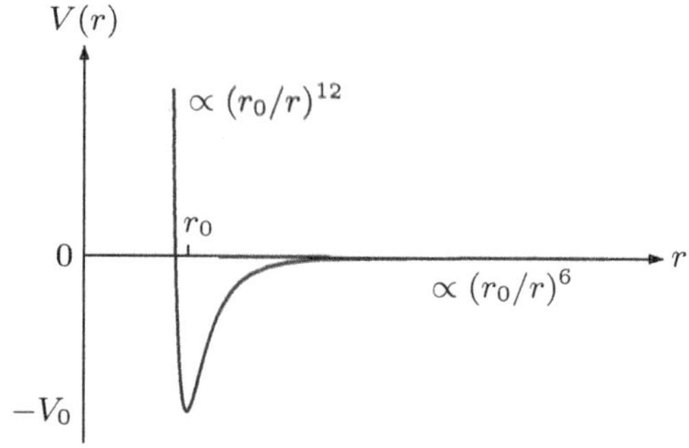

(6.1.5)

The factor $e^{-\beta V(r)}$ does not exceed $e^{\beta V_0}$ for this potential energy, and there is an analogous upper bound for other potentials of this general shape. Then,

$$e^{-\beta V(r)} = 1 + \left(e^{-\beta V(r)} - 1 \right) \equiv 1 + f(r) \tag{6.1.6}$$

introduces the quantity $f(r)$ whose r dependence has a graph like this:

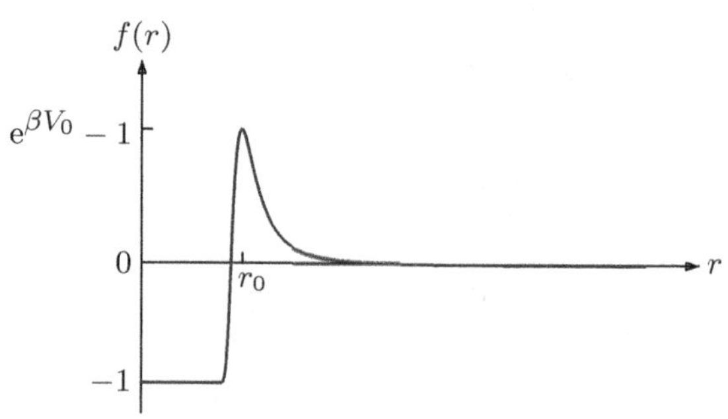

(6.1.7)

*John LENNARD-JONES (1894–1954)

where the most important property is the rapid approach to $f(r) \to 0$ as $r \to \infty$, so that the integral

$$\int (\mathrm{d}\boldsymbol{r})\, f(r) = \int (\mathrm{d}\boldsymbol{r}) \left(\mathrm{e}^{-\beta V(r)} - 1\right) < \infty \tag{6.1.8}$$

converges well for large r.

6.2 Clusters

Now writing f_{jk} for $f(r_{jk})$, we have

$$\begin{aligned} Q(\beta, V, N) &= \frac{1}{N!\,\lambda^{3N}} \int (\mathrm{d}\boldsymbol{r}_1)(\mathrm{d}\boldsymbol{r}_2)\cdots(\mathrm{d}\boldsymbol{r}_N) \prod_{j<k} (1 + f_{jk}) \\ &= \frac{1}{N!\,\lambda^{3N}} \int (\mathrm{d}\boldsymbol{r}_1)\cdots(\mathrm{d}\boldsymbol{r}_N) \left[1 + \sum_{j<k} f_{jk} + \sum_{j<k}\sum_{l<n} f_{jk} f_{ln} + \cdots\right], \end{aligned} \tag{6.2.1}$$

where the ellipsis stands for a hierarchy of contributions with three, four, five, ... f_{jk} factors. It is common to picture the individual terms by graphs that have circles for each atom (or its position variable) and links between those sites when the f_{jk} of the pair is present in the product. For example, the graph

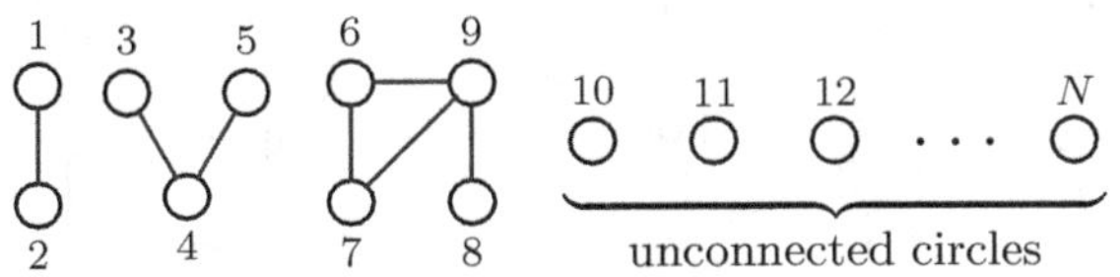

(6.2.2)

represents

$$\begin{aligned} &\int (\mathrm{d}\boldsymbol{r}_1)\cdots(\mathrm{d}\boldsymbol{r}_N) f_{12} f_{34} f_{45} f_{67} f_{69} f_{79} f_{89} \\ &= \left[\int (\mathrm{d}\boldsymbol{r}_1)(\mathrm{d}\boldsymbol{r}_2) f_{12}\right] \left[\int (\mathrm{d}\boldsymbol{r}_3)(\mathrm{d}\boldsymbol{r}_4)(\mathrm{d}\boldsymbol{r}_5) f_{34} f_{45}\right] \\ &\quad \times \left[\int (\mathrm{d}\boldsymbol{r}_6)(\mathrm{d}\boldsymbol{r}_7)(\mathrm{d}\boldsymbol{r}_8)(\mathrm{d}\boldsymbol{r}_9) f_{67} f_{69} f_{79} f_{89}\right] \left[\int (\mathrm{d}\boldsymbol{r})\right]^{N-9}, \end{aligned} \tag{6.2.3}$$

where we have recognized that the integral factors into $1+1+1+(N-9) = N-6$ terms, one for each *cluster*, that is: a group of circles with links between them such that one can reach each circle from any other by traveling along the links. In the example, there is one cluster with two circles and one link, one cluster with three circles and two links, one cluster with four circles and four links, and $N-9$ clusters with a single circle and no links. Between any pair of circles, there can only be one link since a particular factor f_{jk} can only occur once in the product of f factors.

A l-cluster is a cluster with l circles, and any graph is made up of m_1 1-clusters, m_2 2-clusters, m_3 3-clusters, ..., m_l l-clusters, where the counts m_l give the count of particles in accordance with

$$\sum_{l=1}^{N} l\, m_l = N\,. \tag{6.2.4}$$

Since there are different kinds of l-clusters when $l > 2$, knowing the list $m = (m_1, m_2, m_3, \ldots)$ does not specify a unique graph. In the example in (6.2.2), we have $m_1 = N-9$, $m_2 = m_3 = m_4 = 1$, and $m_l = 0$ for $l > 4$, but these counts would also apply to

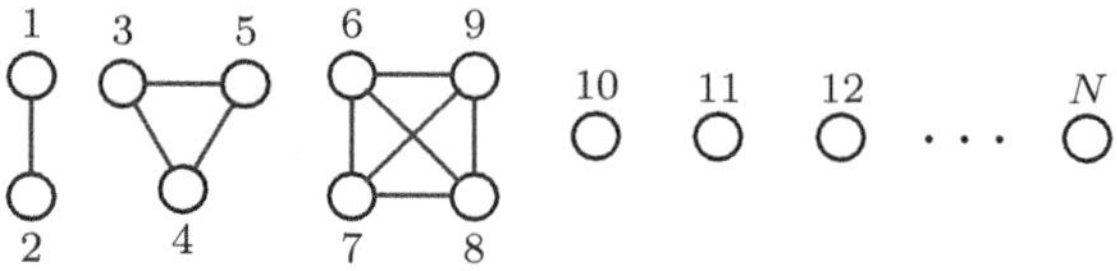

(6.2.5)

with a different 3-cluster and a different 4-cluster.

To help with the counting, we introduce

$$b_l(\beta, V) = \frac{1}{l!\,\lambda^{3(l-1)}\,V} \times \bigl[\text{sum of all possible } l\text{-clusters}\bigr]\,, \tag{6.2.6}$$

illustrated by

$$b_1(\beta, V) = \frac{1}{V}[\bigcirc] = \frac{1}{V}\int (\mathrm{d}\boldsymbol{r}) = 1\,,$$

$$b_2(\beta, V) = \frac{1}{2\lambda^3 V}\Bigl[$$ 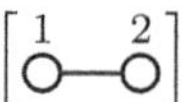 $$\Bigr] = \frac{1}{2\lambda^3 V}\int (\mathrm{d}\boldsymbol{r}_1)(\mathrm{d}\boldsymbol{r}_2)\, f_{12}\,,$$

$$b_3(\beta, V) = \frac{1}{6\lambda^6 V}\Bigl[$$ 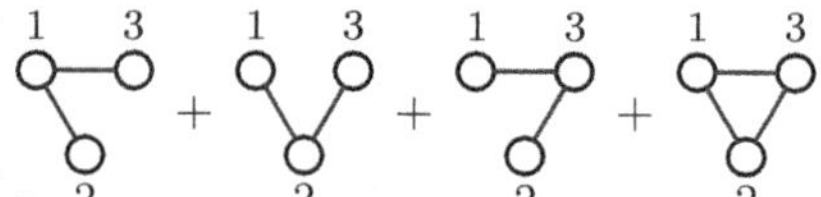 $$\Bigr]$$

$$= \frac{1}{6\lambda^6 V}\int (\mathrm{d}\boldsymbol{r}_1)(\mathrm{d}\boldsymbol{r}_2)(\mathrm{d}\boldsymbol{r}_3)\,\bigl(3f_{12}f_{23} + f_{12}f_{13}f_{23}\bigr)\,. \tag{6.2.7}$$

In $b_3(\beta, V)$, the three graphs with two links must be counted as three different graphs but their values are the same, so that we can just take three times the value of one of them, as we did.

In (6.2.7), each of the integrations over $\boldsymbol{r}_1$, $\boldsymbol{r}_2$, ... covers the whole volume of size V. Owing to the finite extension of this volume, the range $V(\boldsymbol{r}_1)$ covered by the relative vector $\boldsymbol{r}_{12} = \boldsymbol{r}_1 - \boldsymbol{r}_2$ in

$$b_2(\beta, V) = \frac{1}{2\lambda^3 V} \int (\mathrm{d}\boldsymbol{r}_1) \int\limits_{V(\boldsymbol{r}_1)} (\mathrm{d}\boldsymbol{r}_{12})\, f(r_{12}) \tag{6.2.8}$$

depends on $\boldsymbol{r}_1$, and this dependence introduces finite-size corrections. Under typical circumstances, these corrections are negligibly small if the volume is large, because the integral over $\boldsymbol{r}_{12}$ has no substantial contributions from distances r_{12} much larger than the range of the atom-atom interaction force — a few r_0 for the Lennard-Jones potential. Then, the finite-size corrections are essentially independent of the size and shape of the volume V (we are not interested in the complications introduced by considering particularly queer shapes), and can be ignored in the limit of $V \to \infty$.

We shall regard this limit as being performed and replace $b_l(\beta, V)$ by

$$b_l(\beta) = b_l(\beta, V)\Big|_{V \to \infty}. \tag{6.2.9}$$

Then, we integrate the relative vectors over the full three-dimensional space rather than over finite volumes and so arrive at

$$\begin{aligned} b_2(\beta) &= \frac{1}{2\lambda^3} \int (\mathrm{d}\boldsymbol{r}) f(r)\,, \\ b_3(\beta) &= \frac{1}{6\lambda^6}\left[3\left(\int (\mathrm{d}\boldsymbol{r})\, f(r)\right)^2 + \int (\mathrm{d}\boldsymbol{r})(\mathrm{d}\boldsymbol{r}')\, f(r) f(r') f(|\boldsymbol{r} - \boldsymbol{r}'|)\right] \\ &= 2b_2(\beta)^2 + \frac{1}{6\lambda^6} \int (\mathrm{d}\boldsymbol{r})(\mathrm{d}\boldsymbol{r}')\, f(r) f(r') f(|\boldsymbol{r} - \boldsymbol{r}'|) \end{aligned} \tag{6.2.10}$$

and so forth.

In (6.2.6), the factor $\lambda^{-3(l-1)}$ is introduced in $b_l(\beta, V)$, and thus in $b_l(\beta)$, to turn it into a quantity with no metrical dimension, and the division by the factorial $l!$ is another measure of convenience. It will combine with other factorial factors shortly.

Currently, we have

$$\begin{aligned} Q(\beta, V, N) &= \frac{1}{N!\,\lambda^{3N}} \sum_{m_1=1}^{\infty} \sum_{m_2=1}^{\infty} \cdots \sum_{m_N=1}^{\infty} \delta_{N, \sum_l l m_l} S_N(m) \\ &\equiv \frac{1}{N!\,\lambda^{3N}} \sum_m^{(N)} S_N(m)\,, \end{aligned} \tag{6.2.11}$$

where $S_N(m)$ is the sum over all graphs that have m_l l-clusters for $l = 1, 2, \ldots, N$, with a factor of $l!\,\lambda^{3(l-1)} V b_l(\beta)$ for each set of l-clusters to correctly account for the variety of possible l-clusters. Additional factors account for the combinatorial aspects.

First, any assignment of the labels $1, 2, \ldots, N$ to the N sites of a graph, or the N atoms in the gas, is as good as any other, so that we have as many as

$$\frac{N!}{(1!)^{m_1}(2!)^{m_2}\ldots(l!)^{m_l}\ldots} = N!\prod_{l=1}^{N}(l!)^{-m_l} \tag{6.2.12}$$

copies of a graph with given $m = (m_1, m_2, m_3, \ldots)$. Second, if $m_l > 1$, we can exchange *all* labels of one l-cluster with all labels of another without getting a new configuration, such as

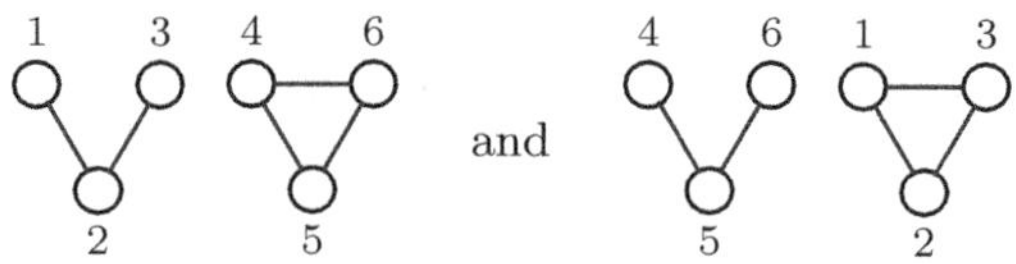

(6.2.13)

are two versions of the same $m_3 = 2$ configuration, which we must not count twice. The factor

$$\prod_{l=1}^{N}\frac{1}{m_l!} \tag{6.2.14}$$

protects us against over-counting of this kind.

Upon collecting all these factors, we have

$$\begin{aligned} S_N(m) &= \left(N!\prod_l (l!)^{-m_l}\right)\left(\prod_l (m_l!)^{-1}\right)\left(\prod_l \left[l!\,\lambda^{3(l-1)} V b_l(\beta)\right]^{m_l}\right) \\ &= N!\,\lambda^{3\sum_l l m_l}\prod_l \frac{1}{m_l!}\left(b_l(\beta)\frac{V}{\lambda^3}\right)^{m_l} \end{aligned} \tag{6.2.15}$$

and, since $\sum_l l m_l = N$, this gives

$$Q(\beta, V, N) = \sum_m^{(N)} \prod_l \frac{1}{m_l!}\left(\frac{V}{\lambda^3} b_l(\beta)\right)^{m_l} \tag{6.2.16}$$

for the canonical partition function. As in (6.2.11), the superscript $^{(N)}$ on the summation sign is a reminder of the constraint obeyed by m.

This constraint makes the evaluation of the sum difficult. So we get rid of it by switching to the grand canonical partition function

$$Z(\beta, V, z) = \sum_{N=0}^{\infty} z^N Q(\beta, V, N) \tag{6.2.17}$$

with fugacity z. The constraint is exploited by writing

$$z^N = z^{\sum_l l m_l} = \prod_l z^{l m_l} = \prod_l (z^l)^{m_l}, \tag{6.2.18}$$

and we get

$$Z(\beta, V, z) = \sum_{N=0}^{\infty} \sum_m^{(N)} \prod_{l=1}^{\infty} \frac{1}{m_l!} \left(\frac{z^l V}{\lambda^3} b_l(\beta) \right)^{m_l}. \tag{6.2.19}$$

We now recall the meaning of $\sum_m^{(N)}$ in (6.2.11) and note that the summation over N removes the Kronecker delta that enforces the constraint of (6.2.4). Therefore, we have

$$\begin{aligned} Z(\beta, V, z) &= \left(\sum_{m_1=0}^{\infty} \sum_{m_2=0}^{\infty} \cdots \right) \prod_{l=1}^{\infty} \frac{1}{m_l!} \left(\frac{z^l V}{\lambda^3} b_l(\beta) \right)^{m_l} \\ &= \prod_{l=1}^{\infty} \sum_{m_l=0}^{\infty} \frac{1}{m_l!} \left(\frac{z^l V}{\lambda^3} b_l(\beta) \right)^{m_l} \\ &= \prod_{l=1}^{\infty} \mathrm{e}^{z^l V \lambda^{-3} b_l(\beta)}, \end{aligned} \tag{6.2.20}$$

or

$$\beta P V = \log\bigl(Z(\beta, V, z)\bigr) = \frac{V}{\lambda^3} \sum_{l=1}^{\infty} z^l b_l(\beta). \tag{6.2.21}$$

This is a remarkably compact final expression.

6.3 Cluster expansion; virial expansion

We supplement (6.2.21) by the statement about the (expected) number of atoms,

$$N = z\frac{\partial}{\partial z}\log\bigl(Z(\beta, V, z)\bigr) = \frac{V}{\lambda^3}\sum_{l=1}^{\infty} l z^l b_l(\beta)\,, \tag{6.3.1}$$

or, after dividing by the volume V,

$$\begin{aligned}\beta P &= \frac{1}{\lambda^3}\sum_{l=1}^{\infty} z^l b_l(\beta)\,,\\ \frac{N}{V} = \frac{1}{v} &= \frac{1}{\lambda^3}\sum_{l=1}^{\infty} l z^l b_l(\beta)\,.\end{aligned} \tag{6.3.2}$$

This pair of equations, which is the pair of (2.9.21) in the present context, defines the so-called *cluster expansion formalism* of Mayer* and Ursell† (1930s). There is also a version for atoms subject to quantum statistics, developed by Kahn‡ and Uhlenbeck§ (late 1930s) and Lee¶ and Yang (late 1950s).

The task is now to eliminate the fugacity z and arrive at a thermal equation of state of the form

$$\beta P v = \sum_{l=1}^{\infty} a_l(\beta)\left(\frac{\lambda^3}{v}\right)^{l-1}, \tag{6.3.3}$$

known as the *virial expansion*, originally introduced by Kamerlingh Onnes,‖ with the virial coefficients $a_l(\beta)$. Here, $\dfrac{\lambda^3}{v} \propto \beta^{\frac{3}{2}} \propto T^{-\frac{3}{2}}$ is a small expansion parameter for high enough temperatures. Accordingly, we anticipate $a_1(\beta) = 1$ for the correct high-temperature limit of the ideal gas.

Let us work out the first few coefficients in (6.3.3). Since $b_1(\beta) = 1$, we have

$$\frac{\lambda^3}{v} \equiv y = z + 2b_2(\beta)z^2 + 3b_3(\beta)z^3 + \cdots \tag{6.3.4}$$

---

*Joseph Edward MAYER (1904–1983) †Harold Douglas URSELL (1907–1969)
‡Boris KAHN (work of 1938) §George Eugene UHLENBECK (1900–1988)
¶Tsung-Dao LEE (b. 1926) ‖Heike KAMERLINGH ONNES (1853–1926)

and

$$\beta P v = \frac{v}{\lambda^3} \sum_{l=1}^{\infty} z^l b_l(\beta) = \frac{1}{y} \left[z + b_2(\beta) z^2 + b_3(\beta) z^3 + \cdots \right], \tag{6.3.5}$$

so that we get $\beta P v$ to order y^2 if we first find

$$z = y + \alpha_2 y^2 + \alpha_3 y^3 + \cdots \tag{6.3.6}$$

to order y^3 from

$$y = \left(y + \alpha_2 y^2 + \alpha_3 y^3\right) + 2b_2(\beta)\left(y^2 + 2\alpha_2 y^3\right) + 3b_3(\beta) y^3 . \tag{6.3.7}$$

This establishes first

$$\alpha_2 = -2b_2(\beta) \tag{6.3.8}$$

and

$$\alpha_3 = -4b_2(\beta)\alpha_2 - 3b_3(\beta) = 8b_2(\beta)^2 - 3b_3(\beta), \tag{6.3.9}$$

and then

$$\begin{aligned} \beta P v &= \frac{1}{y} \left[y + \alpha_2 y^2 + \alpha_3 y^3 + b_2(\beta)\left(y^2 + 2\alpha_2 y^3\right) + b_3(\beta) y^3 \right] + \mathcal{O}(y^3) \\ &= 1 + [\alpha_2 + b_2(\beta)] y + [\alpha_3 + 2\alpha_2 b_2(\beta) + b_3(\beta)] y^2 + \mathcal{O}(y^3) \\ &= 1 - b_2(\beta) y + \left[4b_2(\beta)^2 - 2b_3(\beta)\right] y^2 + \mathcal{O}(y^3) . \end{aligned} \tag{6.3.10}$$

It follows that the first three virial coefficients are

$$\begin{aligned} a_1(\beta) &= 1 , \\ a_2(\beta) &= -b_2(\beta) , \\ a_3(\beta) &= 4b_2(\beta)^2 - 2b_3(\beta) . \end{aligned} \tag{6.3.11}$$

In (6.2.10), we find the expressions for $b_2(\beta)$ and $b_3(\beta)$, which yield

$$\begin{aligned} a_2(\beta) &= -\frac{1}{2\lambda^3} \int (\mathrm{d}\boldsymbol{r})\, f(r) = -\frac{2\pi}{\lambda^3} \int_0^{\infty} \mathrm{d}r\, r^2 f(r) , \\ a_3(\beta) &= -\frac{1}{3\lambda^6} \int (\mathrm{d}\boldsymbol{r})(\mathrm{d}\boldsymbol{r}')\, f(r) f(r') f\left(|\boldsymbol{r} - \boldsymbol{r}'|\right) , \end{aligned} \tag{6.3.12}$$

or graphically

$$a_2(\beta) = -\frac{1}{2V\lambda^3}\left[\ \right], \qquad a_3(\beta) = -\frac{1}{3V\lambda^6}\left[\ \right]. \tag{6.3.13}$$

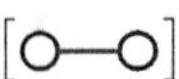

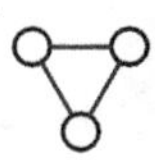

We note that not all 3-clusters contribute to $a_3(\beta)$ but only the *irreducible cluster* . Here, the adjective "irreducible" means that the cluster cannot be decomposed into two clusters by removing a *single* line connecting two circles. The suggestion that also $a_4(\beta), a_5(\beta), \ldots$ involve irreducible clusters only is, indeed, true but the demonstration of this case is technically complicated, so we omit it. The outcome is

$$a_l(\beta) = -\frac{l-1}{l!\,\lambda^{3(l-1)}\,V} \times \big[\text{sum of all irreducible } l\text{-clusters}\big] \tag{6.3.14}$$

for $l > 2$; there is no irreducible 2-cluster.

The simplification achieved by replacing $b_l(\beta, V)$ by its $V \to \infty$ limit term by term made it possible to solve $N = z\frac{\partial}{\partial z}\log Z$ for the fugacity in terms of $y = \frac{\lambda^3}{v}$, and so get the virial coefficients. Some physics is lost in this simplification. In particular, we must limit the application to high enough temperature at which the system is in the gas phase.

6.4 Finite volume, finite number of particles

More generally, if the volume is finite, the number of atoms, with short-range strongly repulsive forces between them, that can be inside the volume is finite — say, at most there can be M atoms. For $N > M$, then, the energy is infinite, $e^{-\beta E}$ vanishes, and so does the canonical partition function,

$$Q(\beta, V, N) = 0 \quad \text{for} \quad N > M\,. \tag{6.4.1}$$

It follows that the grand canonical partition function is a polynomial in the fugacity,

$$Z(\beta, V, z) = \sum_{N=0}^{M} z^N Q(\beta, V, N)\,, \tag{6.4.2}$$

with positive coefficients, so that there is no fugacity value for which $Z(\beta, V, z) = 0$. For the thermal equation of state, we have the pair of (2.9.21),

$$\begin{aligned}\beta P &= \frac{1}{V}\log\big(Z(\beta, V, z)\big)\,,\\ \frac{1}{v} &= \frac{1}{V}z\frac{\partial}{\partial z}\log\big(Z(\beta, V, z)\big)\,,\end{aligned} \tag{6.4.3}$$

where the fugacity is a parameter that needs to be eliminated. Now, for any *finite* volume V, even if it is very large, both βP and $\frac{1}{v}$ are analytical functions of z on and near the real z axis. There is a strip along the z axis inside which the pressure and the specific volume are analytical functions of the fugacity, so that all derived thermodynamical functions are regular, they do not have singularities.

Since $Z > 0$ and $\frac{\partial}{\partial z}Z > 0$ for the physical z values ($z > 0$), we have $P > 0$ and $\langle N\rangle = z\frac{\partial}{\partial z}\log Z > 0$, and also

$$\frac{\partial P}{\partial v} = \frac{\partial P}{\partial(V/\langle N\rangle)} = -\frac{\langle N\rangle^2}{V}\frac{\partial P}{\partial\langle N\rangle} = -\frac{\langle N\rangle^2}{V}\frac{\partial P}{\partial z}\bigg/\frac{\partial\langle N\rangle}{\partial z}\,, \tag{6.4.4}$$

where V and β are constant for these derivatives with respect to the fugacity. Here,

$$\frac{\partial P}{\partial z} = \frac{1}{\beta V}\frac{\partial}{\partial z}\log Z = \frac{1}{\beta V}\frac{\langle N\rangle}{z} \tag{6.4.5}$$

and, as earlier in (2.9.2),

$$\begin{aligned}\frac{\partial\langle N\rangle}{\partial z} &= \frac{1}{z}z\frac{\partial}{\partial z}\frac{1}{Z}z\frac{\partial}{\partial z}Z \\ &= \frac{1}{z}\left[\frac{1}{Z}\left(z\frac{\partial}{\partial z}\right)^2 Z - \left(\frac{1}{Z}z\frac{\partial}{\partial z}Z\right)^2\right] \\ &= \frac{1}{z}\left(\langle N^2\rangle - \langle N\rangle^2\right) = \frac{1}{z}\langle\delta N^2\rangle\,,\end{aligned} \tag{6.4.6}$$

so that

$$\frac{\partial P}{\partial v} = -\frac{\langle N\rangle^2}{V}\frac{\langle N\rangle}{\beta V}\frac{1}{\langle\delta N^2\rangle} = -\frac{k_{\mathrm{B}}T}{v^2}\frac{\langle N\rangle}{\langle\delta N^2\rangle}\,. \tag{6.4.7}$$

Recalling that $\langle\delta N^2\rangle \propto \langle N\rangle$, the right-hand side is composed of intensive variables, as it should, and we observe that

$$\frac{\partial P}{\partial v} \leq 0 \tag{6.4.8}$$

with the equal sign only applying when $\langle\delta N^2\rangle/\langle N\rangle = \infty$. For any finite number of particles, however, this cannot happen. Therefore, we get first-order phase transitions — the pressure stays constant while the volume decreases, see (1.15.15) — only in the *thermodynamical limit*, in which $V \to \infty$ and $\langle N\rangle \to \infty$ with fixed specific volume $v = \frac{V}{\langle N\rangle}$ or, equivalently, fixed particle density $\rho = \frac{\langle N \rangle}{V}$. That is:

$$\begin{aligned}\beta P &= \lim_{V\to\infty} \frac{1}{V} \log\bigl(Z(\beta, V, z)\bigr)\,, \\ \frac{1}{v} &= \lim_{V\to\infty} \frac{1}{V} z \frac{\partial}{\partial z} \log\bigl(Z(\beta, V, z)\bigr)\,,\end{aligned} \tag{6.4.9}$$

where we cannot interchange the limit $V \to \infty$ with the differentiation $\frac{\partial}{\partial z}$ unless the limit is uniform.

Lee and Yang have investigated these issues in the early 1950s and established the following two facts. First, the limit in the expression for βP exists for all $z > 0$ and yields a continuous nondecreasing function of the fugacity z; this limit does not depend on the shape of the volume provided that the volume is not weird, that is: its surface must not grow faster than $V^{2/3}$ as V increases. Second, if we consider a segment of the positive real z axis with a region around it that contains no zeros of $Z(\beta, V, z)$, then $V^{-1} \log Z$ converges uniformly as $V \to \infty$, and the limit is analytical for all z values in this region.

That leaves the possibility of having one zero of $Z(\beta, V, z)$ (or several of them) that converges to a singular value $\overline{z} > 0$ on the real fugacity axis. Then the pressure will be continuous at this fugacity, but its derivative may be discontinuous, so that we get a phase transition there, with one phase for $z < \overline{z}$, another for $z > \overline{z}$. If the derivative is continuous, but the second derivative is not, we have a second-order phase transition, such as the order-disorder transition discussed in the context of the Ising model. Third-order, fourth-order, ... transitions are conceivable as well. Consult specialized advanced textbooks on these matters.

6.5 Second virial coefficient: van der Waals gas

Let us return to the virial approximation in (6.3.3) with (6.3.11) and (6.3.12) that we obtained by letting $V \to \infty$ term by term. In view of the general considerations just now, we know that this is trustworthy only below the last one of those z_0 values on the real z axis, where the temperature is high and we are assuredly in the gas phase. The first virial coefficient is $a_1 = 1$, so

$$a_2(\beta) = -b_2(\beta) = \frac{2\pi}{\lambda^3} \int_0^\infty \mathrm{d}r\, r^2 \Bigl(1 - \mathrm{e}^{-\beta V(r)}\Bigr) \tag{6.5.1}$$

is the most important virial coefficient that we need to calculate. We do this for a simplified version of the Lennard-Jones potential, namely

$$V(r) = \begin{cases} \infty & \text{for } r < r_0\,, \\ -V_0\left(\dfrac{r_0}{r}\right)^6 & \text{for } r > r_0\,. \end{cases} \tag{6.5.2}$$

Then

$$\begin{aligned} a_2(\beta) &= \frac{2\pi}{\lambda^3}\left(\int_0^{r_0} \mathrm{d}r\, r^2 + \int_{r_0}^{\infty} \mathrm{d}r\, r^2\left[1 - \mathrm{e}^{\beta V_0 (r_0/r)^6}\right]\right) \\ &\underset{\beta V_0 \ll 1}{\cong} \frac{2\pi}{\lambda^3}\left(\frac{1}{3}r_0^3 - \int_{r_0}^{\infty} \mathrm{d}r\, r^2 \beta V_0\left(\frac{r_0}{r}\right)^6\right) \\ &= \frac{2\pi}{\lambda^3}\left(\frac{1}{3}r_0^3 - \frac{1}{3}\beta V_0 r_0^3\right) \end{aligned} \tag{6.5.3}$$

or

$$a_2(\beta) \cong \frac{2\pi r_0^3}{3\lambda^3}(1 - \beta V_0)\,, \tag{6.5.4}$$

where we simplified the integration a bit by taking into account that the temperature has to be high.

With just this second virial coefficient, we have

$$\beta P v \cong 1 + a_2(\beta)\frac{\lambda^3}{v} \cong 1 + \frac{2\pi r_0^3}{3v}(1 - \beta V_0)\,. \tag{6.5.5}$$

We rearrange the terms,

$$\beta\left(P + \frac{2\pi r_0^3 V_0}{3v^2}\right)v = 1 + \frac{2\pi r_0^3}{3v} \cong \left(1 - \frac{2\pi r_0^3}{3v}\right)^{-1} \tag{6.5.6}$$

or

$$\beta\left(P + \frac{2\pi r_0^3 V_0}{3v^2}\right)\left(v - \frac{2\pi r_0^3}{3}\right) = 1 \tag{6.5.7}$$

or, finally

$$\left(P + \frac{a}{v^2}\right)(v - b) = \frac{1}{\beta} = k_{\mathrm{B}}T\,, \tag{6.5.8}$$

which is the thermal equation of state of the van der Waals in (1.16.1) for

$$a = \frac{2\pi}{3}r_0^3 V_0 \quad \text{and} \quad b = \frac{2\pi}{3}r_0^3 \tag{6.5.9}$$

with the sole difference that v denotes the volume per particle here, whereas it stands for the molar volume, the volume per mole of particles, in Section 1.16. See Exercise 121 for the corresponding adiabatic equation of state.

The parameter b has a geometrical meaning. Since r_0 is the smallest distance between two particles (atoms), we have this picture for two particles in closest proximity:

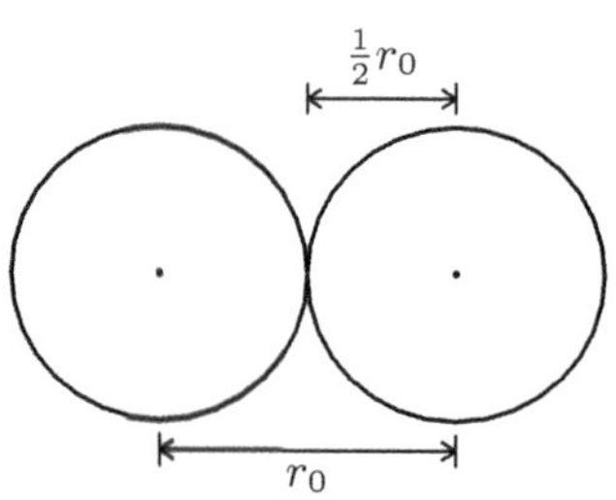

(6.5.10)

The particles themselves are hard balls with radius $\frac{1}{2}r_0$ and volume $\frac{4\pi}{3}\left(\frac{1}{2}r_0\right)^3 = \frac{\pi}{6}r_0^3 = \frac{1}{4}b$, and the volume that is unavailable to other particles because one particle is already present, is the volume of a ball with radius r_0 that is $\frac{4\pi}{3}r_0^3 = 2b$, so that b is a value intermediate between the volume occupied by a particle and the volume made inaccessible by that particle. We leave this matter at that.

When using the Lennard-Jones potential of (6.1.4) in (6.5.1), we get

$$a_2(\beta) = \frac{2\pi}{3}\left(\frac{r_0}{\lambda}\right)^3 \int_0^\infty \frac{dy}{y^2}\left[1 - e^{-\beta V_0(y^4 - 2y^2)}\right] = \frac{2\pi}{3}\left(\frac{r_0}{\lambda}\right)^3 \overline{a}_2(\beta V_0) \tag{6.5.11}$$

for the second virial coefficient, after substituting $y = \left(\frac{r_0}{r}\right)^3$. There is no closed form for $\overline{a}_2(\beta V_0)$, but one can compute this universal function of βV_0 numerically. In the virial expansion of (6.3.3), we then have

$$\beta P v = 1 + \frac{2\pi}{3}\frac{r_0^3}{v}\overline{a}_2(\beta V_0) + \cdots . \tag{6.5.12}$$

One extracts experimental values for $r_0^3\overline{a}_2(\beta V_0)$ by measuring, say, the pressure as a function of temperature $T = (k_B\beta)^{-1}$ and density $\rho = v^{-1}$. A comparison of such experimental values with those for the Lennard-Jones potential shows a very good agreement for many real gases. Notable exceptions are the gases of H_2 and He, where quantum effects are important and the classical-gas approximation that we used is not valid.

Chapter 7

Stochastic processes

We turn our attention to *stochastic processes*, for which Brownian* motion is a prime example. The original observation was about the erratic motion of pollen immersed in water, as viewed under a microscope, a never-ending random walk of the pollen grains resulting — as we now know — from the collision of water molecules with the much larger grains. These collisions transfer momentum between projectile (water) and target (grain) and so lead to sudden changes of the velocity vector and thus to the erratic motion of the pollen grain.

7.1 Random jumps

As a first example of a stochastic process, we consider the situation where an object has random jumps to the left or right along an axis with the probability $p_1(m)$ of jumping by m units in a single jump. More generally, $p_n(m)$ is the probability of jumping by m units in the course of n jumps. Clearly, we have

$$p_0(m) = \delta_{0,m} \qquad \text{(no jump, no change)} \tag{7.1.1}$$

and the normalization

$$\sum_{m=-\infty}^{\infty} p_n(m) = 1\,. \tag{7.1.2}$$

The sum covers negative m values (jumps to the left) as well as positive m values (jumps to the right), and the $m = 0$ term accounts for the probability that no net change occurs after n jumps.

*Robert Brown (1773–1858)

7.1.1 *Recurrence relation; generating function*

Since a displacement by m units in n jumps comes about by having m' and $m-m'$ displacements in fewer jumps, we have the recurrence relation

$$p_n(m)=\sum_{m'=-\infty}^{\infty} p_{n-k}(m-m')p_k(m')\,, \qquad (7.1.3)$$

which must hold for $k=0,1,2,\ldots,n$; the cases $k=0$ and $k=n$ are verified immediately. We deal with the probabilities $p_n(0),p_n(\pm1),p_n(\pm2),\ldots$ as a set, and use the generating function

$$g_n(\phi)=\sum_{m=-\infty}^{\infty} p_n(m)\,\mathrm{e}^{\mathrm{i}m\phi}=\left\langle \mathrm{e}^{\mathrm{i}m\phi}\right\rangle_n \qquad (7.1.4)$$

for this purpose, noting that $g(0)=1$ in virtue of the normalization and $g_0(\phi)=1$ since there is only the $m=0$ term in the sum for $n=0$. Rather than investigating the n-jump probabilities $p_n(m)$ one by one, we turn our attention to the periodic function $g_n(\phi)$ that has $p_n(m)$ as its mth Fourier* coefficient. If we know $g_n(\phi)$, we can extract $p_n(m)$ by means of

$$p_n(m)=\int_{(2\pi)} \frac{\mathrm{d}\phi}{2\pi}\,\mathrm{e}^{-\mathrm{i}m\phi}g_n(\phi)\,, \qquad (7.1.5)$$

where the integration covers any range of 2π.

The recurrence relation in (7.1.3), which is a convolution in m, implies

$$\begin{aligned} g_n(\phi) &= \sum_{m=-\infty}^{\infty}\sum_{m'=-\infty}^{\infty} \mathrm{e}^{\mathrm{i}(m-m')\phi}p_{n-k}(m-m')\,\mathrm{e}^{\mathrm{i}m'\phi}p_k(m') \\ &= \sum_{m'=-\infty}^{\infty} \mathrm{e}^{\mathrm{i}m'\phi}p_k(m') \sum_{m=-\infty}^{\infty} \mathrm{e}^{\mathrm{i}(m-m')\phi}p_{n-k}(m-m') \\ &= \underbrace{\sum_{m'=-\infty}^{\infty} \mathrm{e}^{\mathrm{i}m'\phi}p_k(m')}_{=g_k(\phi)} \underbrace{\sum_{m=-\infty}^{\infty} \mathrm{e}^{\mathrm{i}m\phi}p_{n-k}(m)}_{=g_{n-k}(\phi)} \\ &= g_k(\phi)g_{n-k}(\phi) \end{aligned} \qquad (7.1.6)$$

for $k=0,1,\ldots,n$. In particular, then,

---

*Jean Baptiste Joseph FOURIER (1768–1830)

$$
\begin{aligned}
g_2(\phi) &= g_1(\phi)^2\,,\\
g_3(\phi) &= g_2(\phi)g_1(\phi) = g_1(\phi)^3\,,\\
&\ \,\vdots\\
g_n(\phi) &= g_{n-1}(\phi)g_1(\phi) = g_1(\phi)^n\,,
\end{aligned}
\tag{7.1.7}
$$

and one confirms easily that $g_n(\phi) = g_1(\phi)^n$ solves the recurrence relation. The normalization

$$
g_n(0) = g_1(0)^n = 1 \tag{7.1.8}
$$

is fine, of course, and we can also look at the mean values of m and m^2,

$$
\begin{aligned}
\langle m\rangle_n &= \sum_{m=-\infty}^{\infty} m\,p_n(m) = \frac{1}{\mathrm{i}}\frac{\partial}{\partial\phi}g_n(\phi)\bigg|_{\phi=0}\,,\\
\langle m^2\rangle_n &= \sum_{m=-\infty}^{\infty} m^2\,p_n(m) = \left(\frac{1}{\mathrm{i}}\frac{\partial}{\partial\phi}\right)^2 g_n(\phi)\bigg|_{\phi=0}\,,
\end{aligned}
\tag{7.1.9}
$$

or, more generally,

$$
\begin{aligned}
g_n(\phi) &= \sum_{k=0}^{\infty}\frac{1}{k!}\sum_{m=-\infty}^{\infty} p_n(m)(\mathrm{i}m\phi)^k\\
&= \sum_{k=0}^{\infty}\frac{(\mathrm{i}\phi)^k}{k!}\sum_{m=-\infty}^{\infty} p_n(m)\,m^k = \sum_{k=0}^{\infty}\frac{(\mathrm{i}\phi)^k}{k!}\langle m^k\rangle_n\,.
\end{aligned}
\tag{7.1.10}
$$

Considering the terms up to order ϕ^2, we have

$$
\begin{aligned}
g_n(\phi) &= 1 + \mathrm{i}\phi\langle m\rangle_n + \frac{1}{2}(\mathrm{i}\phi)^2\langle m^2\rangle_n + \cdots\\
&= \left[1 + \mathrm{i}\phi\langle m\rangle_1 + \frac{1}{2}(\mathrm{i}\phi)^2\langle m^2\rangle_1 + \cdots\right]^n\\
&= 1 + \mathrm{i}\phi n\langle m\rangle_1 + \frac{1}{2}(\mathrm{i}\phi)^2\Big(n\langle m^2\rangle_1 + n(n-1)\langle m\rangle_1^2\Big) + \cdots\,,
\end{aligned}
\tag{7.1.11}
$$

and read off that

$$
\begin{aligned}
\langle m\rangle_n &= n\langle m\rangle_1\,,\\
\langle m^2\rangle_n &= n\langle m^2\rangle_1 + n(n-1)\langle m\rangle_1^2\,,
\end{aligned}
\tag{7.1.12}
$$

so that

$$
\langle \delta m^2\rangle_n = \langle m^2\rangle_n - \langle m\rangle_n^2 = n\big(\langle m^2\rangle_1 - \langle m\rangle_1^2\big) = n\langle\delta m^2\rangle_1\,. \tag{7.1.13}
$$

In such a stochastic process, then, the net *drift* $\langle m\rangle_n$ is proportional to n, and so is the *variance* $\left\langle\delta m^2\right\rangle_n$, whereas the *spread* $\sqrt{\left\langle\delta m^2\right\rangle_n}$ grows proportional to $\sqrt{n}$,

$$\sqrt{\left\langle\delta m^2\right\rangle_n}=\sqrt{n}\sqrt{\left\langle\delta m^2\right\rangle_1}\propto\sqrt{n}\,. \tag{7.1.14}$$

These n dependencies are completely independent of the probabilities $p_1(m)$ that characterize the process. The values of $\langle m\rangle_1$ and $\left\langle m^2\right\rangle_1$ do, of course, depend on the choice of $p_1(m)$.

7.1.2 *Example: Symmetric one-unit jumps*

As a more specific example, let us now consider the process characterized by

$$p_1(m)=\frac{1}{2}\delta_{m,1}+\frac{1}{2}\delta_{m,-1}\,,\quad g_1(\phi)=\cos\phi\,, \tag{7.1.15}$$

where the particle jumps one unit to the left or the right with equal probability, but single jumps do not go further and not jumping is not an option. Then

$$\begin{aligned} g_n(\phi)=(\cos\phi)^n&=\frac{1}{2^n}\left(\mathrm{e}^{\mathrm{i}\phi}+\mathrm{e}^{-\mathrm{i}\phi}\right)^n\\ &=\frac{1}{2^n}\sum_{k=0}^{n}\binom{n}{k}\mathrm{e}^{\mathrm{i}k\phi}\,\mathrm{e}^{-\mathrm{i}(n-k)\phi}\,, \end{aligned} \tag{7.1.16}$$

and the nth probability distribution is given by

$$\begin{aligned} p_n(m)&=\int\limits_{(2\pi)}\frac{\mathrm{d}\phi}{2\pi}\,\mathrm{e}^{-\mathrm{i}m\phi}g_n(\phi)\\ &=\frac{1}{2^n}\sum_{k=0}^{n}\binom{n}{k}\underbrace{\int\limits_{(2\pi)}\frac{\mathrm{d}\phi}{2\pi}\,\mathrm{e}^{-\mathrm{i}m\phi}\,\mathrm{e}^{-\mathrm{i}n\phi+2\mathrm{i}k\phi}}_{=\delta_{m+n,2k}}\\ &=\begin{cases}\dfrac{1}{2^n}\dfrac{n!}{\left(\frac{n+m}{2}\right)!\left(\frac{n-m}{2}\right)!} & \text{if } n+m \text{ is even and } |m|\le n\,,\\ 0 & \text{else}\,.\end{cases} \end{aligned} \tag{7.1.17}$$

Here, we have $\langle m\rangle_1 = 0$ and $\langle m^2\rangle_1 = 1$, so that $\langle m\rangle_n = 0$ and $\langle m^2\rangle_n = n$ and $\sqrt{\langle \delta m^2\rangle_n} = \sqrt{n}$, and it follows that for large n (many jumps) the important values of m are quite small compared with n. To understand better the large-n form of $p_n(m)$ we, therefore, keep $|m| \ll n$ in mind when employing Stirling's approximation for the factorials,

$$\begin{aligned} n! &\cong \sqrt{2\pi n}\, n^n \,\mathrm{e}^{-n} = \sqrt{2\pi}\, n^{n+\frac{1}{2}}\,\mathrm{e}^{-n}\,; \\ \left(\frac{n+m}{2}\right)! &\cong \sqrt{2\pi}\left(\frac{n+m}{2}\right)^{\frac{1}{2}(n+m+1)} \mathrm{e}^{-\frac{1}{2}(n+m)} \\ &= \sqrt{2\pi}\,\mathrm{e}^{-\frac{1}{2}(n+m)} 2^{-\frac{1}{2}(n+m+1)}\,\mathrm{e}^{\frac{1}{2}(n+m+1)\log(n+m)}\,, \\ \left(\frac{n-m}{2}\right)! &\cong \sqrt{2\pi}\,\mathrm{e}^{-\frac{1}{2}(n-m)} 2^{-\frac{1}{2}(n-m+1)}\,\mathrm{e}^{\frac{1}{2}(n-m+1)\log(n-m)}\,, \end{aligned} \tag{7.1.18}$$

and combining the latter two into

$$\left(\frac{n+m}{2}\right)!\left(\frac{n-m}{2}\right)! \cong \pi 2^{-n}\,\mathrm{e}^{-n}\,\mathrm{e}^{\frac{1}{2}(n+1)\log(n^2-m^2)+\frac{1}{2}m\log\frac{n+m}{n-m}}\,, \tag{7.1.19}$$

where

$$\begin{aligned} \mathrm{e}^{\frac{1}{2}(n+1)\log(n^2-m^2)} &\cong \mathrm{e}^{\frac{1}{2}(n+1)\left[\log(n^2)-\left(\frac{m}{n}\right)^2\right]} \\ &= n^{(n+1)}\,\mathrm{e}^{-\frac{1}{2}(n+1)\left(\frac{m}{n}\right)^2} \end{aligned} \tag{7.1.20}$$

and

$$\mathrm{e}^{\frac{1}{2}m\log\frac{n+m}{n-m}} \cong \mathrm{e}^{\frac{1}{2}m\left[\frac{m}{n}-\left(-\frac{m}{n}\right)\right]} = \mathrm{e}^{\frac{m^2}{n}} \tag{7.1.21}$$

to second-order of $\dfrac{m}{n}$ in the exponent. After putting the pieces together, we have

$$\begin{aligned} p_n(m) &\cong \frac{1}{2^n}\frac{\sqrt{2\pi}\,n^{n+\frac{1}{2}}\,\mathrm{e}^{-n}}{\pi\, 2^{-n} n^{n+1}\,\mathrm{e}^{-n}\,\mathrm{e}^{\frac{1}{2}(n-1)\left(\frac{m}{n}\right)^2}} \\ &= \sqrt{\frac{2}{\pi n}}\,\mathrm{e}^{-\frac{m^2}{2n}\left(1-\frac{1}{n}\right)} \end{aligned} \tag{7.1.22}$$

or

$$p_n(m) \cong \sqrt{\frac{2}{\pi n}}\,\mathrm{e}^{-\frac{m^2}{2n}}\,, \tag{7.1.23}$$

valid for $n \gg 1$. The summation over m skips every second m value — recall that $m+n$ is even for nonzero values of $p_n(m)$ — and the picture is this:

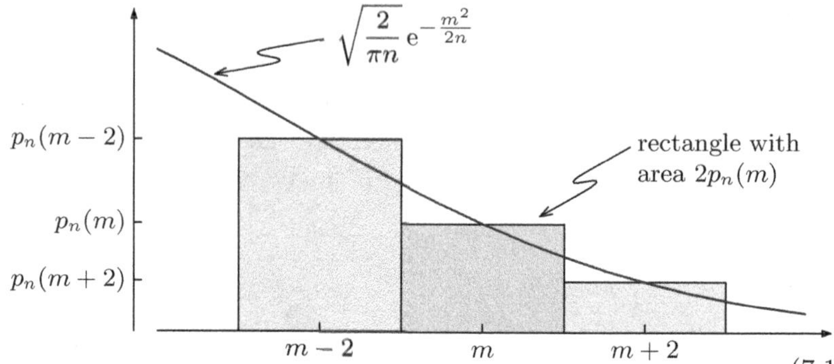

(7.1.24)

When we replace the summation over m (in steps of two) by an integration over m, we get the corresponding approximation of the combined area of the rectangles, which is twice the sum of the probabilities $p_n(m)$ over m. Thus, we should expect

$$\sum_{\substack{m \\ (n+m\,\text{even})}} p_n(m) \cong \frac{1}{2}\int_{-\infty}^{\infty} \mathrm{d}m\,\sqrt{\frac{2}{\pi n}}\,\mathrm{e}^{-\frac{m^2}{2n}} = \frac{1}{2}\sqrt{\frac{2}{\pi n}}\sqrt{\frac{\pi}{\frac{1}{2n}}} = 1\,, \tag{7.1.25}$$

indeed, with no further approximation for large n. That is, the gaussian distribution

$$\mathrm{d}m\,\frac{1}{\sqrt{2\pi n}}\,\mathrm{e}^{-\frac{m^2}{2n}} \tag{7.1.26}$$

is the probability of having a displacement between m and $m+\mathrm{d}m$ units after n jumps if $n \gg 1$. We note that the main contribution to the integral in (7.1.25) stems from $|m| \ll n$, which is consistent with the approximations on the way to (7.1.23).

7.2 Asymptotic gaussian distribution

The *gaussian form* for large n in (7.1.26) illustrates a ubiquitous phenomenon, namely that probability distributions tend to be gaussian when many random processes contribute (many jumps here), whether the individual probability distributions are gaussian or not. All that matters is that the net effect is the result of a large number of random events, usually because a large number of uncontrolled parameters determine the outcome. That the overall description is in terms of a gaussian distribution is known as the *central limit theorem* in statistics, introduced by Pólya* and also studied by Lindeberg† in the early 1920s. Accordingly, we should not be surprised when encountering gaussian probability distributions, they should often be expected.

Often is not the same as always. Consider the *Bernoulli* probability distribution* for n trials with mean value $\langle m\rangle$,

$$p_n(m) = \binom{n}{m}\left(\frac{\langle m\rangle}{n}\right)^m\left(1-\frac{\langle m\rangle}{n}\right)^{n-m}, \tag{7.2.1}$$

which is nonzero for $m = 0, 1, \ldots, n$. Its generating function

$$\begin{aligned} g_n(y) &= \sum_{m=0}^{\infty} y^m p_n(m) = \sum_{m=0}^{\infty}\binom{n}{m}\left(\frac{y\langle m\rangle}{n}\right)^m\left(1-\frac{\langle m\rangle}{n}\right)^{n-m} \\ &= \left[\frac{y\langle m\rangle}{n} + \left(1-\frac{\langle m\rangle}{n}\right)\right]^n = \left[1-\frac{\langle m\rangle}{n}(1-y)\right]^n \end{aligned} \tag{7.2.2}$$

has a simple limiting form for $n\to\infty$,

$$g_n(y) \to \mathrm{e}^{-\langle m\rangle(1-y)} = \sum_{m=0}^{\infty} y^m \frac{\langle m\rangle^m}{m!}\,\mathrm{e}^{-\langle m\rangle} \quad \text{as } n\to\infty\,. \tag{7.2.3}$$

It follows that the asymptotic large-n form of the Bernoulli distribution is the *Poisson† distribution* with the same mean value

$$p_n(m) = \binom{n}{m}\left(\frac{\langle m\rangle}{n}\right)^m\left(1-\frac{\langle m\rangle}{n}\right)^{n-m} \underset{(n\gg 1)}{=} \frac{\langle m\rangle^m}{m!}\,\mathrm{e}^{-\langle m\rangle} \equiv p_\infty(m)\,. \tag{7.2.4}$$

In this large-n limit, we do not get a gaussian distribution.

There is, however, still a gaussian in hiding. When the mean value $\langle m\rangle$ of the Poisson distribution is very large and the spread $\sqrt{\langle\delta m^2\rangle} = \sqrt{\langle m\rangle} \ll \langle m\rangle$ is quite small compared with the large mean value, then, for $m = \langle m\rangle + x$ with $|x| \ll \langle m\rangle$,

$$\begin{aligned} p_\infty(m) = p_\infty(\langle m\rangle + x) &= \frac{\langle m\rangle^{\langle m\rangle + x}}{(\langle m\rangle + x)!}\,\mathrm{e}^{-\langle m\rangle} \\ &\cong \frac{\langle m\rangle^{\langle m\rangle+x}\,\mathrm{e}^{-\langle m\rangle}}{\sqrt{2\pi}(\langle m\rangle + x)^{\langle m\rangle + x + \frac{1}{2}}\,\mathrm{e}^{-(\langle m\rangle + x)}} \\ &\cong \frac{1}{\sqrt{2\pi\langle m\rangle}}\left(1+\frac{x}{\langle m\rangle}\right)^{-(\langle m\rangle + x)}\mathrm{e}^{x}\,, \end{aligned} \tag{7.2.5}$$

---

*George (György) PÓLYA (1887–1985) †Jarl Waldemar LINDEBERG (1876–1932)

*Jakob BERNOULLI (1655–1705) †Siméon Denise POISSON (1781–1840)

and now keeping terms up to order x^2 in

$$\begin{aligned}\left(1+\frac{x}{\langle m\rangle}\right)^{-(\langle m\rangle+x)} &= \mathrm{e}^{-(\langle m\rangle+x)\log\left(1+\frac{x}{\langle m\rangle}\right)} \\ &= \mathrm{e}^{-(\langle m\rangle+x)\left[\frac{x}{\langle m\rangle}-\frac{1}{2}\left(\frac{x}{\langle m\rangle}\right)^2\right]} \\ &= \mathrm{e}^{-x-\frac{1}{2}\frac{x^2}{\langle m\rangle}}\end{aligned} \tag{7.2.6}$$

takes us to the gaussian distribution

$$p_\infty\big(\langle m\rangle + x\big) = \frac{1}{\sqrt{2\pi\langle m\rangle}}\,\mathrm{e}^{-\frac{1}{2}\frac{x^2}{\langle m\rangle}}\,, \tag{7.2.7}$$

which is properly normalized

$$\int_{-\infty}^{\infty} \mathrm{d}x\, p_\infty\big(\langle m\rangle + x\big) = 1 \tag{7.2.8}$$

and has $\langle x\rangle = \big\langle(m-\langle m\rangle)\big\rangle = 0$ and

$$\begin{aligned}\langle x^2\rangle &= \left\langle\big(m-\langle m\rangle\big)^2\right\rangle = \langle\delta m^2\rangle \\ &= \int_{-\infty}^{\infty} \mathrm{d}x\, x^2 p_\infty\big(\langle m\rangle + x\big) = \langle m\rangle\,,\end{aligned} \tag{7.2.9}$$

the correct expected values for $x = m - \langle m\rangle$ and its square.

7.3 Poisson processes

An important kind of a random process is the so-called *Poisson process*, in which events occur independently of each other at a certain rate r. The probability of having an event between t and $t+\mathrm{d}t$ is $r\,\mathrm{d}t$, irrespective of whether and when there were earlier events.

We denote by $p_m(t)$ the probability that there are exactly m events during the lapse of time t. These probabilities have unit sum,

$$\sum_{m=0}^{\infty} p_m(t) = 1 \quad \text{for} \quad t > 0\,, \tag{7.3.1}$$

and obey the recurrence relation

$$p_{m+1}(t) = r\int_0^t \mathrm{d}t'\, p_{m-m'}(t-t')p_{m'}(t') \qquad (7.3.2)$$

for $m' = 0, 1, 2, \ldots, m$. It states that we get $m+1$ events in time t by having one event at any intermediate time t' (factor $r\,\mathrm{d}t'$), preceded by m' events and succeeded by $m-m'$ events.

The convolution in (7.3.2) invites us to look at the Laplace* transform,

$$\begin{aligned}
\widetilde{p}_{m+1}(\gamma) &= \int_0^\infty \mathrm{d}t\ \mathrm{e}^{-\gamma t} p_{m+1}(t) \\
&= r\int_0^\infty \mathrm{d}t \int_0^t \mathrm{d}t'\ \mathrm{e}^{-\gamma(t-t')} p_{m-m'}(t-t')\,\mathrm{e}^{-\gamma t'} p_{m'}(t') \\
&= r\int_0^\infty \mathrm{d}t'\ \mathrm{e}^{-\gamma t'} p_{m'}(t') \underbrace{\int_{t'}^\infty \mathrm{d}t\ \mathrm{e}^{-\gamma(t-t')} p_{m-m'}(t-t')}_{=\widetilde{p}_{m-m'}(\gamma)} \\
&= r\widetilde{p}_{m-m'}(\gamma)\widetilde{p}_{m'}(\gamma)
\end{aligned} \qquad (7.3.3)$$

or

$$r\widetilde{p}_{m+1}(\gamma) = \bigl[r\widetilde{p}_{m-m'}(\gamma)\bigr]\bigl[r\widetilde{p}_{m'}(\gamma)\bigr]\,. \qquad (7.3.4)$$

This recurrence relation for $\widetilde{p}_{m+1}(\gamma)$ is solved immediately,

$$r\widetilde{p}_m(\gamma) = \bigl[r\widetilde{p}_0(\gamma)\bigr]^{m+1} \quad \text{or} \quad \widetilde{p}_m(\gamma) = r^m\,\widetilde{p}_0(\gamma)^{m+1}\,. \qquad (7.3.5)$$

We determine $\widetilde{p}_0(\gamma)$ from the sum rule in (7.3.1), which translates into

$$\sum_{m=0}^\infty \widetilde{p}_m(\gamma) = \int_0^\infty \mathrm{d}t\ \mathrm{e}^{-\gamma t} \underbrace{\sum_{m=0}^\infty p_m(t)}_{=1} = \frac{1}{\gamma}\,. \qquad (7.3.6)$$

According to (7.3.5), we have a geometric series for the sum of the $\widetilde{p}_m(\gamma)$s,

$$\sum_{m=0}^\infty \widetilde{p}_m(\gamma) = \widetilde{p}_0(\gamma)\sum_{m=0}^\infty \bigl[r\widetilde{p}_0(\gamma)\bigr]^m = \frac{\widetilde{p}_0(\gamma)}{1-r\widetilde{p}_0(\gamma)}\,, \qquad (7.3.7)$$

---

*Marquis de Pierre Simon LAPLACE (1749–1827)

so that

$$\widetilde{p}_0(\gamma) = \frac{1}{\gamma + r} \,. \tag{7.3.8}$$

It follows that

$$\widetilde{p}_m(\gamma) = \frac{r^m}{(\gamma + r)^{m+1}} = \frac{r^m}{m!} \int_0^\infty \mathrm{d}t\, t^m \,\mathrm{e}^{-(\gamma + r)t} \,, \tag{7.3.9}$$

where Euler's factorial integral of (3.6.16) is used in recognizing that $\widetilde{p}_m(\gamma)$ is the Laplace transform of

$$p_m(t) = \frac{(rt)^m}{m!}\,\mathrm{e}^{-rt} \,. \tag{7.3.10}$$

This is the Poisson distribution of (7.2.4) with $\langle m \rangle = rt$, which tells us that, on average, rt events occur in the lapse of t, fully consistent with the meaning of r as the rate at which independent random events happen.

Let us note that we could have found $p_0(t)$ by another line of reasoning that is exhibited in

$$p_0(t) = \mathrm{e}^{-rt} = (1 - r\,\mathrm{d}t)^{\frac{t}{\mathrm{d}t}} \,. \tag{7.3.11}$$

Each factor $1 - r\,\mathrm{d}t$ is the probability of no event in the very short time step $\mathrm{d}t$, so short that the probability of more than one event can be safely ignored, and there are $\dfrac{t}{\mathrm{d}t} \gg 1$ such steps in the time interval of duration t.

As a little application, let us ask this question: If we keep observing until exactly m events have occurred, what is the probability that the observation is over between time t and time $t + \mathrm{d}t$? This is, of course, just asking for the probability of observing the mth event between t and $t + \mathrm{d}t$, which is $p_{m-1}(t) r\,\mathrm{d}t$, namely the product of the probability $p_{m-1}(t)$ of $m - 1$ events before time t and the probability $r\,\mathrm{d}t$ of another event between t and $t+\mathrm{d}t$. Since the mth event does occur eventually, this probability distribution must be normalized to unit integral, and it is:

$$\int_0^\infty \mathrm{d}t\, r p_{m-1}(t) = \int_0^\infty \mathrm{d}(rt)\, \frac{(rt)^{m-1}}{(m-1)!}\,\mathrm{e}^{-rt} = 1 \,. \tag{7.3.12}$$

On average, then, we must wait for the time

$$\int_0^\infty \mathrm{d}t\, r p_{m-1}(t)\, t = \frac{m}{r} \tag{7.3.13}$$

until we observe the mth event. This result is rather intuitive: Since the events occur randomly and independently at rate r, we have on average rt events in the lapse of t and, therefore, it will on average take time $\frac{m}{r}$ to observe m events.

7.4 Brownian motion: Langevin's random force

Let us now consider the situation of Brownian motion as a dynamical problem, as described by Newton's equation of motion for a pollen grain,

$$M\frac{\mathrm{d}}{\mathrm{d}t}\boldsymbol{v}(t) = \boldsymbol{F}_{\mathrm{random}}(t) = -\frac{1}{B}\boldsymbol{v}(t) + \boldsymbol{F}(t)\,, \qquad (7.4.1)$$

where $\boldsymbol{F}_{\mathrm{random}}(t)$ is the random force that act, at time t, on the grain of mass M that results from the collisions with the much smaller water molecules. Following Langevin,* we decompose the random force into a drift part $-\frac{1}{B}\boldsymbol{v}(t)$, with the *mobility* B, and a fluctuating part $\boldsymbol{F}(t)$ that has a vanishing average,

$$\overline{\boldsymbol{F}(t)} = \big\langle \boldsymbol{F}(t)\big\rangle = \langle \boldsymbol{F}\rangle_t = 0\,. \qquad (7.4.2)$$

Here $\overline{\cdots}$ means the long-time average over time, and $\langle\cdots\rangle$ stands for an ensemble average, and we take for granted that the physics is ergodic, so that the two averages agree with each other. Then,

$$M\frac{\mathrm{d}}{\mathrm{d}t}\boldsymbol{v}(t) = -\frac{1}{B}\boldsymbol{v}(t) + \boldsymbol{F}(t) \qquad (7.4.3)$$

is the equation of motion for the detailed trajectory of a pollen grain, and its ensemble average reads

$$M\frac{\mathrm{d}}{\mathrm{d}t}\langle\boldsymbol{v}\rangle_t = -\frac{1}{B}\langle\boldsymbol{v}\rangle_t + \underbrace{\langle\boldsymbol{F}\rangle_t}_{=0} = -\frac{1}{B}\langle\boldsymbol{v}\rangle_t\,. \qquad (7.4.4)$$

This has the immediate solution

$$\langle\boldsymbol{v}\rangle_{=}\langle\boldsymbol{v}\rangle_0\,\mathrm{e}^{-t/\tau} \quad \text{with} \quad \tau = BM\,, \qquad (7.4.5)$$

and

$$\langle\boldsymbol{r}\rangle_t = \langle\boldsymbol{r}\rangle_0 + \langle\boldsymbol{v}\rangle_0\tau\Big(1-\mathrm{e}^{-t/\tau}\Big) \qquad (7.4.6)$$

---

*Paul LANGEVIN (1872–1946)

is the ensemble-averaged position in time. The acceleration contribution proportional to $-\frac{1}{\tau}\boldsymbol{v}(t)$ accounts for the viscous force by the water on the pollen, it brings the pollen grains to rest *on average*. This relaxation happens on the time scale set by τ; after a few τ, a stationary situation is reached, in which there are fluctuations but no more overall drift.

We return to (7.4.3),

$$\frac{\mathrm{d}}{\mathrm{d}t}\boldsymbol{v} = -\frac{1}{\tau}\boldsymbol{v} + \frac{1}{M}\boldsymbol{F}(t)\,, \tag{7.4.7}$$

and multiply by $\boldsymbol{r}$,

$$\boldsymbol{r}\cdot\frac{\mathrm{d}}{\mathrm{d}t}\boldsymbol{v} = -\frac{1}{\tau}\boldsymbol{r}\cdot\boldsymbol{v} + \frac{1}{M}\boldsymbol{r}\cdot\boldsymbol{F}\,, \tag{7.4.8}$$

where

$$\boldsymbol{r}\cdot\boldsymbol{v} = \boldsymbol{r}\cdot\frac{\mathrm{d}}{\mathrm{d}t}\boldsymbol{r} = \frac{1}{2}\frac{\mathrm{d}}{\mathrm{d}t}\boldsymbol{r}^2 = \frac{1}{2}\frac{\mathrm{d}}{\mathrm{d}t}r^2 \tag{7.4.9}$$

and

$$\boldsymbol{r}\cdot\frac{\mathrm{d}}{\mathrm{d}t}\boldsymbol{v} = \frac{\mathrm{d}}{\mathrm{d}t}(\boldsymbol{r}\cdot\boldsymbol{v}) - \boldsymbol{v}^2 = \frac{1}{2}\left(\frac{\mathrm{d}}{\mathrm{d}t}\right)^2 r^2 - v^2\,. \tag{7.4.10}$$

Therefore, we have

$$\left(\frac{\mathrm{d}}{\mathrm{d}t}\right)^2 r^2 + \frac{1}{\tau}\frac{\mathrm{d}}{\mathrm{d}t}r^2 = 2v^2 + \frac{2}{M}\boldsymbol{r}\cdot\boldsymbol{F}\,, \tag{7.4.11}$$

and taking ensemble averages gives

$$\left(\frac{\mathrm{d}}{\mathrm{d}t}\right)^2 \langle r^2\rangle_t + \frac{1}{\tau}\frac{\mathrm{d}}{\mathrm{d}t}\langle r^2\rangle_t = 2\langle v^2\rangle_t\,, \tag{7.4.12}$$

since $\langle\boldsymbol{r}\cdot\boldsymbol{F}\rangle = 0$, as the randomly fluctuating force $\boldsymbol{F}(t)$ should be the same irrespective of what is the position vector $\boldsymbol{r}(t)$ of the pollen grain.

When stationarity is reached, $\langle v^2\rangle_t$ no longer changes in time and its value is that in the canonical ensemble with the temperature T of the water in which the pollen grains are immersed, see Exercises 125 and 126,

$$\langle v^2\rangle = \frac{2}{M}\left\langle\frac{M}{2}v^2\right\rangle = \frac{2}{M}\frac{3}{2}k_{\mathrm{B}}T = \frac{3k_{\mathrm{B}}T}{M}\,, \tag{7.4.13}$$

and so we have

$$\frac{\mathrm{d}}{\mathrm{d}t}\left(\frac{\mathrm{d}}{\mathrm{d}t} + \frac{1}{\tau}\right)\langle r^2\rangle_t = \frac{6k_{\mathrm{B}}T}{M}\,. \tag{7.4.14}$$

This is solved by

$$\langle r^2\rangle_t = \langle r^2\rangle_0 + \tau\left(\frac{\mathrm{d}\langle r^2\rangle}{\mathrm{d}t}\right)_0\left(1 - \mathrm{e}^{-t/\tau}\right) + \frac{6k_\mathrm{B}T}{M}\tau^2\left(\frac{t}{\tau} - 1 + \mathrm{e}^{-t/\tau}\right), \tag{7.4.15}$$

and we note that the term that is proportional to t dominates at late times:

$$\langle r^2\rangle_t \cong \frac{6k_\mathrm{B}T}{M}\tau t \quad \text{for} \quad t \gg \tau\,. \tag{7.4.16}$$

In accordance with $\langle r^2\rangle_t = 6Dt$ for late times, this identifies the *diffusion coefficient*

$$D = \frac{k_\mathrm{B}T}{M}\tau = Bk_\mathrm{B}T\,, \tag{7.4.17}$$

a quantity that can be measured. At very early times, $t \ll \tau$, we have

$$\begin{aligned}\langle r^2\rangle_t &= \langle r^2\rangle_0 + \left(\frac{\mathrm{d}\langle r^2\rangle}{\mathrm{d}t}\right)_0 t + \frac{3k_\mathrm{B}T}{M}t^2\\ &= \langle r^2\rangle_0 + \left(\frac{\mathrm{d}\langle r^2\rangle}{\mathrm{d}t}\right)_0 t + \langle v^2\rangle_0 t^2\,,\end{aligned} \tag{7.4.18}$$

fully consistent with the average of $\boldsymbol{r}(t)^2 = \left[\boldsymbol{r}(0) + \boldsymbol{v}(0)t\right]^2$ for an ensemble of particles in force-free motion. This force-free motion, however, lasts only for a period that is short compared with τ, the time constant of the frictional viscous drag force, the contribution proportional to $-\dfrac{M}{\tau}\boldsymbol{v}(t)$.

7.5 Autocorrelation function

For an account of the approach to stationarity, we solve the equation for $\boldsymbol{v}(t)$ in (7.4.7) for an arbitrary fluctuating force $\boldsymbol{F}(t)$,

$$\boldsymbol{v}(t) = \boldsymbol{v}(0)\,\mathrm{e}^{-t/\tau} + \frac{1}{M}\int_0^t \mathrm{d}t'\;\mathrm{e}^{-(t-t')/\tau}\boldsymbol{F}(t')\,, \tag{7.5.1}$$

and square this expression before taking the ensemble average,

$$\begin{aligned}\langle v^2\rangle_t = \langle v^2\rangle_0\,\mathrm{e}^{-2t/\tau} &+ \frac{2}{M}\int_0^t \mathrm{d}t'\;\mathrm{e}^{-(2t-t')/\tau}\big\langle \boldsymbol{v}(0)\cdot\boldsymbol{F}(t')\big\rangle \qquad (7.5.2)\\ &+ \frac{1}{M^2}\int_0^t \mathrm{d}t'\;\mathrm{e}^{-(t-t')/\tau}\int_0^t \mathrm{d}t''\;\mathrm{e}^{-(t-t'')/\tau}\big\langle \boldsymbol{F}(t')\cdot\boldsymbol{F}(t'')\big\rangle\,.\end{aligned}$$

Here we have

$$\big\langle \boldsymbol{v}(0)\cdot\boldsymbol{F}(t')\big\rangle = 0 \tag{7.5.3}$$

since the fluctuating force at t' is not correlated with the initial velocity, and we meet the *autocorrelation function*

$$\frac{1}{M^2}\big\langle \boldsymbol{F}(t')\cdot\boldsymbol{F}(t'')\big\rangle = K(t'-t'') \tag{7.5.4}$$

of the fluctuating acceleration $\frac{1}{M}\boldsymbol{F}(t)$.The physical situation is the same irrespective of when we observe the pollen immersed in the water and, therefore, only the time difference $t'-t''$ matters for the autocorrelation function, not the absolute time. This is already indicated by writing K as a function of $t'-t''$. Further, the left-hand side is the same when t' and t'' are interchanged, so that

$$K(t'-t'') = K(t''-t') \tag{7.5.5}$$

is an even function of $t'-t''$. We also note that $K(0)>0$, since it is the average of a square,

$$K(0) = \frac{1}{M^2}\big\langle \boldsymbol{F}(t')^2\big\rangle > 0\,, \tag{7.5.6}$$

and all other values of $K(t'-t'')$ are bounded by $K(0)$,

$$|K(t'-t'')| \le K(0) \quad \text{for all } t'-t''\,. \tag{7.5.7}$$

One sees this by considering the positive quantities

$$\begin{aligned} 0 < \Big\langle \big(\boldsymbol{F}(t') \pm \boldsymbol{F}(t'')\big)^2 \Big\rangle &= \big\langle \boldsymbol{F}(t')^2\big\rangle + \big\langle \boldsymbol{F}(t'')^2\big\rangle \pm 2\big\langle \boldsymbol{F}(t')\cdot\boldsymbol{F}(t'')\big\rangle \\ &= 2M^2K(0) \pm 2M^2K(t'-t'')\,, \end{aligned} \tag{7.5.8}$$

which requires $K(t'-t'')<K(0)$ and also $-K(t'-t'')<K(0)$.

Yet another property of the autocorrelation function is that it will be markedly different from zero ($\equiv$ no correlations) only for finite differences $t'-t''$ that are short compared with the characteristic time scales of the fluctuations or comparable with them, while $K(t'-t'')\cong 0$ for time differences $|t'-t''|$ large compared with the characteristic times of the fluctuations, which are rapid on the timescale $\sim\tau$ of the damping by the viscous frictional force.

We introduce the Fourier transform $\widetilde{K}(\omega)$,

$$K(t'-t'') = \int_{-\infty}^{\infty} \frac{d\omega}{2\pi}\, e^{i\omega(t'-t'')}\,\widetilde{K}(\omega)\,,$$

$$\widetilde{K}(\omega) = \int_{-\infty}^{\infty} d(t'-t'')\, e^{-i\omega(t'-t'')}\,K(t'-t'')\,, \tag{7.5.9}$$

which is real and even,

$$\widetilde{K}(\omega) = \widetilde{K}(-\omega) = \widetilde{K}(\omega)^*\,, \tag{7.5.10}$$

since $K(t'-t'')$ is real and even. Further, $K(t'-t'')$ is peaked at $t'-t''=0$ with a width that is much smaller than τ, and, therefore, $\widetilde{K}(\omega)$ will be peaked at $\omega = 0$ with a peak much broader than $\frac{1}{\tau}$. For the double integral in

$$\langle v^2\rangle_t = \langle v^2\rangle_0\, e^{-2t/\tau} + \int_0^t dt' \int_0^t dt''\, e^{-(2t-t'-t'')/\tau}\,K(t'-t'')\,, \tag{7.5.11}$$

this means that the replacement

$$\widetilde{K}(\omega) \cong \widetilde{K}(\omega=0) = \int_{-\infty}^{\infty} d(t'-t'')\,K(t'-t'') \tag{7.5.12}$$

is permissible in

$$\begin{aligned}
&\int_0^t dt' \int_0^t dt''\, e^{-(2t-t'-t'')/\tau} \int_{-\infty}^{\infty} \frac{d\omega}{2\pi}\,\widetilde{K}(\omega)\, e^{i\omega(t'-t'')} \\
&= \int_{-\infty}^{\infty} \frac{d\omega}{2\pi}\,\widetilde{K}(\omega) \left| \int_0^t dt'\, e^{-(t-t')/\tau + i\omega t'} \right|^2 \\
&= \int_{-\infty}^{\infty} \frac{d\omega}{2\pi}\,\widetilde{K}(\omega) \left| \frac{e^{i\omega t} - e^{-t/\tau}}{i\omega + 1/\tau} \right|^2 \\
&= \int_{-\infty}^{\infty} \frac{d\omega}{2\pi}\,\widetilde{K}(\omega)\, \underbrace{\frac{1 - 2\,e^{-t/\tau}\cos(\omega t) + e^{-2t/\tau}}{\omega^2 + (1/\tau)^2}}_{\cong 0 \text{ when } |\omega\tau| \gtrsim \mathcal{O}(1)}\,,
\end{aligned} \tag{7.5.13}$$

where the factor multiplying $\widetilde{K}(\omega)$ becomes very small before $\widetilde{K}(\omega)$ starts to deviate much from $\widetilde{K}(\omega=0)$. Now, if we replace $\widetilde{K}(\omega) \to \widetilde{K}(0)$ in the first line, the ω integration gives a delta function,

$$\int_{-\infty}^{\infty} \frac{\mathrm{d}\omega}{2\pi}\, \mathrm{e}^{\mathrm{i}\omega(t'-t'')} = \delta(t'-t''), \tag{7.5.14}$$

and we get

$$\widetilde{K}(0)\int_{0}^{t} \mathrm{d}t'\, \mathrm{e}^{-2t/\tau}\, \mathrm{e}^{2t'/\tau} = \widetilde{K}(0)\frac{\tau}{2}\left(1-\mathrm{e}^{-2t/\tau}\right) \tag{7.5.15}$$

and so arrive at

$$\begin{aligned} \left\langle v^2\right\rangle_t &= \left\langle v^2\right\rangle_0 \mathrm{e}^{-2t/\tau} + \frac{\tau}{2}\left(1-\mathrm{e}^{-2t/\tau}\right)\widetilde{K}(0) \\ &= \frac{\tau}{2}\widetilde{K}(0) + \left[\left\langle v^2\right\rangle_0 - \frac{\tau}{2}\widetilde{K}(0)\right]\mathrm{e}^{-2t/\tau}. \end{aligned} \tag{7.5.16}$$

Since the stationary value of $\left\langle v^2\right\rangle$ is $\dfrac{3k_{\mathrm{B}}T}{M}$, see (7.4.13), we must have

$$\widetilde{K}(0) = \int_{-\infty}^{\infty} \mathrm{d}t\, K(t) = \frac{6k_{\mathrm{B}}T}{M\tau}. \tag{7.5.17}$$

Finally, then,

$$\left\langle v^2\right\rangle_t = \frac{3k_{\mathrm{B}}T}{M} + \left(\left\langle v^2\right\rangle_0 - \frac{3k_{\mathrm{B}}T}{M}\right)\mathrm{e}^{-2t/\tau}. \tag{7.5.18}$$

We note that this approximate answer is reliable for $t \gtrsim \tau$, but not for $t \ll \tau$. For these very short times, we have

$$1 - 2\,\mathrm{e}^{-t/\tau}\cos(\omega t) + \mathrm{e}^{-2t/\tau} = \left(\omega^2 + \frac{1}{\tau^2}\right)t^2 \tag{7.5.19}$$

and get

$$t^2 \int_{-\infty}^{\infty} \frac{\mathrm{d}\omega}{2\pi}\, \widetilde{K}(\omega) = t^2 K(t'-t''=0) \tag{7.5.20}$$

for the double integral in (7.5.11) rather than $\widetilde{K}(0)t$ as in the approximate expression. Nevertheless, we shall ignore this deviation at very short times and use the approximation for $\langle v^2\rangle_t$ on the right-hand side of (7.4.12),

$$\frac{\mathrm{d}}{\mathrm{d}t}\left(\frac{\mathrm{d}}{\mathrm{d}t}+\frac{1}{\tau}\right)\langle r^2\rangle_t=\frac{6k_{\mathrm{B}}T}{M}+\left(2\langle v^2\rangle_0-\frac{6k_{\mathrm{B}}T}{M}\right)\mathrm{e}^{-2t/\tau}\,. \tag{7.5.21}$$

The solution with $\langle v^2\rangle_0=\dfrac{3k_{\mathrm{B}}T}{M}$ is given in (7.4.15), to which we need to add the solution of

$$\left(\frac{\mathrm{d}}{\mathrm{d}t}+\frac{1}{\tau}\right)\frac{\mathrm{d}}{\mathrm{d}t}\langle r^2\rangle_t=\mathrm{e}^{-t/\tau}\frac{\mathrm{d}}{\mathrm{d}t}\,\mathrm{e}^{t/\tau}\frac{\mathrm{d}}{\mathrm{d}t}\langle r^2\rangle_t=\left(2\langle v^2\rangle_0-\frac{6k_{\mathrm{B}}T}{M}\right)\mathrm{e}^{-2t/\tau} \tag{7.5.22}$$

with the initial conditions $\langle r^2\rangle_0=0$ and $\left(\dfrac{\mathrm{d}\langle r^2\rangle}{\mathrm{d}t}\right)_0=0$, since the nonzero initial values are already accounted for in (7.4.15). This additional term is

$$\tau^2\left(\langle v^2\rangle_0-\frac{3k_{\mathrm{B}}T}{M}\right)\left(1-\mathrm{e}^{-t/\tau}\right)^2 \tag{7.5.23}$$

as one verifies by differentiation:

$$\mathrm{e}^{-t/\tau}\frac{\mathrm{d}}{\mathrm{d}t}\,\mathrm{e}^{t/\tau}\frac{\mathrm{d}}{\mathrm{d}t}\left(1-\mathrm{e}^{-t/\tau}\right)^2=\mathrm{e}^{-t/\tau}\frac{\mathrm{d}}{\mathrm{d}t}\frac{2}{\tau}\left(1-\mathrm{e}^{-t/\tau}\right)=\frac{2}{\tau^2}\,\mathrm{e}^{-2t/\tau}\,. \tag{7.5.24}$$

Accordingly, the approach to stationarity is described by

$$\begin{aligned}\langle r^2\rangle_t&=\langle r^2\rangle_0+\tau\left(\frac{\mathrm{d}\langle r^2\rangle}{\mathrm{d}t}\right)_0\left(1-\mathrm{e}^{-t/\tau}\right)\\&\quad+\frac{6k_{\mathrm{B}}T}{M}\tau^2\left(\frac{t}{\tau}-1+\mathrm{e}^{-t/\tau}\right)\\&\quad+\tau^2\left(\langle v^2\rangle_0-\frac{3k_{\mathrm{B}}T}{M}\right)\left(1-\mathrm{e}^{-t/\tau}\right)^2.\end{aligned} \tag{7.5.25}$$

For times late enough that $\mathrm{e}^{-t/\tau}\cong 0$, this simplifies to

$$\langle r^2\rangle_t=\langle r^2\rangle_0+\tau\left(\frac{\mathrm{d}\langle r^2\rangle}{\mathrm{d}t}\right)_0+\tau^2\left(\langle v^2\rangle_0-\frac{9k_{\mathrm{B}}T}{M}\right)+\underbrace{\frac{6k_{\mathrm{B}}T}{M}\tau t}_{=6Dt} \tag{7.5.26}$$

with a constant term determined by the intial conditions and a term linear in time t, which — as it should be — is exactly that of (7.4.16) with the diffusion coefficient D of (7.4.17).

7.6 Fluctuations and dissipation

In (7.5.17), we established that

$$\widetilde{K}(0) = \int_{-\infty}^{\infty} \mathrm{d}t\, K(t) = \frac{6k_{\mathrm{B}}T}{M\tau} \tag{7.6.1}$$

or (any value for t_0 is as good as any other)

$$\int_{-\infty}^{\infty} \mathrm{d}t\, \frac{1}{M^2}\big\langle \boldsymbol{F}(t_0+t)\cdot\boldsymbol{F}(t_0)\big\rangle = \frac{6k_{\mathrm{B}}T}{M^2B}, \tag{7.6.2}$$

after recalling $\tau = MB$ from (7.4.5) and the definition of $K(t)$ in (7.5.4). We further recall that the mobility B is given by the drift term of the random force,

$$\big\langle \boldsymbol{F}_{\mathrm{random}}(t)\big\rangle = -\frac{1}{B}\boldsymbol{v}(t), \tag{7.6.3}$$

which is the dissipative, viscous, frictional force in the equation of motion in (7.4.3). We can thus write

$$\frac{1}{B} = \frac{1}{6k_{\mathrm{B}}T}\int \mathrm{d}t\, \big\langle \boldsymbol{F}(t_0+t)\cdot\boldsymbol{F}(t_0)\big\rangle\,. \tag{7.6.4}$$

This is an example of a *fluctuation-dissipation theorem*: On the right-hand side, we have the autocorrelation function of the *fluctuating force*; on the left-hand side, we have the coefficient of the *dissipative force*.

Relations of this kind between fluctuating forces and dissipative forces that originate in the same physical mechanism (collisions of the pollen grains with the much smaller water molecules in the case of Brownian motion) occur in many contexts (in electrical circuits, for example, or in laser physics) and they always establish a surprising connection: The dissipative forces take us from nonequilibrium situations to equilibrium, where matters are stationary but fluctuating. We can, therefore, learn about nonequilibrium properties by studying fluctuations in equilibrium.

7.7 Dynamics in phase space

Another way of looking at the time evolution of an ensemble (pollen, say, in Brownian motion) relies on a suitable extension of the Liouville* equation

*Joseph LIOUVILLE (1809–1882)

of classical mechanics, a linear first-order partial differential equation for the phase-space density of an ensemble. For simplicity, we consider the one-dimensional case and denote the phase-space density by $f(x,p,t)$, the generalization to three dimensions will be quite immediate. What we are looking for are the additional terms in

$$\frac{\partial}{\partial t}f(x,p,t) = -\frac{p}{M}\frac{\partial}{\partial x}f(x,p,t) - F(x)\frac{\partial}{\partial p}f(x,p,t) + \left[\frac{\partial}{\partial t}f(x,p,t)\right]_{\text{random}} \tag{7.7.1}$$

that supplement the Liouville equation and account for the dissipation and the fluctuations. So, let us put the velocity term proportional to $\frac{p}{M}$ and the force term proportional to $F(x)$ aside and focus on the additional terms symbolized by $\left[\cdots\right]_{\text{random}}$.

If there are random forces that alter the probability $\mathrm{d}x\,\mathrm{d}p\,f(x,p,t)$ of finding the (representative) particle between x and $x+\mathrm{d}x$ as well as p and $p+\mathrm{d}p$ at time t, then we can describe them by the random change in momentum that the fluctuating forces bring about. For this purpose, we denote the probability that a particle with momentum p will change the momentum by an amount $p'\cdots p'+\mathrm{d}p'$ in the lapse of $\mathrm{d}t$ by $\mathrm{d}t\,\mathrm{d}p'\,w(p,p')$. The resulting change of $f(x,p,t)$,

$$f(x,p,t+\mathrm{d}t) - f(x,p,t) = \mathrm{d}t\int \mathrm{d}p'\,\big[f(x,p-p',t)w(p-p',p') - f(x,p,t)w(p,p')\big]\,, \tag{7.7.2}$$

has two terms: one for the jumps from $p-p'$ to p, which add to the probability at momentum p, the other for the jumps from p to $p+p'$, which deplete the probability at p.

Now, it is typically the case that the random forces are weak, so that $w(p,p')$ is substantially nonzero only for small p' values and, therefore, we can expand $f(x,p-p',t)w(p-p',p')$ in powers of the p' that is subtracted from p,

$$\begin{aligned} f(x,p-p',t)w(p-p',p') = {}& f(x,p,t)w(p,p') \\ & - p'\frac{\partial}{\partial p}\big[f(x,p,t)w(p,p')\big] \\ & + \frac{1}{2}p'^2\frac{\partial^2}{\partial p^2}\big[f(x,p,t)w(p,p')\big]\,, \end{aligned} \tag{7.7.3}$$

and stop at this second-order term. Then

$$f(x,p,t+\mathrm{d}t)-f(x,p,t)=\mathrm{d}t\left[-\frac{\partial}{\partial p}\left(f(x,p,t)\int \mathrm{d}p'\,p'w(p,p')\right)\right.\qquad(7.7.4)$$
$$\left.+\frac{1}{2}\frac{\partial^2}{\partial p^2}\left(f(x,p,t)\int \mathrm{d}p'\,{p'}^2w(p,p')\right)\right],$$

where $\int \mathrm{d}p'\,p'w(p,p')$ is the average change in momentum p per time interval $\mathrm{d}t$, which we identify with the effect of the frictional force,

$$\int \mathrm{d}p'\,p'w(p,p')=-\frac{1}{\tau}p\,. \qquad (7.7.5)$$

Regarding $\frac{1}{2}\int \mathrm{d}p\,{p'}^2w(p,p')$, we expect it to be related to the fluctuations of p'^2 in equilibrium, and approximate it with a p-independent constant C whose value we'll determine shortly,

$$\frac{1}{2}\int \mathrm{d}p'\,{p'}^2w(p,p')\equiv C \quad \text{with} \quad \frac{\partial}{\partial p}C=0\,. \qquad (7.7.6)$$

Then, the contribution of the random forces is

$$\begin{aligned}\left[\frac{\partial}{\partial t}f(x,p,t)\right]_{\text{random}} &= \frac{1}{\tau}\frac{\partial}{\partial p}(pf(x,p,t))+C\frac{\partial^2}{\partial p^2}f(x,p,t)\\ &= \frac{\partial}{\partial p}\left(\frac{p}{\tau}+C\frac{\partial}{\partial p}\right)f(x,p,t)\\ &= C\frac{\partial}{\partial p}\left[\mathrm{e}^{-\frac{p^2}{2\tau C}}\frac{\partial}{\partial p}\left(\mathrm{e}^{\frac{p^2}{2\tau C}}f(x,p,t)\right)\right].\end{aligned} \qquad (7.7.7)$$

Now, when stationary conditions are reached, the probability distribution should be that of Maxwell and Boltzmann,

$$f(x,p,t\to\infty)\propto \mathrm{e}^{-\frac{1}{k_{\mathrm{B}}T}H(x,p)}=\mathrm{e}^{-\frac{1}{k_{\mathrm{B}}T}\left[\frac{1}{2M}p^2+V(x)\right]} \qquad (7.7.8)$$

with the Hamilton function $H(x,p)=\frac{1}{2M}p^2+V(x)$, where $V(x)$ is the potential energy associated with the force $F(x)=-\frac{\partial}{\partial x}V(x)$ in the Liouville equation. The random-force contribution vanishes for this stationary $f(x,p,t\to\infty)$ if $\mathrm{e}^{\frac{p^2}{2\tau C}}f(x,p,t\to\infty)$ does not depend on p, which instructs us to choose $C=Mk_{\mathrm{B}}T/\tau$.

In summary, then, the modified Liouville equation is

$$\begin{aligned}\frac{\partial}{\partial t} f(x,p,t) &= -\frac{p}{M}\frac{\partial}{\partial x} f(x,p,t) - F(x)\frac{\partial}{\partial p} f(x,p,t)\\ &\quad + \frac{1}{\tau}\frac{\partial}{\partial p}\bigl(pf(x,p,t)\bigr) + \frac{Mk_{\mathrm{B}}T}{\tau}\frac{\partial^2}{\partial p^2} f(x,p,t)\,, \qquad (7.7.9)\end{aligned}$$

which is known as the *Fokker*–Planck equation.* Its stationary solution is $f \propto \mathrm{e}^{-H(x,p)/(k_{\mathrm{B}}T)}$ of (7.7.8), as one confirms immediately upon recalling that

$$-\frac{p}{M}\frac{\partial f}{\partial x} - F(x)\frac{\partial f}{\partial p} = -\frac{\partial H}{\partial p}\frac{\partial f}{\partial x} + \frac{\partial H}{\partial x}\frac{\partial p}{\partial x} = \{H, f\}\,, \qquad (7.7.10)$$

and the Poisson bracket $\bigl\{H, g(H)\bigr\}$ of H with any function $g(H)$ of H vanishes. The approach to stationarity is the essence of the Minus-First Law in (1.1.1).

Note that in the Fokker–Planck equation we have another link between fluctuations and dissipation. The two additional terms in the second line of (7.7.9) govern the approach to equilibrium (dissipation) and the spread of momenta in equilibrium (fluctuations), respectively. This will be illustrated by the Brownian motion example to which we turn shortly.

As introduced in (7.7.1) and used in (7.7.2), the phase-space density $f(x,p,t)$ is a probability density, and $\mathrm{d}x\,\mathrm{d}p\, f(x,p,t)$ is the probability of finding the representative particle in the phase-space volume element of size $\mathrm{d}x\,\mathrm{d}p$ at the phase (x,p) at time t. Accordingly, $f(x,p,t)$ is normalized to unit phase-space integral,

$$\int_{-\infty}^{\infty} \mathrm{d}x \int_{-\infty}^{\infty} \mathrm{d}p\, f(x,p,t) = 1\,. \qquad (7.7.11)$$

Of course, the Fokker–Planck equation is consistent with this constraint,

$$\begin{aligned}\frac{\mathrm{d}}{\mathrm{d}t}\int_{-\infty}^{\infty} \mathrm{d}x \int_{-\infty}^{\infty} \mathrm{d}p\, f(x,p,t) &= \int_{-\infty}^{\infty} \mathrm{d}x \int_{-\infty}^{\infty} \mathrm{d}p\, \frac{\partial}{\partial t} f(x,p,t)\\ = \int_{-\infty}^{\infty} \mathrm{d}p \left[-\frac{p}{M} f(x,p,t)\right]\Bigg|_{x=-\infty}^{\infty} & \qquad (7.7.12)\\ + \int_{-\infty}^{\infty} \mathrm{d}x \left[\left(-F(x) + \frac{1}{\tau}p + \frac{Mk_{\mathrm{B}}T}{\tau}\frac{\partial}{\partial p}\right) f(x,p,t)\right]\Bigg|_{p=-\infty}^{\infty} &= 0\,,\end{aligned}$$

*Adriaan Daniël FOKKER (1887–1972)

where we recognize that all terms on the right-hand side of (7.7.9) have anti-derivatives and so get evaluated at infinitely large values of x or p, where $f(x,p,t)$ vanishes thoroughly.

7.8 Brownian motion: Fokker–Planck equation

The Fokker–Planck equation (7.7.9) tells us how the system approaches the stationary state $f(x,p,t\to\infty)\propto \mathrm{e}^{-H(x,p)/(k_\mathrm{B}T)}$ from the initial state specified by $f(x,p,t=0)=f_0(x,p)$. For an illustration of this approach to equilibrium, let us consider the simplified situation in which there are no external forces, $F(x)=0$. Then the phase-space density $f(x,p,t)$ is the solution of the Fokker–Planck equation

$$\tau\frac{\partial}{\partial t}f(x,p,t)=-\frac{p\tau}{M}\frac{\partial}{\partial x}f(x,p,t)+\frac{\partial}{\partial p}\bigl(pf(x,p,t)\bigr)+Q^2\frac{\partial^2}{\partial p^2}f(x,p,t) \quad (7.8.1)$$

with $Q=\sqrt{Mk_\mathrm{B}T}$. One can solve this equation and express $f(x,p,t)$ for $t>0$ as a linear map of $f_0(x,p)$ with the aid of a suitable Green's* function. We shall get to that eventually, after first following a different route, along which we turn our attention to the time dependence of expected values.

7.8.1 *Expected values*

The expected value, at time t, of the physical quantity described by the phase-space function $g(x,p,t)$ — $g=\frac{1}{2M}p^2$ for the kinetic energy is an example — is obtained by averaging $g(x,p,t)$ with the weight $\mathrm{d}x\,\mathrm{d}p\, f(x,p,t)$,

$$\langle g(x,p,t)\rangle_t=\int \mathrm{d}x\,\mathrm{d}p\; g(x,p,t)f(x,p,t)\,. \quad (7.8.2)$$

In this ensemble average, the integration covers the whole phase space as it does in the normalization integral (7.7.11). The time derivative of $\langle g\rangle_t$ has two contributions, one from the parametric time dependence in $g(x,p,t)$, if there is one, the other from the evolution of $f(x,p,t)$ in accordance with (7.8.1),

$$\begin{aligned}\tau\frac{\partial}{\partial t}\langle g(x,p,t)\rangle_t&=\left\langle \tau\frac{\partial}{\partial t}g(x,p,t)\right\rangle_t \qquad (7.8.3)\\ &\quad+\int \mathrm{d}x\,\mathrm{d}p\; g(x,p,t)\left(-\frac{p\tau}{M}\frac{\partial}{\partial x}+\frac{\partial}{\partial p}p+Q^2\frac{\partial^2}{\partial p^2}\right)f(x,p,t)\,.\end{aligned}$$

*George GREEN (1793–1841)

We perform integrations by parts, with no contributions from the boundary terms at infinitely large values of x and p, and so convert the derivatives of $f(x,p,t)$ into derivatives of $g(x,p,t)$,

$$\tau\frac{\partial}{\partial t}\langle g(x,p,t)\rangle_t = \left\langle\left(\tau\frac{\partial}{\partial t}+\frac{p\tau}{M}\frac{\partial}{\partial x}-p\frac{\partial}{\partial p}+Q^2\frac{\partial^2}{\partial p^2}\right)g(x,p,t)\right\rangle_t . \quad (7.8.4)$$

For $g(x,p,t)=1$, this repeats (7.7.12) for $F(x)=0$, and for $g(x,p,t)=x^jp^k$ we obtain

$$\tau\frac{\partial}{\partial t}\left\langle x^jp^k\right\rangle_t = \frac{j\tau}{M}\left\langle x^{j-1}p^{k+1}\right\rangle_t - k\left\langle x^jp^k\right\rangle_t + k(k-1)Q^2\left\langle x^jp^{k-2}\right\rangle_t . \quad (7.8.5)$$

This is a set of coupled linear differential equations for the $n+1$ expected values $\left\langle x^jp^k\right\rangle_t$ with the same sum $j+k=n$ whereby the expected values for $j+k=n-2$ act as the source terms — a recurrence relation that takes us from $j+k=n-2$ to $j+k=n$.

For $j+k=1$, then, we have

$$\begin{aligned}\frac{\partial}{\partial t}\langle p\rangle_t &= -\frac{1}{\tau}\langle p\rangle_t\,,\\ \frac{\partial}{\partial t}\langle x\rangle_t &= \frac{1}{M}\langle p\rangle_t\,,\end{aligned} \quad (7.8.6)$$

where the first equation is the one-dimensional version of (7.4.4), and find

$$\begin{aligned}\langle p\rangle_t &= \langle p\rangle_0\,\mathrm{e}^{-t/\tau}\,,\\ \langle x\rangle_t &= \langle x\rangle_0 + \frac{\tau}{M}\langle p\rangle_0\left(1-\mathrm{e}^{-t/\tau}\right).\end{aligned} \quad (7.8.7)$$

The three equations for $j+k=2$,

$$\begin{aligned}\frac{\partial}{\partial t}\left\langle p^2\right\rangle_t &= -\frac{2}{\tau}\left(\left\langle p^2\right\rangle_t - Q^2\right),\\ \frac{\partial}{\partial t}\langle xp\rangle_t &= -\frac{1}{\tau}\langle xp\rangle_t + \frac{1}{M}\left\langle p^2\right\rangle_t\,,\\ \frac{\partial}{\partial t}\left\langle x^2\right\rangle_t &= \frac{2}{M}\langle xp\rangle_t\,,\end{aligned} \quad (7.8.8)$$

are solved by

$$
\begin{aligned}
\langle p^2\rangle_t &= Q^2 + \Bigl(\langle p^2\rangle_0 - Q^2\Bigr)\,\mathrm{e}^{-2t/\tau}\,,\\
\langle xp\rangle_t &= \frac{\tau}{M}Q^2 + \frac{\tau}{M}\Bigl(\langle p^2\rangle_0 + \frac{M}{\tau}\langle xp\rangle_0 - 2Q^2\Bigr)\,\mathrm{e}^{-t/\tau}\\
&\quad - \frac{\tau}{M}\Bigl(\langle p^2\rangle_0 - Q^2\Bigr)\,\mathrm{e}^{-2t/\tau}\,,\\
\langle x^2\rangle_t &= \frac{2\tau t}{M^2}Q^2 + \frac{\tau^2}{M^2}\Bigl(\langle p^2\rangle_0 + \frac{2M}{\tau}\langle xp\rangle_0 + \frac{M^2}{\tau^2}\langle x^2\rangle_0 - 3Q^2\Bigr)\\
&\quad - \frac{2\tau^2}{M^2}\Bigl(\langle p^2\rangle_0 + \frac{M}{\tau}\langle xp\rangle_0 - 2Q^2\Bigr)\,\mathrm{e}^{-t/\tau}\\
&\quad + \frac{\tau^2}{M^2}\Bigl(\langle p^2\rangle_0 - Q^2\Bigr)\,\mathrm{e}^{-2t/\tau}\,.
\end{aligned}
\tag{7.8.9}
$$

The terms are here grouped in accordance with their time dependence; for later reference, we recast these expressions in a grouping that separates the expected values at $t = 0$ from the terms that are independent of the particular initial phase-space density $f_0(x, p)$,

$$
\begin{aligned}
\langle p^2\rangle_t &= Q^2\Bigl(1 - \mathrm{e}^{-2t/\tau}\Bigr) + \Bigl\langle \Bigl(p\,\mathrm{e}^{-t/\tau}\Bigr)^2\Bigr\rangle_0\,,\\
\langle xp\rangle_t &= \frac{\tau}{M}Q^2\Bigl(1 - \mathrm{e}^{-t/\tau}\Bigr)^2 + \Bigl\langle \Bigl[x + \frac{p\tau}{M}\Bigl(1 - \mathrm{e}^{-t/\tau}\Bigr)\Bigr]p\,\mathrm{e}^{-t/\tau}\Bigr\rangle_0\,,\\
\langle x^2\rangle_t &= \Bigl(\frac{\tau Q}{M}\Bigr)^2\Bigl(\frac{2t}{\tau} + 1 - \Bigl(2 - \mathrm{e}^{-t/\tau}\Bigr)^2\Bigr)\\
&\quad + \Bigl\langle \Bigl[x + \frac{p\tau}{M}\Bigl(1 - \mathrm{e}^{-t/\tau}\Bigr)\Bigr]^2\Bigr\rangle_0\,.
\end{aligned}
\tag{7.8.10}
$$

The late-time expressions, valid when $t \gg \tau$,

$$
\langle p^2\rangle_t = Q^2 = Mk_\mathrm{B}T\,,\quad \langle x^2\rangle_t = \frac{2\tau t}{M^2}Q^2 = \frac{2k_\mathrm{B}T}{M}\tau t\,,
\tag{7.8.11}
$$

are the one-dimensional versions of (7.4.13) and (7.4.16), respectively.

7.8.2 *Generating function*

We observe that the expressions for $j+k = 2$ in (7.8.9) or (7.8.10) are rather involved, and it is clear that the recursive evaluation for $j + k = 3, 4, 5, \ldots$ is error-prone and impractical. Instead we use the exponential function

$$
g(x,p,t) = \mathrm{e}^{\mathrm{i}ax\,+\,\mathrm{i}bp} = \sum_{j=0}^{\infty}\sum_{k=0}^{\infty} = \frac{(\mathrm{i}a)^j\,(\mathrm{i}b)^k}{j!\;k!}x^j p^k
\tag{7.8.12}
$$

as the generating function for all $x^j p^k$ and find the expected value of

$$G_t(a,b) = \left\langle \mathrm{e}^{\mathrm{i}ax + \mathrm{i}bp} \right\rangle_t \tag{7.8.13}$$

from the differential equation that results from (7.8.4),

$$\begin{aligned} \tau\frac{\partial}{\partial t}G_t(a,b) &= \left\langle \left(\frac{p\tau}{M}\mathrm{i}a - p\,\mathrm{i}b + Q(\mathrm{i}b)^2\right)\mathrm{e}^{\mathrm{i}ax + \mathrm{i}bp} \right\rangle_t \\ &= \left\langle \left(\frac{\tau a}{M}\frac{\partial}{\partial b} - b\frac{\partial}{\partial b} - Q^2 b^2\right)\mathrm{e}^{\mathrm{i}ax + \mathrm{i}bp} \right\rangle_t , \end{aligned} \tag{7.8.14}$$

that is

$$\left[\tau\frac{\partial}{\partial t} + \left(b - \frac{\tau a}{M}\right)\frac{\partial}{\partial b} + Q^2 b^2\right] G_t(a,b) = 0\,. \tag{7.8.15}$$

For each value of a, this is a linear partial differential equation in the two variables t and b.

Let us first consider the case of $a = 0$, for which (7.8.15) is equivalent to

$$\left(\tau\frac{\partial}{\partial t} + b\frac{\partial}{\partial b}\right)\mathrm{e}^{\frac{1}{2}Q^2b^2} G_t(0,b) = 0\,. \tag{7.8.16}$$

It follows that

$$\mathrm{e}^{\frac{1}{2}Q^2b^2} G_t(0,b) = \mathrm{func}\left(b\,\mathrm{e}^{-t/\tau}\right), \tag{7.8.17}$$

and the initial conditions at $t = 0$ determine the single-argument function,

$$\mathrm{e}^{\frac{1}{2}Q^2b^2} G_0(0,b) = \mathrm{func}(b)\,, \tag{7.8.18}$$

so that

$$\begin{aligned} G_t(0,b) &= \mathrm{e}^{-\frac{1}{2}Q^2b^2}\left[\mathrm{e}^{\frac{1}{2}Q^2b^2} G_0(0,b)\right]_{b\,\to\,b\,\mathrm{e}^{-t/\tau}} \\ &= \mathrm{e}^{-\frac{1}{2}Q(t)^2b^2} G_0\left(0, b\,\mathrm{e}^{-t/\tau}\right) \\ \text{with}\quad Q(t) &= Q\sqrt{1 - \mathrm{e}^{-2t/\tau}}\,. \end{aligned} \tag{7.8.19}$$

With the definition of $G_t(a,b)$ in (7.8.13) this says

$$\left\langle \mathrm{e}^{\mathrm{i}bp} \right\rangle_t = \left\langle \mathrm{e}^{-\frac{1}{2}Q(t)^2b^2}\, \mathrm{e}^{\mathrm{i}bp\,\mathrm{e}^{-t/\tau}} \right\rangle_0 . \tag{7.8.20}$$

We extract $\langle p^k\rangle_t$ by expanding both sides in powers of b. On the right-hand side this expansion is provided by the generating function of the Hermite polynomials $\mathrm{H}_k(\ )$,

$$\mathrm{e}^{2uv-v^2}=\sum_{k=0}^{\infty}\frac{v^k}{k!}\mathrm{H}_k(u)\,,\tag{7.8.21}$$

which we exploit for

$$v^2=\frac{1}{2}Q(t)^2b^2\,,\quad 2uv=\mathrm{i}bp\,\mathrm{e}^{-t/\tau}\tag{7.8.22}$$

or

$$v=\frac{1}{\sqrt{2}}Q(t)b\,,\quad u=\frac{\mathrm{i}}{\sqrt{2}}\frac{p}{Q(t)}\,\mathrm{e}^{-t/\tau}\,.\tag{7.8.23}$$

Then,

$$\sum_{k=0}^{\infty}\frac{(\mathrm{i}b)^k}{k!}\langle p^k\rangle_t=\sum_{k=0}^{\infty}\frac{[2^{-\frac{1}{2}}Q(t)b]^k}{k!}\left\langle\mathrm{H}_k\left(\frac{\mathrm{i}}{\sqrt{2}}\frac{p}{Q(t)}\,\mathrm{e}^{-t/\tau}\right)\right\rangle_0\tag{7.8.24}$$

and, therefore,

$$\langle p^k\rangle_t=\left(\frac{Q(t)}{\mathrm{i}\sqrt{2}}\right)^k\left\langle\mathrm{H}_k\left(\frac{\mathrm{i}}{\sqrt{2}}\frac{p}{Q(t)}\,\mathrm{e}^{-t/\tau}\right)\right\rangle_0\,.\tag{7.8.25}$$

For $k=0$, 1, and 2, we have $\mathrm{H}_0(u)=1$, $\mathrm{H}_1(u)=2u$, $\mathrm{H}_2(u)=(2u)^2-2$, and get $\langle p^0\rangle_t=\langle 1\rangle_t=1$ and

$$\langle p\rangle_t=\langle p\rangle_0\,\mathrm{e}^{-t/\tau}\,,\quad \langle p^2\rangle_t=\langle p^2\rangle_0\,\mathrm{e}^{-2t/\tau}+Q(t)^2\,,\tag{7.8.26}$$

which repeat, as they should, the corresponding statements in (7.8.7) and (7.8.9). The late-time ($t\gg\tau$) values of all $\langle p^k\rangle_t$s can also be determined from (7.8.25); see Exercise 129.

Before proceeding to the $a\neq 0$ case, we take another look at (7.8.17) and (7.8.18), which we now combine into the statement

$$\mathrm{e}^{\frac{1}{2}Q^2b^2}\left\langle\mathrm{e}^{\mathrm{i}bp}\right\rangle_t=\left\langle\mathrm{e}^{\mathrm{i}bp\,+\,\frac{1}{2}Q^2b^2}\right\rangle_t=\left\langle\mathrm{e}^{\mathrm{i}bp\,+\,\frac{1}{2}Q^2b^2}\right\rangle_0\Big|_{b\,\to\,b\,\mathrm{e}^{-t/\tau}}\,.\tag{7.8.27}$$

The expansion in powers of b — we exploit once more the generating function in (7.8.21), here for $u=\dfrac{p}{\sqrt{2}\,Q}$ and $v=\dfrac{\mathrm{i}Qb}{\sqrt{2}}$ — gives

$$\sum_{k=0}^{\infty}\frac{(\mathrm{i}Qb)^k}{2^{\frac{1}{2}k}\,k!}\left\langle\mathrm{H}_k\left(\frac{p}{\sqrt{2}\,Q}\right)\right\rangle_t=\sum_{k=0}^{\infty}\frac{(\mathrm{i}Qb\,\mathrm{e}^{-t/\tau})^k}{2^{\frac{1}{2}k}\,k!}\left\langle\mathrm{H}_k\left(\frac{p}{\sqrt{2}\,Q}\right)\right\rangle_0\,,\tag{7.8.28}$$

from which we learn that

$$\left\langle \mathrm{H}_k\left(\frac{p}{\sqrt{2}\,Q}\right)\right\rangle_t = \mathrm{e}^{-kt/\tau}\left\langle \mathrm{H}_k\left(\frac{p}{\sqrt{2}\,Q}\right)\right\rangle_0 . \qquad (7.8.29)$$

It's not powers of p that have a simple exponential time dependence, but these Hermite polynomials of $p/(\sqrt{2}\,Q)$; with the exception of $k = 0$, all these expected values vanish at late times. In the examples for $k = 1$ and $k = 2$,

$$\begin{aligned} \left\langle \sqrt{2}\frac{p}{Q}\right\rangle_t &= \mathrm{e}^{-t/\tau}\left\langle \sqrt{2}\frac{p}{Q}\right\rangle_0 , \\ \left\langle 2\left(\frac{p^2}{Q^2}-1\right)\right\rangle_t &= \mathrm{e}^{-2t/\tau}\left\langle 2\left(\frac{p^2}{Q^2}-1\right)\right\rangle_0 , \end{aligned} \qquad (7.8.30)$$

we again recognize the corresponding statements in (7.8.7) and (7.8.9). In view of (7.8.25) and (7.8.29), the recurrence relation (7.8.5) for $j = 0$ is a recurrence relation for the Hermite polynomials in disguise; we leave the details as an exercise to the reader.

When $a \neq 0$ in (7.8.15), now, we read it as an inhomogeneous partial differential equation for $\log G_t = \log\bigl(G_t(a,b)\bigr)$,

$$\tau\frac{\partial \log G_t}{\partial t} + \left(b - \frac{a\tau}{M}\right)\frac{\partial \log G_t}{\partial b} + Q^2b^2 = 0 . \qquad (7.8.31)$$

The general solution of the homogeneous equation is a variation of (7.8.17),

$$\log G_t = \mathrm{func}\Bigl((b - a\tau/M)\,\mathrm{e}^{-t/\tau}\Bigr) , \qquad (7.8.32)$$

to which we add a particular solution of the inhomogeneous equation,

$$\log G_t = -\frac{1}{2}Q^2(b + a\tau/M)^2 - (Qa/M)^2\tau t . \qquad (7.8.33)$$

Accordingly, we have

$$\begin{aligned} G_t(a,b) = {}& \mathrm{e}^{-(Qa/M)^2\tau t}\,\mathrm{e}^{-\frac{1}{2}Q^2(b+a\tau/M)^2} \\ & \times \mathrm{func}\Bigl((b - a\tau/M)\,\mathrm{e}^{-t/\tau}\Bigr) . \end{aligned} \qquad (7.8.34)$$

After ensuring that func(b) is such that $G_0(a,b)$ is recovered for $t = 0$, this becomes

$$\begin{aligned} G_t(a,b) = {}& \mathrm{e}^{-\frac{1}{2}\left(a\tau/M\;\; b\right)\mathsf{Q}(t)^2\begin{pmatrix} a\tau/M \\ b\end{pmatrix}} \\ & \times G_0\Bigl(a, \frac{a\tau}{M} + \Bigl(b - \frac{a\tau}{M}\Bigr)\,\mathrm{e}^{-t/\tau}\Bigr) \end{aligned} \qquad (7.8.35)$$

with the 2×2 matrix

$$\mathsf{Q}(t)^2 = Q^2 \begin{pmatrix} 1 + 2t/\tau - \left(2 - \mathrm{e}^{-t/\tau}\right)^2 & \left(1 - \mathrm{e}^{-t/\tau}\right)^2 \\ \left(1 - \mathrm{e}^{-t/\tau}\right)^2 & 1 - \mathrm{e}^{-2t/\tau} \end{pmatrix}, \qquad (7.8.36)$$

and we arrive at

$$\left\langle \mathrm{e}^{\mathrm{i}ax + \mathrm{i}bp} \right\rangle_t = \mathrm{e}^{-\frac{1}{2}(a\tau/M\ b)\mathsf{Q}(t)^2 \binom{a\tau/M}{b}} \left\langle \mathrm{e}^{\mathrm{i}ax(t) + \mathrm{i}bp(t)} \right\rangle_0, \qquad (7.8.37)$$

where

$$x(t) = x + \frac{p\tau}{M}\left(1 - \mathrm{e}^{-t/\tau}\right) \quad \text{and} \quad p(t) = p\,\mathrm{e}^{-t/\tau} \qquad (7.8.38)$$

are ingredients in (7.8.10) and also in (7.8.7), inasmuch as

$$\langle x \rangle_t = \langle x(t) \rangle_0 \quad \text{and} \quad \langle p \rangle_t = \langle p(t) \rangle_0 . \qquad (7.8.39)$$

This confirms that the terms linear in a and b give the correct values of $\langle x \rangle_t$ and $\langle p \rangle_t$, when we expand both sides in (7.8.37) in powers of a and b. As a further check, we verify that the expected values of x^2, xp, and p^2 are also correct. On the left-hand side, we have the second-order term

$$\left\langle \frac{1}{2}(\mathrm{i}ax + \mathrm{i}bp)^2 \right\rangle_t = -\frac{1}{2}(a\ b)\left\langle \begin{pmatrix} x^2 & xp \\ px & p^2 \end{pmatrix} \right\rangle_t \binom{a}{b}, \qquad (7.8.40)$$

and on the right-hand side we have

$$\begin{aligned} &-\frac{1}{2}(a\tau/M\ b)\mathsf{Q}(t)^2 \binom{a\tau/M}{b} + \left\langle \frac{1}{2}(\mathrm{i}ax(t) + \mathrm{i}bp(t))^2 \right\rangle_0 \\ &= -\frac{1}{2}(a\ b)\left[\begin{pmatrix} a\tau/M & 0 \\ 0 & 1 \end{pmatrix} \mathsf{Q}(t)^2 \begin{pmatrix} a\tau/M & 0 \\ 0 & 1 \end{pmatrix} \right. \\ &\qquad \left. + \left\langle \begin{pmatrix} x(t)^2 & x(t)p(t) \\ p(t)x(t) & p(t)^2 \end{pmatrix} \right\rangle_0 \right] \binom{a}{b}. \end{aligned} \qquad (7.8.41)$$

The resulting equality

$$\begin{aligned} \left\langle \begin{pmatrix} x^2 & xp \\ px & p^2 \end{pmatrix} \right\rangle_t &= \begin{pmatrix} a\tau/M & 0 \\ 0 & 1 \end{pmatrix} \mathsf{Q}(t)^2 \begin{pmatrix} a\tau/M & 0 \\ 0 & 1 \end{pmatrix} \\ &\quad + \left\langle \begin{pmatrix} x(t)^2 & x(t)p(t) \\ p(t)x(t) & p(t)^2 \end{pmatrix} \right\rangle_0 \end{aligned} \qquad (7.8.42)$$

is a compact way of writing (7.8.10) as one confirms by inspection.

7.8.3 *Green's function*

Rather than extracting higher-order expectation values of the $\langle x^j p^k \rangle_t$ kind from the generating function in (7.8.37), we now use it for a different purpose, namely for constructing the Green's function $\mathcal{G}(x,p,t;x',p')$ that we need in

$$f(x,p,t) = \int \mathrm{d}x'\,\mathrm{d}p'\;\mathcal{G}(x,p,t;x',p')f_0(x',p')\,, \tag{7.8.43}$$

which states the solution of the Fokker–Planck equation (7.8.1) for the specified initial phase-space density. We proceed from

$$f(x,p,t) = \int \mathrm{d}x'\,\mathrm{d}p'\;f(x',p',t)\delta(x'-x)\delta(p'-p)\,, \tag{7.8.44}$$

employ the Fourier integral for Dirac's delta function twice,

$$\delta(x'-x)\delta(p'-p) = \int \frac{\mathrm{d}a}{2\pi}\frac{\mathrm{d}b}{2\pi}\;\mathrm{e}^{\mathrm{i}a(x'-x)+\mathrm{i}b(p'-p)}\,, \tag{7.8.45}$$

and recognize $G_t(a,b)$ of (7.8.13) in the integral over x' and p',

$$f(x,p,t) = \int \frac{\mathrm{d}a}{2\pi}\frac{\mathrm{d}b}{2\pi}\;\mathrm{e}^{-\mathrm{i}ax-\mathrm{i}bp}G_t(a,b)\,. \tag{7.8.46}$$

With the expected value at $t=0$ in (7.8.37), then, this identifies

$$\begin{aligned}\mathcal{G}(x,p,t;x',p') = \int \frac{\mathrm{d}a}{2\pi}\frac{\mathrm{d}b}{2\pi}\;&\mathrm{e}^{-\frac{1}{2}\left(a\tau/M\;\; b\right)\mathsf{Q}(t)^2\begin{pmatrix}a\tau/M\\ b\end{pmatrix}}\\ &\times\mathrm{e}^{-\mathrm{i}a(x-x'(t))-\mathrm{i}b(p-p'(t))}\,,\end{aligned} \tag{7.8.47}$$

where $x'(t)$ and $p'(t)$ are as in (7.8.38) with x' and p' on the right-hand sides instead of x and p. The evaluation of this two-dimensional gaussian integral gives us the explicit expression for the Green's function,

$$\mathcal{G}(x,p,t;x',p') = \frac{M}{2\pi\tau\det\{\mathsf{Q}(t)\}}\;\mathrm{e}^{-\frac{1}{2}\left((M/\tau)\Delta x\;\;\Delta p\right)\mathsf{Q}(t)^{-2}\begin{pmatrix}(M/\tau)\Delta x\\ \Delta p\end{pmatrix}} \tag{7.8.48}$$

with $\mathsf{Q}(t)$ the positive square root of $\mathsf{Q}(t)^2$ and

$$\begin{aligned}\Delta x &= x - x'(t) = x - x' - \frac{p'\tau}{M}\left(1-\mathrm{e}^{-t/\tau}\right),\\ \Delta p &= p - p'(t) = p - p'\,\mathrm{e}^{-t/\tau}\,.\end{aligned} \tag{7.8.49}$$

A given initial phase-space density $f_0(x, p)$ evolves into a later density $f(x, p, t)$, which we can now obtain from (7.8.43) by integration. Since the Green's function does not just involve the differences $x - x'$ and $p - p'$, this integration is not a two-fold convolution, but we could rewrite it to make it look like one; see Exercise 130.

We note, however, that as far as x and x' are concerned, (7.8.43) is a convolution and, therefore, when we integrate $\mathcal{G}(x, p, t; x', p')$ over x, the x' dependence disappears,

$$\int \mathrm{d}x\, \mathcal{G}(x, p, t; x', p') = \frac{1}{\sqrt{2\pi}\, Q(t)}\, \mathrm{e}^{-\frac{1}{2}Q(t)^{-2}(p - \mathrm{e}^{-t/\tau}p')^2} . \qquad (7.8.50)$$

As a consequence, the marginal probability density in p at a later time t, which we denote by $f(\cdot, p, t)$, is completely determined by the marginal density at the initial time,

$$\begin{aligned} f(\cdot, p, t) &= \int \mathrm{d}x\, f(x, p, t) \\ &= \int \mathrm{d}p'\, \frac{1}{\sqrt{2\pi}\, Q(t)}\, \mathrm{e}^{-\frac{1}{2}Q(t)^{-2}(p - \mathrm{e}^{-t/\tau}p')^2} f_0(\cdot, p') . \end{aligned} \qquad (7.8.51)$$

At late times, when $\mathrm{e}^{-t/\tau} \cong 0$ and $Q(t) \cong Q$, we get a universal final probability density for the momentum,

$$f(\cdot, p, t)\Big|_{t \to \infty} = \frac{1}{\sqrt{2\pi}\, Q}\, \mathrm{e}^{-\frac{1}{2}\frac{p^2}{Q^2}} , \qquad (7.8.52)$$

which is the kinetic energy contribution in (7.7.8), now properly normalized to unit integral. This final density makes no reference to the initial one, it is the same for all $f_0(\cdot, p)$. All memory of the initial momentum distribution is lost in the course of time.

In the Brownian motion situation without a force-term in the Fokker–Planck equation (7.8.1), the distribution in x does not become stationary. There is, in particular, the linear-in-t growth of $\langle x^2 \rangle_t$ when $t \gg \tau$ that we have in (7.8.9) and, earlier, in (7.5.26). We integrate $\mathcal{G}(x, p, t; x', p')$ over p and find the marginal distribution in x at time t in terms of the initial phase-space density,

$$f(x, \cdot, t) = \int \mathrm{d}x'\, \mathrm{d}p'\, \frac{M}{\tau} \sqrt{\frac{2\pi}{\left[\mathsf{Q}(t)^2\right]_{\mathrm{UL}}}}\, \mathrm{e}^{-\frac{1}{2}\left[\mathsf{Q}(t)^2\right]_{\mathrm{UL}}^{-1}\left(\frac{M}{\tau}\Delta x\right)^2} f_0(x', p') , \qquad (7.8.53)$$

where $\left[\mathsf{Q}(t)^2\right]_{\mathrm{UL}}$ is the upper-left entry in the 2×2 matrix of (7.8.36). We could use this to find explicit expressions for the $\left\langle x^k \right\rangle_t$s, and the most efficient way is to evaluate expected value of $\mathrm{e}^{\mathrm{i}ax}$ and then expand it in powers of a, which takes us back to (7.8.37) for $b = 0$; see Exercise 131.

At very later times, we have

$$\Delta x = x - x' - \frac{p'\tau}{M} \quad \text{and} \quad \left[\mathsf{Q}(t)^2\right]_{\mathrm{UL}} = \frac{2t}{\tau} Q^2 \,, \tag{7.8.54}$$

so that

$$\begin{aligned} f(x,\cdot,t)\Big|_{t\,\gg\,\tau} &= \int \mathrm{d}x'\,\mathrm{d}p'\,\frac{M}{Q}\sqrt{\frac{\pi}{\tau t}}\,\mathrm{e}^{-\left(\frac{M}{2Q}\right)^2 \frac{1}{\tau t}(x\,-\,x'\,-\,p'\tau/M)^2} f_0(x',p') \\ &= \int \mathrm{d}x'\,\frac{M}{Q}\sqrt{\frac{\pi}{\tau t}}\,\mathrm{e}^{-\left(\frac{M}{2Q}\right)^2 \frac{1}{\tau t}(x\,-\,x')^2}\,\tilde{f}_0(x')\,, \end{aligned} \tag{7.8.55}$$

where

$$\tilde{f}_0(x') = \int \mathrm{d}p'\, f_0\Big(x' - \frac{p'\tau}{M}, p'\Big) \tag{7.8.56}$$

is a skewed marginal of $f_0(x,p)$. In terms of this skewed marginal, the late-time expected values of powers of x can be extracted from

$$\begin{aligned} \left\langle \mathrm{e}^{\mathrm{i}ax} \right\rangle_{t\,\gg\,\tau} &= \mathrm{e}^{-(Qa/M)^2\tau t} \int \mathrm{d}x'\,\mathrm{e}^{\mathrm{i}ax'}\,\tilde{f}_0(x') \\ &= \mathrm{e}^{-(Qa/M)^2\tau t} \left\langle \mathrm{e}^{\mathrm{i}a(x\,+\,p\tau/M)} \right\rangle_0 \,, \end{aligned} \tag{7.8.57}$$

where the dominant contribution to $\left\langle x^k \right\rangle_{t\gg\tau}$ comes from the t-dependent factor. For the even powers of a, the factor $\langle\cdots\rangle_0$ can be ignored; for the odd powers we need the linear-in-a contribution, so that

$$\left\langle \mathrm{e}^{\mathrm{i}ax} \right\rangle_{t\,\gg\,\tau} = \mathrm{e}^{-(Qa/M)^2\tau t}\left[1 + \mathrm{i}\,a\Big(\langle x\rangle_0 + \frac{\tau}{M}\langle p\rangle_0\Big)\right] \tag{7.8.58}$$

and, finally,

$$\begin{aligned} \left\langle x^{2k} \right\rangle_{t\,\gg\,\tau} &= \frac{(2k)!}{k!}\left[\left(\frac{Q}{M}\right)^2 \tau t\right]^k \\ \left\langle x^{2k+1} \right\rangle_{t\,\gg\,\tau} &= \frac{(2k+1)!}{k!}\left[\left(\frac{Q}{M}\right)^2 \tau t\right]^k \Big(\langle x\rangle_0 + \frac{\tau}{M}\langle p\rangle_0\Big) \end{aligned} \tag{7.8.59}$$

for $k = 0, 1, 2, \ldots$; accordingly, we refine the statement after (7.8.52): All memory of the initial probability density is lost, except for the expected values of x and p.

Except for the $k = 0$ cases in (7.8.59), all these late-time expectation values increase with time. Such limitless growth is unrealistic, of course, and identifies a shortcoming of the model system specified by the Fokker–Planck equation (7.8.1) — there is no confining force that keeps the particles within a finite volume. This can be incorporated by reintroducing the force term $-F(x)\dfrac{\partial}{\partial p}f(x,p,t)$ of (7.7.9). A tractable example is that of a linear restoring force, $F(x) = -M\omega^2 x$, associated with the potential energy of a harmonic oscillator, $V(x) = \frac{1}{2}M\omega^2 x^2$. Some aspects thereof are addressed in Exercises 132–135.

Exercises with Hints

Chapter 1

1 Why is it $-\boldsymbol{E}(\boldsymbol{r})\cdot\mathrm{d}\boldsymbol{P}(\boldsymbol{r})$ and $-\boldsymbol{H}(\boldsymbol{r})\cdot\mathrm{d}\boldsymbol{M}(\boldsymbol{r})$ in (1.1.7) and (1.1.8) rather than $-\boldsymbol{P}(\boldsymbol{r})\cdot\mathrm{d}\boldsymbol{E}(\boldsymbol{r})$ and $-\boldsymbol{M}(\boldsymbol{r})\cdot\mathrm{d}\boldsymbol{H}(\boldsymbol{r})$?

2 Complement Section 1.2 by stating the chemical potential $\mu(S,V,n)$ for a one-component ideal gas.

3 For a certain rubber band of length L that consists of n moles of rubber, the differential of the internal energy U is

$$\mathrm{d}U = T\,\mathrm{d}S + \tau\,\mathrm{d}L + \mu\,\mathrm{d}n\,,$$

where T, S, τ, μ are the temperature, entropy, tension, and chemical potential, respectively. The equation of state $TS = 3\tau L$ holds for this rubber band. It is also known that $\tau L^{1/2}$ does not change if both S and n are kept constant (no heat transfer, no particle exchange). Find $U(S,L,n)$ up to a multiplicative constant.

4 The thermodynamical equilibrium states of a certain single-substance gas are fully specified by T, V, and n — the temperature, the volume, and the number of moles, respectively. The thermal equation of state of the gas under consideration is

$$P^2V = \text{constant when } T \text{ and } n \text{ are constant,}$$

and its adiabatic equation of state is

$$PV^2 = \text{constant when } S \text{ and } n \text{ are constant,}$$

where P is the pressure and S is the entropy. Explain why there must be functions $a(S/n) > 0$ and $b(T) > 0$ such that

$$P = \frac{a(S/n)}{(V/n)^2} = \frac{b(T)}{\sqrt{V/n}},$$

and conclude that $a(s) = \dfrac{s^3}{3w}$ and $b(T) = \dfrac{1}{3}\sqrt{wT^3}$ with a material constant w. What is the SI unit of w? What is the relation between PV and TS?

5 Two portions of an ideal gas with adiabatic index γ occupy two parts of an isolated container, separated by a thin wall that cannot be penetrated by the gas atoms:

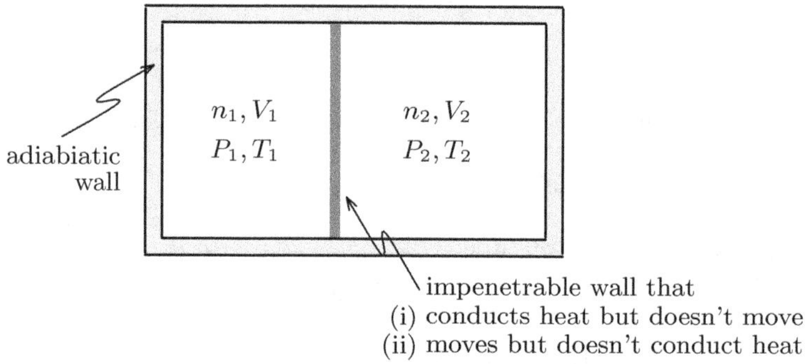

Initially, there are n_1 moles of gas in the left compartment of volume V_1, with pressure P_1 and temperature T_1; and there are n_2 moles of gas in the right compartment of volume V_2, with pressure P_2 and temperature T_2. Consider these two scenarios: (i) The wall cannot move but conducts heat; (ii) the wall can move but does not conduct heat. For both scenarios, then, what are the final temperatures and pressures on both sides? And by how much does the total entropy change from the initial to the final state?

6 In Section 1.5 we found — by an argument that exploits the maximum property of the entropy — that $T_1 = T_2$ for two systems in contact and in an equilibrium state for the combined system. Establish the same fact by an argument based on the minimum property of the internal energy.

7 Modify the argument that established $T_1 = T_2$ in Section 1.5 to demonstrate that also $P_1 = P_2$ (equal pressure) and $\mu_1 = \mu_2$ (equal chemical potentials).

8 In Section 1.5 we found that, if we have different temperatures on two sides of a partition, heat flows from the region of higher temperature to that of lower temperature. What is the corresponding statement about mass flow between regions with different chemical potentials?

9 Supplement Section 1.7 by the enthalpy $\mathrm{H}(S, P, n)$ for the ideal gas, and then find T, V, and μ as functions of S, P, and n.

10 A thermodynamical system undergoes a reversible cycle: From state A to state B by an expansion at constant pressure; from state B to state C by a pressure reduction at constant volume; then from state C back to state A at constant temperature:

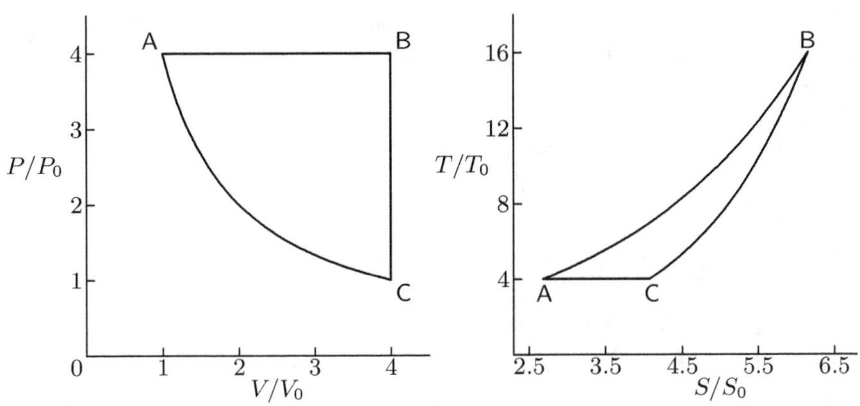

On the left the cycle is depicted in a V, P diagram, on the right in a S, T diagram. The figures are not sketched, but drawn accurately for an ideal gas with $\gamma\frac{5}{3}$, so that the same area is enclosed in both diagrams, provided that the reference volume V_0, the reference pressure P_0, the reference entropy S_0, and the reference temperature T_0 are such that $V_0 P_0 = S_0 T_0$. What is the physical significance of the two areas? Explain why the two areas are the same for any reversible cycle, not just in this particular example.

11 Next, identify in the S, T diagram the areas that represent the heat absorbed by the system ($\mathrm{d}S > 0$) and the heat extracted from the system ($\mathrm{d}S < 0$). The efficiency of the heat engine specified by a cycle is the net work done by the system on the environment divided by the heat absorbed, that is: the fraction of the absorbed heat that is converted into net work. Conclude that of all cycles with the permissible temperatures T from a given interval, $T_1 < T < T_2$, the cycle with the highest efficiency encloses a rectangular area in the S, T diagram: isothermal processes alternating with isentropic processes — the Carnot* cycle. Determine its efficiency.

12 Clearly, the cycle depicted in Exercise 10 is not a Carnot cycle, but one with a much smaller efficiency. How much smaller?

13 According to Stefan and Boltzmann, see Section 3.4, the internal energy of electromagnetic radiation at temperature T ("photon gas") is

$$U = \frac{\pi^2}{15}\frac{(k_\mathrm{B}T)^4 V}{(\hbar c)^3}\,.$$

Note that there is no dependence on the amount of substance (the number of photons). Convert this into expressions for $U(S,V)$, the free energy $F(T,V)$, the enthalpy $\mathrm{H}(S,P)$, and the free enthalpy $G(T,P)$ as functions of their natural variables.

14 For the photon gas of Exercise 13, which curves identify isothermal processes in a V, P diagram? Which curves identify isentropic processes? Explain why there is no meaningful value of C_P, the heat capacity for constant pressure, for a photon gas.

15 For a system that can be characterized by entropy S, volume V, and mole number n, show that

$$\left(\frac{\partial F}{\partial V}\right)_{S,n} = S\left(\frac{\partial P}{\partial S}\right)_{V,n} - P$$

and

$$\left(\frac{\partial U}{\partial V}\right)_{T,n} = T\left(\frac{\partial P}{\partial T}\right)_{V,n} - P$$

and also

$$n\left(\frac{\partial \mu}{\partial V}\right)_{T,n} = V\left(\frac{\partial P}{\partial V}\right)_{T,n}.$$

How are these three statements modified for the photon gas? For the free energy of the photon gas found in Exercise 13, verify explicitly that the modified statements are true.

*Nicolas Léonard Sadi CARNOT (1796–1832)

16 Show that the constant-volume heat capacity of (1.6.4) can also be calculated in accordance with

$$C_V = \left(\frac{\partial U}{\partial T}\right)_{V,n},$$

but note that T is not a natural variable of U. What is the analogous statement for the constant-pressure heat capacity C_P?

17 Show that the so-called "$T\,\mathrm{d}S$ equations" hold:

$$\text{(a)}\quad T\,\mathrm{d}S = C_V\,\mathrm{d}T + \frac{\alpha T}{K_T}\,\mathrm{d}V\,,$$
$$\text{(b)}\quad T\,\mathrm{d}S = C_P\,\mathrm{d}T - \alpha TV\,\mathrm{d}P\,.$$

Derive $C_P - C_V = \dfrac{\alpha^2}{K_T}VT$ [= (1.12.10)] from (a); from (b).

18 In Sections 1.4 and 1.8 we established the minimum property of the internal energy and the Helmholtz free energy. The corresponding minimum properties of the enthalpy and the Gibbs free energy are stated at the end of Section 1.8. Confirm that these statements are correct.

19 Find the values of $\dfrac{C_P}{C_V}$ and $\dfrac{C_P - C_V}{PV/T}$ for the gas of Exercise 4.

20 As discussed in Section 1.11, the Gibbs–Duhem relation tells us that a thermodynamic potential with T, P, and μ as its natural variables is meaningless. What happens if you try, nevertheless, to construct such a potential?

21 What is the analog of the Gibbs–Duhem relation for the rubber band of Exercise 3?

22 For the van der Waals gas, we have the coexistence pressure $\overline{P}(T)$ for temperatures just below the critical temperature, $0 < T_{\mathrm{c}} - T \ll T_{\mathrm{c}}$, in (1.16.25). Supplement this with the corresponding expression for temperatures far below the critical temperature,

$$\overline{P}(T) = 27P_{\mathrm{c}}\,\mathrm{e}^{-\frac{27}{8}\frac{T_{\mathrm{c}}}{T}} \quad \text{for} \quad 0 < T \ll T_{\mathrm{c}}\,.$$

Then confirm the low-temperature part of the dotted curve in (1.16.31).

23 For the van der Waals gas, show that

$$\left.\frac{\mathrm{d}^2\overline{P}}{\mathrm{d}T^2}\right|_{0<T_\mathrm{c}-T\to 0} = \frac{48}{5}\frac{P_\mathrm{c}}{T_\mathrm{c}^2}\,,$$

and then confirm the high-temperature part of the dotted curve in (1.16.31).

24 For the van der Waals gas, write the latent heat $q(T)$ for subcritical temperature T in the form $q(T) = RT_\mathrm{c}\mathrm{func}(x)$ and determine the function of the parameter x of (1.16.26). What is $q(T)$ for temperatures just below the critical temperature, or very far below?

25 Consider a thermal equation of state such as that for the van der Waals gas in Section 1.16, that is: for temperatures T below a critical temperature T_c, the derivative of $P(T,v)$ with respect to the molar volume v has the wrong sign in some v interval. For temperatures $T < T_\mathrm{c}$, we must use the Maxwell construction to identify the coexistence range $v^{(2)}(T) < v < v^{(1)}(T)$ where the liquid phase and the gas phase coexist, and $\overline{P}(T) = P\bigl(T, v^{(2)}(T)\bigr) = P\bigl(T, v^{(1)}(T)\bigr)$ determines the coexistence pressure for the subcritical temperature T. Show that

$$\left.\frac{\mathrm{d}\overline{P}(T)}{\mathrm{d}T}\right|_{T=T_\mathrm{c}} = \left.\frac{\partial P(T,v)}{\partial T}\right|_{T=T_\mathrm{c},\; v=v_\mathrm{c}}.$$

Use this to confirm (1.16.25) for the van der Waals gas.

26 For the situation of Exercise 25, show that

$$T\frac{\mathrm{d}\overline{P}(T)}{\mathrm{d}T} = \overline{P}(T) + \frac{1}{\Delta v}\int\limits_{v^{(2)}(T)}^{v^{(1)}(T)} \mathrm{d}v\left[\left(T\frac{\partial P}{\partial T}\right)_v (T,v) - P(T,v)\right].$$

Then confirm that this is (1.16.23) in the case of the van der Waals gas.

27 The Gibbs surfaces in (1.13.10) are drawn to scale for the two phases of the van der Waals gas at subcritical temperatures. Anticipate the lesson of Exercise 121, namely that $c_V = \dfrac{3}{2}R$ outside the coexistence region, and infer $f_0(T)$ from (1.16.20). Then find the chemical potential as a function of T and v. How would you determine $\mu^{(1)}(T,P)$ for the gas phase and $\mu^{(2)}(T,P)$ for the liquid phase?

28 The Clausius model of a real gas is specified by the equation of state

$$P(T,v) = \frac{RT}{v-b} - \frac{a}{(v+c)^2 T}$$

with positive material constants a, b, and c. Just like the van der Waals gas, the Clausius gas has a gas-to-liquid phase transition for temperatures below the critical temperature T_c. Express the critical temperature T_c and also the critical values of the molar volume (v_c) and the pressure (P_c) in terms of a, b, c, and the gas constant R. Then find the value of $\frac{P_c v_c}{RT_c}$ and the coexistence pressure $P(T)$ for temperatures just below the critical temperature, $0 \lesssim T_c - T \ll T_c$.

29 The thermal equation of state of the so-called Berthelot* gas is

$$P(T,v) = \frac{RT}{v-b} - \frac{a}{v^2 RT} \quad \text{with } a > 0 \text{ and } b > 0\,.$$

Answer the questions analogous to those in Exercise 28 about the Clausius gas.

30 The Dieterici† model of a real gas is specified by the thermal equation of state

$$P(T,v) = \frac{RT}{v-b} \exp\left(-\frac{a}{vRT}\right) \quad \text{with } a > 0 \text{ and } b > 0\,.$$

Answer the questions analogous to those in Exercise 28 about the Clausius gas.

31 Explain, why

$$c_V(T,v) = vT\frac{\mathrm{d}^2\overline{P}}{\mathrm{d}T^2}(T) - T\frac{\mathrm{d}^2\overline{\mu}}{\mathrm{d}T^2}(T)$$

holds for the coexisting phases in Section 1.17, and then conclude that

$$\frac{\mathrm{d}^2\overline{P}}{\mathrm{d}T^2} = \frac{\Delta c_P}{T\Delta v} - 2\frac{\Delta(\alpha v)}{\Delta v}\frac{\mathrm{d}\overline{P}}{\mathrm{d}T} + \frac{\Delta(K_T v)}{\Delta v}\left(\frac{\mathrm{d}\overline{P}}{\mathrm{d}T}\right)^2.$$

Find also the corresponding expression for $\frac{\mathrm{d}^2\overline{\mu}}{\mathrm{d}T^2}$.

---

*Pierre Eugène Marcellin BERTHELOT (1827–1907) †Conrad DIETERICI (1858–1929)

Chapter 2

32 Section 2.1: Find $\langle N_1\rangle$ and δN_1 if the probabilities for the two partial volumes are p and $1-p$.

33 In Section 2.1 and Exercise 32, we tacitly assume that the atoms are distinguishable. What do you get for $\langle N_1\rangle$ and δN_1 if the atoms are indistinguishable?

34 Invent a simple model system that is not ergodic.

35 The volume of a D-dimensional ball of radius R is

$$V_D(R)=\int(\mathrm{d}\boldsymbol{x})\,\eta\biggl(R^2-\sum_{j=1}^{D}x_j^2\biggr)\quad\text{with}\quad(\mathrm{d}\boldsymbol{x})=\mathrm{d}x_1\,\mathrm{d}x_2\,\cdots\,\mathrm{d}x_D\,.$$

Establish first

$$2\int_0^\infty \mathrm{d}R\,R\;\mathrm{e}^{-\lambda^2R^2}\,V_D(R)=\frac{\pi^{\frac{1}{2}D}}{\lambda^{D+2}}=\frac{\pi^{\frac{1}{2}D}}{(\frac{1}{2}D)!}\int_0^\infty \mathrm{d}t\,t^{\frac{1}{2}D}\,\mathrm{e}^{-\lambda^2 t}$$

and infer then that $V_D(R)=\dfrac{\pi^{\frac{1}{2}D}}{(\frac{1}{2}D)!}R^D$.

36 For $x\gg 1$, justify the steps in

$$\begin{aligned}x!=\int_0^\infty \mathrm{d}t\,t^x\,\mathrm{e}^{-t}&=x^x\,\mathrm{e}^{-x}\int_{-x}^\infty \mathrm{d}t\,\Bigl(1+\frac{t}{x}\Bigr)^x\,\mathrm{e}^{-t}\\&\cong x^x\,\mathrm{e}^{-x}\int_{-\infty}^\infty \mathrm{d}t\;\mathrm{e}^{-t^2/(2x)}\end{aligned}$$

and then get Stirling's approximation (2.3.17) from the final integral.

37 Improve Stirling's approximation (2.3.17) by including a factor $\mathrm{e}^{\lambda/x}$ on the right-hand side and choose the number λ such that $(x+1)!/x!$ approximates $x+1$ as precisely as possible if you use this modified Stirling approximation for $(x+1)!$ and $x!$.

38 In the microcanonical ensemble, we have the entropy $S(E,V,N)$ and can solve for the internal energy $E=U(S,V,N)$. In the canonical ensemble, we have the free energy $F(\beta, V, N) = -\beta^{-1} \log\bigl(Q(\beta, V, N)\bigr)$ and the expected energy $\langle E\rangle = -\left(\dfrac{\partial Q}{\partial \beta}\right)_{V,n}$. Show that $\langle E\rangle = F + \beta\left(\dfrac{\partial F}{\partial \beta}\right)_{V,n}$ and then infer that $U = \langle E\rangle$.

39 Consider the D-dimensional analog of a cube and the largest ball that fits into the cube. How large is the ratio of their volumes? What about the smallest ball into which the cube fits?

40 Consider the canonical ensemble for the situation of (2.3.14)–(2.4.7). Find the partition function, and then rederive the Sackur–Tetrode equation.

41 You add a milli-calorie of heat to a certain substance at a temperature of 300 K (water, perhaps, but that is not relevant). What is the corresponding change in the number of microstates available to the system?

42 The example of (2.5.22)–(2.5.34): Use the results for the microcanonical ensemble to express the entropy in terms of βE_0 and N. Do you get what you expect in the low-temperature limit ($\beta E_0 \gg 1$) and the high-temperature limit ($\beta E_0 \ll 1$)?

43 Next, use $S = \dfrac{1}{T}\bigl(\langle E\rangle - F\bigr)$ to find the entropy from the free energy of the canonical ensemble. Then compare with the entropy as found for the microcanonical ensemble.

44 We denote the energy and the particle number in the kth microstate by $E_k(V)$ and N_k, respectively. Then the partition functions for the canonical and grand canonical ensembles are

$$Q(\beta, V, N) = \sum_k \mathrm{e}^{-\beta E_k}\, \delta_{N,N_k} \quad \text{and} \quad Z(\beta, V, z) = \sum_k \mathrm{e}^{-\beta(E_k \,-\, \mu N_k)}$$

with the fugacity $z = \mathrm{e}^{\beta\mu}$. Show that these partition functions are related to each other by

$$Z(\beta, V, z) = \sum_{N=0}^{\infty} z^N\, Q(\beta, V, N)\,.$$

This states that $Z(\beta, V, z)$ is a generating function for the $Q(\beta, V, N)$s. How, then, do you find $Q(\beta, V, N)$ when you know $Z(\beta, V, z)$?

45 We also have the microcanonical ensemble, for which the count of microstates with energy less than E distributed over N particles is

$$\Omega(E, V, N) = \sum_k \eta(E - E_k)\, \delta_{N,N_k}\,.$$

How are $\Omega(E, V, N)$ and $Q(\beta, V, N)$ related to each other?

46 In addition to the microcanonical, the canonical, and the grand canonical ensembles, there is also the so-called Maxwell's demon ensemble — introduced by Rzążewski* and others — for situations in which particles can be exchanged with a bath without, however, exchanging energy. The variables of the corresponding partition function $\Xi(E, V, \mu)$ are the energy E, the volume V, and the chemical potential μ. How is $\Xi(E, V, \mu)$ related to the grand canonical partition function $Z(\beta, V, z)$?

47 A milli-mole of ideal gas occupies volume V while exchanging heat and particles with a bath at 300 K. How large are the fluctuations in energy and density as compared with their mean values?

48 In the situation of Exercise 47, what is the probability that an energy fluctuation of the size $10^{-8}\langle E\rangle$ is observed?

49 Consider a model system of N constituents distributed over M sites with $M \geq N$. Each constituent has two internal states with energies $\pm\frac{1}{2}E_0$, and there is no interaction between the constituents. Find the canonical partition function $Q^{(M)}(\beta, N)$ and determine the expected value of the energy and its spread.

50 Next, find the grand canonical partition function $Z^{(M)}(\beta, z)$. In the grand canonical ensemble, then, what is the expected number of constituents, and what is its spread?

51 Use the Gibbs entropy formula in (2.7.4) for another derivation of (2.8.2).

52 The grand canonical partition function $Z(\beta, V, z)$ of a dilute gas is given by $\log Z = (k_{\mathrm{B}}T_0\beta)^{-\kappa}\dfrac{V}{V_0}z$, where T_0, V_0, and κ are positive material constants. Confirm that PV = constant for isothermal changes. What is the adiabatic equation of state? Confirm that you get the expected answer when $\kappa = \dfrac{3}{2}$.

*Kazimierz RZĄŻEWSKI (b. 1943)

53 Next, determine the free energy $F(T,V,N)$ and the internal energy $U(S,V,N)$. What are the heat capacities for constant volume and constant pressure?

54 Supplement the three ensembles discussed in Section 2.10 by the Maxwell's demon ensemble of Exercise 46. What is p_k, the probability of the kth microstate, in this case?

Chapter 3

55 In Exercise 40, you found the canonical partition function for the classical ideal gas confined to volume V. Now infer the grand canonical partition function and confirm that it is of the form stated in Exercise 52 with $\kappa = \frac{3}{2}$.

56 The single-particle energy in Exercise 55 is the nonrelativistic kinetic energy $\frac{1}{2m}\boldsymbol{p}^2$. Now imagine an ideal gas of "luxons" — that is: massless, conserved particles with kinetic energy $c|\boldsymbol{p}|$, where c is the speed of light — confined to volume V. Find the canonical partition function, the grand canonical partition functions, and the expected energy per particle.

57 Next, find the Helmholtz free energy $F(T,V,N)$ and the Gibbs free energy $G(T,P,N)$ as functions of their natural variables for the ideal gas of luxons. Then, determine the heat capacitances C_V and C_P for constant volume and constant pressure, respectively.

58 In the so-called *Jüttner* gas*, the particles have mass m and the relativistic kinetic energy $c\sqrt{\boldsymbol{p}^2 + (mc)^2} - mc^2$. For an ideal Jüttner gas confined to volume V, find the canonical partition function, the grand canonical partition function, and the expected energy per particle. Verify that you get the results of Exercise 55 when $\boldsymbol{p}^2 \ll (mc)^2$ (formally: the nonrelativistic limit $c \to \infty$) and the results of Exercise 56 in the $m \to 0$ limit of the ideal luxon gas.

59 What are the canonical partition function and the expected energy per particle for the ideal Jüttner gas when $k_{\mathrm{B}}T \ll mc^2$ or $k_{\mathrm{B}}T \gg mc^2$? Do you find what you expect?

---

*Ferencz JÜTTNER (1878–1958)

60 In Exercises 55–59, the particles of the classical ideal gas are confined to a finite volume and only have kinetic energy. Now, consider nonrelativistic particles confined by a force that derives from a single-particle potential energy of the form $\kappa|\boldsymbol{r}|^{\nu}$ with $\kappa > 0$ and $\nu > 0$. Rather than the simple volume factor V in $q(\beta, V)$, we now have

$$q(\beta) = \underbrace{\int (\mathrm{d}\boldsymbol{r})\, \mathrm{e}^{-\beta\kappa|\boldsymbol{r}|^{\nu}}}_{\text{replaces factor } V} \underbrace{\int \frac{(\mathrm{d}\boldsymbol{p})}{(2\pi\hbar)^3}\, \mathrm{e}^{-\beta\frac{1}{2m}\boldsymbol{p}^2}}_{\text{as before}} .$$

Is the strength κ of the confining force an extensive or an intensive variable? Find the canonical partition function and determine the average energy per particle in units of $k_{\mathrm{B}}T$. Explain why there is a ν-independent average kinetic energy.

61 Next, consider the case of a harmonic force, that is: $\kappa|\boldsymbol{r}|^{\nu} = \frac{1}{2}m\omega^2\boldsymbol{r}^2$, and find $q(\beta)$ by summing over the energies of the quantum-mechanical harmonic oscillator. Confirm that there is no difference between this $q(\beta)$ and the one found in Exercise 60 when the temperature is high.

62 Consider a classical ideal gas of molecules that rotate, vibrate, and have electronic excitations, such that the single-molecule energies are a sum $\varepsilon_k = \varepsilon_k^{(\mathrm{rot})} + \varepsilon_k^{(\mathrm{vib})} + \varepsilon_k^{(\mathrm{el})}$ of the respective contributions. Show that the heat capacity C_V is a corresponding sum: $C_V = C_V^{(\mathrm{rot})} + C_V^{(\mathrm{vib})} + C_V^{(\mathrm{el})}$.

63 According to Section 3.3, the average occupation number is

$$\langle n_j \rangle = \frac{1}{\mathrm{e}^{\beta(\varepsilon_j - \mu)} \pm 1} \quad \left\{ \begin{array}{l} \text{for fermions} \\ \text{for bosons} \end{array} \right\},$$

where ε_j is the jth single-particle energy and μ is the chemical potential. In the boson case, assume that the temperature is sufficiently high that there is no Bose–Einstein condensation. Now find the corresponding expression for the correlation $\langle n_j n_{j'} \rangle - \langle n_j \rangle \langle n_{j'} \rangle$ and determine the corresponding fluctuation $\left\langle \delta N^2 \right\rangle$ in the total number of particles. Verify that it is consistent with (2.9.2).

64 In (3.4.9), we have the average energy $\langle E \rangle$ of a photon gas at temperature T. What is the corresponding expression for $\langle N \rangle$, the average number of photons?

65 Show that the integrals of the kind appearing in (3.4.10) can be evaluated in terms of the zeta function in (3.6.13),

$$\int_0^\infty \frac{\mathrm{d}x\, x^\alpha}{\mathrm{e}^x - 1} = \alpha!\,\zeta(\alpha+1) \quad \text{for} \quad \alpha > 0\,,$$

and so confirm the value stated in (3.4.10).

66 In Section 3.5, we discuss the Debye model. There is also the Einstein model, in which one assumes that a single frequency ω_0 is dominating,

$$g(\omega) = 3N\delta(\omega - \omega_0)\,.$$

Find the low-temperature behavior in this case.

67 For the ideal boson gas of Section 3.6, show that the Helmholtz free energy F, the Gibbs free energy G, and the entropy S are

$$-\frac{\beta F}{N} = \frac{v}{\lambda^3} g_{5/2}(z) - \log z\,,$$

$$\frac{\beta G}{N} = \log z\,,$$

$$\frac{S}{k_\mathrm{B}N} = \frac{5}{2}\frac{v}{\lambda^3} g_{5/2}(z) - \log z$$

for $T > T_\mathrm{c}$. What are the corresponding expressions for $T < T_\mathrm{c}$.?

68 In the ideal boson gas of Section 3.6, we have all bosons in the gas phase when $v > v_\mathrm{c}$ for given temperature $T = 1/(k_\mathrm{B}\beta)$. Now imagine that the specific volume is reduced until the critical volume v_c is reached. Further reduction of the volume occurs at constant pressure $\overline{P}(T)$, the coexistence pressure for temperature T, and we have bosons in the gas phase coexisting with bosons in the condensate phase for $0 < v < v_\mathrm{c}$. At $v = 0$, finally, all bosons are in the condensate phase. What is the change in the entropy as v is reduced from $v = v_\mathrm{c}$ to $v = 0$? Determine $\dfrac{\mathrm{d}\overline{P}(T)}{\mathrm{d}T}$ and verify that the Clausius–Clapeyron equation is obeyed.

69 Figure (3.6.45) shows the constant-volume heat capacity of the ideal boson gas as a function of temperature T; the plot is valid for all values of the specific volume v. We know, however, that C_V is a function of T and v.

Where, then, is the v dependence in this plot? Produce the corresponding plot of $C_V/(k_B N)$ as a function of v.

70 Next, recall the discussion of the heat capacity of coexisting phases in Section 1.17. What can you say about the term that is added to $\lceil c_V \rfloor$ in (1.17.17)?

71 In (3.6.34), we have the heat capacity C_V for the ideal boson gas. Now, establish that

$$\frac{C_P}{C_V} = \frac{5}{3}\frac{g_{5/2}(z)g_{1/2}(z)}{g_{3/2}(z)^2}$$

for supercritical temperatures. What happens in the limit $T_c < T \to T_c$?

72 Bose–Einstein condensation revisited: Rather than just putting $z = 1$ for subcritical temperature, let us express the fugacity z in terms of N_0 and so arrive at

$$\frac{N_0}{N} = 1 - \left(\frac{T}{T_c}\right)^{\frac{3}{2}} \frac{g_{\frac{3}{2}}(z)}{g_{\frac{3}{2}}(1)} \quad \text{with } z = \frac{N_0}{N_0 + 1}\,.$$

Confirm this more precise version of (3.6.25). Then use the expansion

$$g_\alpha\left(z = e^{\beta\mu}\right) = (-\alpha)!(-\beta\mu)^{\alpha-1} + \sum_{j=0}^{\infty} \zeta\,(\alpha - j)\,\frac{(\beta\mu)^j}{j!}$$

in (3.6.18) to determine an approximation for $0 < -\beta\mu = -\log z \ll 1$. How does this modify the graph of $\dfrac{N_0}{N}$ as a function of T in (3.6.26)?

73 Section 3.6 deals with an ideal gas of bosons confined to volume V. What about the one-dimensional and two-dimensional analogs, in which an ideal gas of bosons is confined to a line of length L or an area of size A, respectively? Does Bose–Einstein condensation occur in such a one-dimensional or two-dimensional boson gas? — If you answer "yes" in one or both cases, what is the respective critical temperature?

74 As in Exercises 60 and 61, the ideal gas of bosons can be confined by a force that derives from a potential energy, rather than being boxed into volume V. As an example, consider now confinement by a harmonic force, for which the single-particle energies are integer multiples of $\hbar\omega$. Is there Bose–Einstein condensation under these circumstances? In the spirit of Exercise 73, answer this question for one-dimensional, two-dimensional, and three-dimensional boson gases.

75 We can view the bosons in the excited states of the ideal boson gas as a thermodynamical system with N_{ex} particles, in contact with the bosons in the condensate phase, which we now regard as a bath that exchanges particles with the system. The focus is here on the thermodynamical properties of the bosons in the excited states — the "thermal cloud." Since the bath particles are all in the ground state to which we assign the single-particle energy $\varepsilon = 0$ by convention, this is the situation of the Maxwell's demon ensemble of Exercises 46 and 54. We keep track of energy/temperature and particle number/fugacity while suppressing all other variables in the notation. With N bosons in total, explain why

$$\Omega(E, N) = \sum_{N_{\text{ex}}=0}^{N} \Omega_{\text{ex}}(E, N_{\text{ex}})$$

holds for the total count of microstates Ω and the count of excited microstates Ω_{ex}, and why

$$Q(\beta, N) = \sum_{N_{\text{ex}}=0}^{N} Q_{\text{ex}}(\beta, N_{\text{ex}})$$

holds for the corresponding canonical partition functions.

76 Next, we assume that there are so many particles in the condensate that it acts as a particle reservoir — any exchange of relatively few particles with the thermal cloud does not affect the count of particles in the condensate noticeably. Then, the partition function for the Maxwell's demon ensemble,

$$\Xi_{\text{ex}}(E, z) = \sum_{N_{\text{ex}}=0}^{\infty} z^{N_{\text{ex}}} \Omega_{\text{ex}}(E, N_{\text{ex}}),$$

is the generating function for the $\Omega_{\text{ex}}(E, N_{\text{ex}})$s. Show that the expected number of excited particles in the microcanonical ensemble is

$$\langle N_{\text{ex}} \rangle = z \frac{\partial\, \Xi_{\text{ex}}(E, z)}{\partial z} \bigg|_{z=1}.$$

How large is the corresponding spread?

77 In Section 3.7, we have $\beta P\lambda^3 = f_{5/2}(z)$ and $\rho\lambda^3 = f_{3/2}(z)$ for the ideal gas of fermions, and inferred that the leading correction to the thermal equation of state $\beta P = \rho$ (ideal classical gas) is $\beta P \cong \rho + 2^{-5/2}\rho^2\lambda^3$. What are the corresponding expressions for an ideal gas of bosons?

78 For the ideal gas of fermions, and also for the ideal gas of bosons above the critical temperature, we have $\log\bigl(Z(\beta, V, z)\bigr) = \dfrac{V}{\lambda^3}h(z)$ with the appropriate function $h(z)$ of the fugacity z. We know that the adiabatic equation of state $P^3V^5 =$ constant holds in the dilute-gas limit when $h(z) = z$. What is the adiabatic equation of state for an arbitrary $h(z)$?

79 By a suitable contour integral, show that

$$\int_{-\infty}^{\infty} \mathrm{d}\vartheta \, \frac{\mathrm{e}^{\mathrm{i}t\vartheta}}{2(\cosh\vartheta)^2} = \frac{\frac{1}{2}\pi t}{\sinh\left(\frac{1}{2}\pi t\right)}$$

and then use this to confirm the integrals in (3.7.31).

80 Show that the canonical partition function for the ideal fermion gas is given by

$$\log\Bigl(Q(\beta, V, N)\Bigr) = \frac{V}{\lambda^3} f_{\frac{5}{2}}\Bigl(f_{\frac{3}{2}}^{-1}(N\lambda^3/V)\Bigr) - N\log\Bigl(f_{\frac{3}{2}}^{-1}(N\lambda^3/V)\Bigr).$$

Then verify that you obtain the correct answer in the dilute-gas limit of $N\lambda^3 \ll V$, that is: $\log Q = N - N\log(N\lambda^3/V)$, and determine the leading correction to this expression.

81 Equation (3.7.4) expresses $f_\alpha(z)$ in terms of $g_\alpha(-z)$. Show that one can alternatively use

$$f_\alpha(z) = g_\alpha(z) - 2^{1-\alpha} g_\alpha(z^2)$$

if negative arguments are undesired.

Chapter 4

82 The TF functional $E_{\mathrm{TF}}[\rho]$ in (4.1.5) is stationary at the TF density ρ_{TF}. Show that $E_{\mathrm{TF}}[\rho] > E_{\mathrm{TF}}[\rho_{\mathrm{TF}}]$ for all other permissible ρ.

83 Explain why $\left.\dfrac{\partial}{\partial\lambda}E_{\mathrm{TF}}[\rho^{(\lambda)}]\right|_{\lambda=1}=0$ for $\rho^{(\lambda)}(\boldsymbol{r})=\lambda^3\rho_{\mathrm{TF}}(\lambda\boldsymbol{r})$ for $E_{\mathrm{TF}}[\rho]$ in (4.1.5). What relation is implied among the three terms in $E_{\mathrm{TF}}[\rho]$?

84 With the aid of suitable integrations by parts, first show that these two identities hold for the TF function $F(x)$:

$$\int_0^\infty \mathrm{d}x\,\frac{F(x)^{5/2}}{x^{1/2}}+\int_0^\infty \mathrm{d}x\,F'(x)^2=B\,,$$
$$\frac{1}{5}\int_0^\infty \mathrm{d}x\,\frac{F(x)^{5/2}}{x^{1/2}}-\frac{1}{2}\int_0^\infty \mathrm{d}x\,F'(x)^2=0\,,$$

then express the two integrals as multiples of Baker's constant B. Then, find the values of the three terms in the TF functional $E_{\mathrm{TF}}[\rho]$ in (4.1.5) for the stationary density ρ_{TF}, and verify that the relation found in Exercise 83 is correct.

85 Consider the TF model for a positively charged ion, that is: the number N of electrons is less than the atomic number Z. Then, the corresponding solution of the TF equation (4.1.22) changes sign at a finite distance x_0. How is x_0 determined by N and Z? Confirm that $x_0\to\infty$ as $Z>N\to Z$.

86 As a very rough approximation of a neutral atom with Z electrons, consider the TF model without the repulsive electron-electron interaction. Show that the electron density is of the form

$$\rho(r)=\frac{1}{3\pi^2}\left[\frac{2m}{\hbar^2}\left(\frac{Ze^2}{r}-\frac{Ze^2}{r_0}\right)\right]_+^{3/2}$$

with $r_0>0$, and determine the value of r_0 as a function of Z.

87 Next, confirm that the energy is proportional to $Z^{7/3}$ and determine the positive proportionality factor in $-E(Z)\propto Z^{7/3}\dfrac{e^2}{a_0}$. Is it larger or smaller than the value 0.7687 of the standard TF model? Explain why.

88 An ultracold (temperature $T=0$) gas of N spin-$\frac{1}{2}$ atoms of mass m is trapped by a harmonic force with potential energy $\frac{1}{2}m\omega^2\boldsymbol{r}^2$. The atoms interact with a repulsive contact force, so that the potential energy is

$W\delta(\boldsymbol{r}_1 - \boldsymbol{r}_2)$ for one atom at $\boldsymbol{r}_1$ and another at $\boldsymbol{r}_2$, where $W > 0$. Explain why the energy functional of the atom density $\rho(\boldsymbol{r})$ is

$$E[\rho] = \int (\mathrm{d}\boldsymbol{r})\, \frac{\hbar^2}{10\pi^2 m} \left[3\pi^2 \rho(\boldsymbol{r})\right]^{5/3} + \int (\mathrm{d}\boldsymbol{r})\, \frac{1}{2} m\omega^2 \boldsymbol{r}^2 \rho(\boldsymbol{r})$$

$$+ \frac{1}{2} \int (\mathrm{d}\boldsymbol{r})\, W\rho(\boldsymbol{r})^2 \equiv E_{\mathrm{kin}}[\rho] + E_{\mathrm{trap}}[\rho] + E_{\mathrm{int}}[\rho]$$

in the TF approximation. Which equations are obeyed by the density $\rho_{\mathrm{TF}}(\boldsymbol{r})$ for which $E[\rho]$ is minimal?

89 Next, show that $2E_{\mathrm{kin}}[\rho_{\mathrm{TF}}] - 2E_{\mathrm{trap}}[\rho_{\mathrm{TF}}] + 3E_{\mathrm{int}}[\rho_{\mathrm{TF}}] = 0$.

90 For the standard one-dimensional Ising chain with on-site excitation energy E_0 and next-neighbor interaction energy J, find the heat capacity for low and high temperatures, and show that the Ising chain obeys the Third Law.

91 Consider the standard one-dimensional Ising chain (or ring) with N sites, on-site energy E_0, and next-neighbor interaction strength J. Denote by N_+ and N_- the number of sites with $s_j = +1$ or $s_j = -1$, respectively. Likewise, $N_+^{(\mathrm{nn})}$ is the number of next neighbors with $s_j s_{j+1} = +1$, and $N_-^{(\mathrm{nn})}$ is the number of next neighbors with $s_j s_{j+1} = -1$. What are the expected values of N_+, N_-, $N_+^{(\mathrm{nn})}$, and $N_-^{(\mathrm{nn})}$ in terms of N, βE_0, and βJ?

92 A modified version of the one-dimensional Ising model of Section 4.2 has the usual on-site energy $\frac{1}{2}E_0 s_j$ and a next-neighbor interaction energy with alternating strength, namely $-J_1 s_j s_{j+1}$ when j is odd, and $-J_2 s_j s_{j+1}$ when j is even, with $J_1 \neq J_2$. Find the free energy $F(\beta, E_0, J_1, J_2, N)$.

93 Consider a one-dimensional Ising chain (or ring) with N links between next neighbors. There is no on-site energy, and the next-neighbor interaction energy for sites j and $j+1$ is

$$\left\{\begin{array}{l} +J \text{ if } s_j s_{j+1} = -1 \\ -J_+ \text{ if } s_j = s_{j+1} = +1 \\ -J_- \text{ if } s_j = s_{j+1} = -1 \end{array}\right\} = J\frac{1 - s_j s_{j+1}}{2} - J_+ \frac{(1+s_j)(1+s_{j+1})}{4}$$
$$- J_- \frac{(1-s_j)(1-s_{j+1})}{4}.$$

We have the standard Ising model for $J_+ = J_- = J$ and a modified Ising model otherwise. Show that such a modified Ising model is equivalent to a standard Ising model with a certain on-site energy E_0' and a certain interaction energy J' plus an energy off-set $\mathcal{E}$, in the sense that the energy of the kth microstate can be written as

$$E_k = N\mathcal{E} + \frac{1}{2}E_0'\sum_j s_j - J'\sum_j s_j s_{j+1} \,.$$

State $\mathcal{E}$, E_0', and J' in terms of J, J_+, and J_-. Then, find the canonical partition function of this modified Ising model. When $J_\pm = J \pm \epsilon$, what is the free energy to first order in ϵ?

94 Consider a one-dimensional chain (or ring) of particles with N next-neighbor links and no on-site energy. The energy of the kth microstate is

$$E_k = -J\sum_j s_j s_{j+1} \quad \text{with } s_j = 0 \text{ or } +1 \text{ or } -1 \,.$$

Note that we have the additional option of $s_j = 0$ here, while there is only $s_j = \pm 1$ in the standard Ising model. Show that the canonical partition function is

$$Q(K,N) = \left(\cosh(K) + \tfrac{1}{2} + \sqrt{[\cosh(K) - \tfrac{1}{2}]^2 + 2}\right)^N$$

where $K = \beta J$. What is the free energy per site?

95 Next, determine the heat capacity per site at low temperatures ($K \gg 1$) and high temperatures ($K \ll 1$); in both cases, state the leading term. Confirm that this system obeys the Third Law.

96

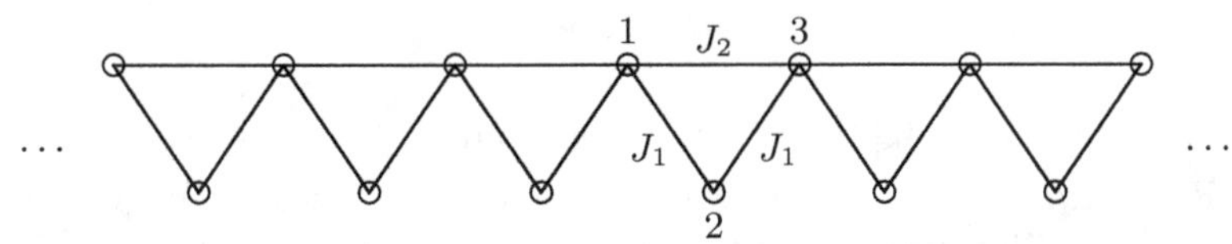

A modified Ising model has N sites in a zigzag with links of strength J_1 between all pairs of next-neighbor sites and links of strength J_2 between every second pair of next-next-neighbor sites, so that the triangle 1,2,3 indicated in the picture contributes

$$-J_1(s_1 s_2 + s_2 s_3) - J_2 s_1 s_3 \quad \text{with } s_j = \pm 1$$

to the energy. For $K_1 = \beta J_1$ and $K_2 = \beta J_2$, we denote the canonical partition function by $Q(K_1, K_2, N)$. Show that $Q(K_1, 0, N) = 2^N \cosh(K_1)^N$ is the partition function for $J_2 = 0$, and that $Q(0, K_2, N) = 2^N \cosh(K_2)^{\frac{1}{2}N}$ is the partition function for $J_1 = 0$.

97 Next, explain why a relation of the form

$$Q(K_1, K_2, N) = g^{\frac{1}{2}N} Q(\widetilde{K}, 0, \tfrac{1}{2}N)$$

must hold and determine g and $\widetilde{K}$ as functions of K_1 and K_2; find $Q(K_1, K_2, N)$. As a check, verify that the partition function is

$$Q(K, K, N) = \left(\mathrm{e}^{3K} + 3\,\mathrm{e}^{-K}\right)^{\frac{1}{2}N}$$

when $K_1 = K_2 = K$; find the heat capacity in this case.

98 Renormalization group method of Section 4.3: Show that

$$q(K_2) = \frac{1}{2}q(K_1) + \frac{1}{2}K_1 + \frac{1}{2}\log 2 \quad \text{with} \quad \cosh(2K_2) = \mathrm{e}^{2K_1}$$

holds for $q(K) = \log\bigl(2\cosh(K)\bigr)$.

99 For the one-dimensional Ising model with $E_0 \neq 0$, set $K = \beta J$ and $L = \frac{1}{2}\beta E_0$, so that the canonical partition function is

$$Q(L, K, N) = \sum_{s_1, s_2, \ldots, s_N} \mathrm{e}^{-L\sum_j s_j + K\sum_j s_j s_{j+1}}\,.$$

Sum over $s_2, s_4, s_6, \ldots$ to establish the renormalization-group relation

$$Q(L, K, N) = g(L, K)^{\frac{1}{2}N} Q\bigl(L', K', \tfrac{1}{2}N\bigr)\,.$$

Find L', K', and $g(L, K)$ in terms of L and K.

100 To show that (4.4.29) has only one solution for $0 < K_0 < \infty$, set $x_0 = \mathrm{e}^{-\frac{4}{3}K_0}$, find the polynomial equation obeyed by x_0, and examine its properties. What are $\cosh(4K_0)$, $\sinh(4K_0)$, and $\tanh(4K_0)$ in terms of x_0? Confirm that, if x is an approximation for x_0, then $\frac{1}{2}(1 + x^6)$ is a better approximation. Starting with $x = \frac{1}{2} \cong x_0$, what are the successive approximations you get?

101 Confirm the small-K and large-K approximations for $q(K)$ mentioned after (4.5.3).

102 Mean-field approximation of Section 4.8: Find the internal energy $\langle E\rangle$ and verify that it has the correct value when $E_0 = 0$ and $T = 0$. What do you get for $T > T_c$? Is this value correct?

Chapter 5

103 Consider the minimum property of the free energy of (5.4.1). What is the entropy $S(\beta)$ in terms of ρ_β of (5.4.5)? With the eigenkets $|E_k\rangle$ of H in (5.2.12), the probability of finding the system in the kth state is $p_k = \langle E_k|\rho_\beta|E_k\rangle$. Express $S(\beta)$ in terms of the p_ks and comment on what you find.

104 Next, if a constant energy E_0 is added to H, that is: $H \to H + E_0$, what are the resulting changes in F, ρ_β, and S?

105 For a classical system with energy E_k in the kth configuration, the analog of the minimal principle for the free energy in (5.4.1) states

$$F(\beta) \leq \sum_k \left(w_k E_k + \frac{1}{\beta} w_k \log w_k\right)$$

where $w_k \geq 0$ for all k and $\sum\limits_k w_k = 1$. For which choice of weights w_k does the equal sign hold? Show that any other choice yields a truly larger value for the right-hand side.

106 Verify that

$$\log A = \int_0^\infty \frac{\mathrm{d}t}{t}\left(\mathrm{e}^{-t} - \mathrm{e}^{-At}\right) \qquad \text{for } A > 0$$

and then use this for a derivation of (5.4.14).

107 Modified Ising model of Section 5.5: Show that the free energy, with the optimal value for ϵ or the sub-optimal value, agrees in first order of K' with the expression for $\epsilon = 0$. Find the heat capacity to first order in J'.

108 Next, consider the other extreme: What is $F(\beta, 0, J')$, and what is $F(\beta, J, J')$ for $J \ll J'$?

109 Recognize that the example in (5.5.3) is equivalent to a standard Ising chain and use this equivalence to find the canonical partition function and then the free energy when $K = \beta J \neq 0$ and $K' = \beta J \neq 0$. Confirm the small-K' approximation in (5.5.10) and the small-K approximation found in Exercise 108.

110 Next, determine the heat capacity to the leading order in K' when $0 < K' \ll 1$.

111

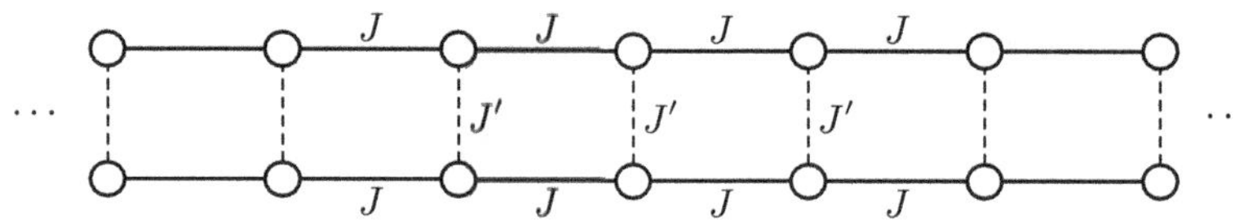

We have two long parallel Ising chains, each made up of $\frac{1}{2}N$ particles with no on-site energy and a next-neighbor interaction of strength J (solid connecting lines). There is also an interaction of strength J' between each particle of one chain and the nearest particle of the other chain (dashed connecting lines). For $J' = 0$, what is the free energy $F(T, J, J' = 0)$? For $J = 0$, what is the free energy $F(T, J = 0, J')$? For $0 < J' \ll J$, what is the free energy $F(T, J, J')$ to first order in J'?

112 Standard one-dimensional Ising model: Supplement the expected values of $s_j s_{j+1}$ and $s_j s_{j+2}$, given in Section 5.5, by the expected values of $s_j s_{j+m}$ for $m > 2$. For large m, one expects an exponential decay, $\langle s_j s_{j+m} \rangle \propto \mathrm{e}^{-m/\ell}$ with the *correlation length* ℓ, which one calculates as the limit

$$\ell = -\lim_{m\to\infty} \frac{m}{\log \langle s_j s_{j+m} \rangle}.$$

What is the value of ℓ here?

113 If A and B do not commute — that is $AB \neq BA$; they could be square matrices or differential operators or quantum-mechanical operators or ... — then $\mathrm{e}^{\epsilon(A+B)}$ and $\mathrm{e}^{\frac{1}{2}\epsilon A}\,\mathrm{e}^{\epsilon B}\,\mathrm{e}^{\frac{1}{2}\epsilon A}$ are not the same. Show that the two expressions agree to order ϵ^2. What is the error of order ϵ^3? How does the error in the approximation

$$\mathrm{e}^{A+B} \cong \left[\mathrm{e}^{\frac{1}{2N}A}\,\mathrm{e}^{\frac{1}{N}B}\,\mathrm{e}^{\frac{1}{2N}A}\right]^N \quad \text{with integer } N \gg 1$$

scale with N?

114 The example of Section 5.6: Show that the correct result is obtained for $N \to \infty$.

Chapter 6

115 As an illustration of the argument about negligible boundary contributions at (6.2.8), consider $f(|x-y|) = \mathrm{e}^{-|x-y|/a}$ and $-\frac{1}{2}L < x, y < \frac{1}{2}L$ with $a, L > 0$. Show that

$$\frac{1}{L} \int_{-\frac{1}{2}L}^{\frac{1}{2}L} \mathrm{d}x \int_{-\frac{1}{2}L}^{\frac{1}{2}L} \mathrm{d}y\, f(|x-y|) \cong \int_{-\infty}^{\infty} \mathrm{d}(x-y)\, f(|x-y|)$$

when $L \gg a$.

116 Section 6.3: We already found the first three virial coefficients $a_1(\beta)$, $a_2(\beta)$, and $a_3(\beta)$ in terms of $b_1(\beta)$, $b_2(\beta)$, and $b_3(\beta)$. Now find also $a_4(\beta)$. Which clusters contribute to the fourth virial coefficient?

117 What are the first few virial coefficients $a_1(\beta)$, $a_2(\beta)$, $a_3(\beta)$ for the Clausius gas, the Berthelot gas, and the Dieterici gas of Exercises 28–30?

118 Find the virial coefficients $a_2(\beta)$ and $a_3(\beta)$ for the extreme situation of impenetrable balls, for which

$$V(r) = \begin{cases} \infty \text{ for } r < r_0 \\ 0 \text{ for } r > r_0 \end{cases}$$

is the potential energy of the particle-particle interaction.

119 For the noble gases, the critical temperatures and pressures are these:

	T_c	P_c
He	5.19 K	227 kPa
Ne	44.40 K	2,760 kPa
Ar	150.8 K	4,870 kPa
Kr	209.3 K	5,500 kPa
Xe	289.8 K	5,840 kPa

What are the corresponding values of the parameters a and b of the van der Waals gas, and what are the values of V_0 and r_0 in the approximation of Section 6.5?

120 Consider the approximation to the cluster expansion in Section 6.3 of the notes, in which we only take the 2-clusters into account but neglect the contributions from 3-clusters, 4-clusters, etc., similar to the discussion in Section 6.5. Show that, under these circumstances, the free energy is approximated by

$$F(\beta, V, N) = -\frac{N}{\beta}\left(1 + \log\frac{V}{N\lambda^3} + b_2(\beta)\frac{N\lambda^3}{V}\right).$$

121 For the van der Waals gas we have $b_2(\beta) = \dfrac{\beta a - b}{\lambda^3}$ in Section 6.5. Use this to find the function $f_0(T)$ that remained unspecified in (1.16.17)–(1.16.20) and confirm that $c_V = \frac{3}{2}R$. Then determine the adiabatic equation of state obeyed by the van der Waals gas. Make sure that you get $PV^{\frac{5}{3}} = \text{constant}$ when $a = 0$ and $b = 0$.

Chapter 7

122 Rather than the symmetric jumps in (7.1.15), consider the more general situation of random jumps with $p_1(m) = q\delta_{m,1} + (1-q)\delta_{m,-1}$ where $0 < q < 1$. Find $\langle m\rangle_n$ and $\langle \delta m^2\rangle_n$.

123 Now, consider random jumps with $p_1(m) \propto a^{|m|}$ where $0 < a < 1$. Find $\left\langle e^{im\phi}\right\rangle_n$ as well as $\langle m\rangle_n$ and $\left\langle \delta m^2\right\rangle_n$.

124 In a Poisson process of the kind discussed in Section 7.3, the events are random jumps of a clock hand by angle φ either clockwise or anticlockwise, which happen with the probabilities q and $1-q$, respectively. If we specify the position of the hand by azimuthal angle ϕ, with $\phi = 0$ at $t = 0$, what is the expected value of $e^{ik\phi}$ at time t for $k = 0, \pm 1, \pm 2, \ldots$?

125 Consider a classical gas of N structureless particles with a Hamilton function $H(\boldsymbol{x}) = \sum_{k=1}^{N} \dfrac{p_k^2}{2m_k} + V(\boldsymbol{r}_1, \boldsymbol{r}_2, \ldots, \boldsymbol{r}_N)$, where the potential energy

$V(\ )$ is any reasonable function of the position vectors, and we assign "∞" to it if one or more $\boldsymbol{r}_k$s are outside the volume within which the gas is contained. The symbol $\boldsymbol{x}$ stands for all $6N$ phase-space variables, the $3N$ momentum-vector components plus the $3N$ position coordinates, and we denote the $6N$-dimensional phase-space volume element by $(\mathrm{d}\boldsymbol{x})$. At temperature $T = (k_\mathrm{B}\beta)^{-1}$, then, the expected value of a function $f(\boldsymbol{x})$ of the phase-space variables is

$$\langle f(\boldsymbol{x})\rangle = \frac{\displaystyle\int (\mathrm{d}\boldsymbol{x})\ \mathrm{e}^{-\beta H(\boldsymbol{x})}\, f(\boldsymbol{x})}{\displaystyle\int (\mathrm{d}\boldsymbol{x})\ \mathrm{e}^{-\beta H(\boldsymbol{x})}}\,.$$

Show that $\left\langle x_j \dfrac{\partial H(\boldsymbol{x})}{\partial x_k}\right\rangle = \delta_{jk} k_\mathrm{B} T$ for any pair (x_j, x_k) of the phase-space variables.

126 What do you obtain for the expected values of $\sum_{k=1}^{N} \boldsymbol{p}_k \cdot \boldsymbol{v}_k$ and $\sum_{k=1}^{N} \boldsymbol{r}_k \cdot \boldsymbol{F}_k$? Here, $\boldsymbol{v}_k = \dfrac{\mathrm{d}}{\mathrm{d}t}\boldsymbol{r}_k$ is the velocity of the kth particle and $\boldsymbol{F}_k = \dfrac{\mathrm{d}}{\mathrm{d}t}\boldsymbol{p}_k$ is the force acting on the kth particle.

127 An atom in a gas has the velocity vector $\boldsymbol{v} = \boldsymbol{p}/m$ and moves at the speed $v = |\boldsymbol{v}|$. Find the average speed $\langle v\rangle$ and also the average reciprocal speed $\langle v^{-1}\rangle$ for the atoms of an ideal classical gas at temperature T. Confirm that $\langle v\rangle\langle v^{-1}\rangle \geq 1$.

128 Demonstrate that, quite generally, the inequality $\langle X\rangle\langle X^{-1}\rangle \geq 1$ holds for any positive quantity X.

129 Find the limiting value of $\langle p^k\rangle_t$ as $t \gg \tau$ from (7.8.25) or (7.8.20). Alternatively, determine these limiting values directly from the $j = 0$ terms in (7.8.5).

130 Introduce new integration variables $\tilde{x}$ and $\tilde{p}$ by setting $p' = \tilde{p}\,\mathrm{e}^{t/\tau}$ and $x' = \tilde{x} - \dfrac{\tilde{p}\tau}{M}\left(\mathrm{e}^{t/\tau} - 1\right)$, and so rewrite (7.8.43) in the form of a convolution,

$$f(x,p,t) = \int \mathrm{d}\tilde{x}\,\mathrm{d}\tilde{p}\ \tilde{G}_t(x-\tilde{x}, p-\tilde{p})\tilde{f}_t(\tilde{x},\tilde{p})\,,$$

with $\tilde{f}_t(\tilde{x},\tilde{p})$ suitably related to $f_0(x',p')$.

131 Use the $b = 0$ version of (7.8.37) to find an explicit expression for $\langle x^k \rangle_t$. For $t \gg \tau$, compare with (7.8.59).

132 Include a force term $-F(x)\frac{\partial}{\partial p}f(x,p,t) = M\omega^2 x\frac{\partial}{\partial p}f(x,p,t)$ into the Fokker–Planck equation (7.8.1). What are the resulting changes in (7.8.4), (7.8.5), and (7.8.15)?

133 Next, show that the modified version of (7.8.5) has a stationary solution, reached at very late times, with $\langle x^j p^k \rangle_t = \langle x^j \rangle_t \langle p^k \rangle_t$ and vanishing expectation values for odd powers, and find the even-power values. Do these late-time values have any memory of the initial probability distribution?

134 Further, confirm that the modified version of (7.8.15) has a stationary solution of the form $G_t(a,b) = \mathrm{e}^{-Aa^2 - Bb^2}$ and determine A and B. Confirm that what you get is consistent with your expectations in view of (7.7.8).

135 Finally, find the Green's function for this situation.

Hints

1 Distinguish between extensive and intensive quantities.

2 No hint needed.

3 Follow the procedure of Section 1.2: Read the equation of state as a differential equation for $U(S, L, n)$, then proceed.

4 Recognize two differential equations, one for $U(S, V, n)$, the other for $F(T, V, n)$, and recall how T and S are related to one another; ensure that $a(S/n)$ is consistent with $b(T)$.

5 Begin with determining the quantities that do not change as the system progresses from the initial non-equilibrium situation to the final equilibrium situation.

6 First apply (1.4.4) to the situation of (1.5.1) and conclude that

$$U_1(S_1 + \delta S, X_1) + U_2(S_2 - \delta S, X_2) \geq U_1(S_1, X_1) + U_2(S_2, X_2),$$

then proceed.

7, 8 Rather than letting the systems exchange heat, consider trading volume or substance.

9 No hint needed.

10 The area enclosed by the cycle in the V, P diagram is the net work done by the system; the area in the S, T diagram is the net heat transferred to the system. Invoke energy conservation ($\mathrm{d}U = T\,\mathrm{d}S - P\,\mathrm{d}V$ if nothing else changes) to show that the areas are always the same.

11 You should, of course, find the familiar result for the efficiency of the Carnot cycle: $\dfrac{T_2 - T_1}{T_2}$.

12 Note that $(V, P) = (V_0, 4P_0)$ in state A and $(V, P) = (4V_0, P_0)$ in state C. From statements in Section 1.2 infer that $\dfrac{T}{T_0} = \dfrac{PV}{P_0V_0}$ and $\dfrac{S}{S_0} = \dfrac{5}{2}\log\dfrac{V}{V_0} + \dfrac{3}{2}\log\dfrac{3P}{2P_0}$ with $P_0V_0 = S_0T_0$; then proceed. The reference volume V_0 here need not be the same as that in (1.2.16).

13 Recognize a relation like $\dfrac{\partial U}{\partial S} \propto U^{\frac{1}{4}}$. Alternatively, proceed from the ansatz $F = \mathrm{func}(T)\,V$ and determine the function of T such that you recover the Stefan–Boltzmann expression for U. You should find

$$F(T,V) = -\frac{\pi^2}{45}\frac{(k_\mathrm{B}T)^4 V}{(\hbar c)^3},$$

for example.

14 Is it possible to change the temperature of the photon gas while keeping the pressure constant?

15 Combine the fundamental statements about $\mathrm{d}U$ and $\mathrm{d}F$, or the Gibbs–Duhem relation, with suitable Maxwell relations. For the photon gas, n is not a variable and the chemical potential vanishes (what did you find for the free enthalpy in Exercise 13?).

16 Recall the fundamental statements about $\mathrm{d}U$ and dH.

17 Pay close attention to the precise definitions of C_V, C_P, α, and K_T.

18 No further hints needed.

19 The thermal equation of state tells you the difference $C_P - C_V$; see (1.10.14). You also need the adiabatic equation of state in (1.10.22).

20 Recall (1.11.5).

21 No hint needed.

22 Supplement (1.16.39) with the corresponding approximation for $x \gg 1$ and then use this in (1.16.13) and (1.16.14). Alternatively, the ansatz $v^{(2)} = (1+\epsilon)b$ with $0 < \epsilon \ll 1$ in (1.16.15) yields $v^{(1)} = b\epsilon\,\mathrm{e}^{1/\epsilon}$ and the rest follows.

23 First use (1.16.39) and related statements to establish that the leading term of the "terms of order Δv" in (1.17.25) is $-\dfrac{2}{5}\dfrac{\Delta v}{v_\mathrm{c}}$; then proceed.

24 Since low temperatures correspond to large values of x, you need the large-x approximation of Exercise 22.

25 Remember that $\left(\dfrac{\partial P}{\partial v}\right)_T = 0$ when $T = T_\mathrm{c}$ and $v = v_\mathrm{c}$.

26 You could try an integration by parts.

27 There are two integration constants, when solving $f_0''(T) = -\dfrac{3R}{2T}$. The natural requirement that $f(T,v)$ turns into the ideal-gas expression, see (1.7.6), when $v \gg b$ and $T \gg T_\mathrm{c}$ determines the value of one constant; the value of the other is a matter of convention. You should find that

$$\mu(T,v) = -RT\log\left(\frac{v-b}{2b}\left(\frac{T}{T_\mathrm{c}}\right)^{3/2}\right) - \frac{2a}{v} + RT\frac{v}{v-b} + \frac{3}{4}RT$$

if you adopt the convention that $\mu(T_\mathrm{c}, v_\mathrm{c}) = 0$. For the elimination of v in favor of P, needed in the Legendre transformation from the free energy to the free enthalpy, you have to find the roots of a cubic polynomial; it is expedient to write it as a polynomial in $1 - \dfrac{3b}{v}$.

28–30 Follow the procedure applied to the van der Waals gas in Section 1.16. The lesson of Exercise 25 is useful. For the value of $\dfrac{P_\mathrm{c}v_\mathrm{c}}{RT_\mathrm{c}}$, you should find $\dfrac{8b+8c}{3b+2c}$, $\dfrac{8}{3}$, and $\dfrac{1}{2}\,\mathrm{e}^2 \cong \dfrac{11}{3}$, respectively.

31 Combine the statements in (1.17.3), (1.17.7), and (1.17.22). For $\overline{\mu}(T)$, you should find

$$\frac{\mathrm{d}^2\overline{\mu}}{\mathrm{d}T^2} = -\frac{\Delta(c_P/v)}{T\Delta(1/v)} + 2\frac{\Delta\alpha}{\Delta(1/v)}\frac{\mathrm{d}\overline{P}}{\mathrm{d}T} - \frac{\Delta K_T}{\Delta(1/v)}\left(\frac{\mathrm{d}\overline{P}}{\mathrm{d}T}\right)^2 .$$

32 The factor $\dfrac{1}{2^N}$ in (2.1.1) is $\left(\frac{1}{2}\right)^{N_1}\left(\frac{1}{2}\right)^{N-N_1}$ and gets replaced. You should find p^k in (2.1.3) instead of $\dfrac{1}{2^k}$.

33 The binomial factor in (2.1.1) is the count of configurations with N_1 distinguishable particles in the first half. When the particles are indistinguishable, there is only one such configuration.

34 You could try a system of noninteracting harmonic oscillators.

35 Note that $2R\,\mathrm{e}^{-\lambda^2R^2} = -\dfrac{1}{\lambda^2}\dfrac{\partial}{\partial R}\,\mathrm{e}^{-\lambda^2R^2}$ invites an integration by parts; recall that the derivative of Heaviside's unit step function is Dirac's delta function, $\dfrac{\mathrm{d}}{\mathrm{d}x}\eta(x) = \delta(x)$; recognize Euler's factorial integral; and exploit that two functions are the same if their Laplace transforms are.

36 Since $\left(1+t/x\right)^x \to \mathrm{e}^t$ as $x \to \infty$, you get a large-x approximation for $\left(1+t/x\right)^x$ from

$$x\log\left(1+\frac{t}{x}\right) = t - \frac{1}{2}\frac{t^2}{x} + \cdots\,,$$

where you keep as many terms as you need.

37 Write $\dfrac{(x+1)!}{x!} = (x+1)\bigl(1+\{\text{powers of } x^{-1}\}\bigr)$ and choose λ such that the leading correction has a power as high as possible. You should find $\lambda = \dfrac{1}{12}$. As another exercise, confirm that the choice of $b = \frac{1}{2}(a+1)$ is best in $\dfrac{(x+a)!}{x!} \cong (x+b)^a$ when $x \gg |a|$.

38 Observe that $\beta\left(\dfrac{\partial F}{\partial \beta}\right)_{V,n} = TS$, then proceed.

39 The D-dimensional analog of a cube with edge length a are all points $\boldsymbol{x}$ with $-\frac{1}{2}a \le x_j \le \frac{1}{2}a$ for $j = 1, 2, \ldots, D$, where $\boldsymbol{x}$ has the same meaning as in Exercise 35.

40 Note that the integral of $\mathrm{e}^{-\beta\sum_j \frac{1}{2m}\boldsymbol{p}_j^2}$ is the product of N single-particle integrals; remember to ensure the correct Boltzmann counting.

41 Combine (1.1.22) and (2.3.8).

42, 43 In a first step establish

$$\frac{N}{2} \pm \frac{E}{E_0} = \frac{N}{1+\mathrm{e}^{\pm\beta E_0}} = \frac{N}{2}\,\frac{\mathrm{e}^{\mp\beta E_0}}{\cosh\left(\frac{1}{2}\beta E_0\right)}\,,$$

then proceed from (2.5.25).

44 No hint needed.

45 Remember the lesson of Exercise 38.

46 The relation between $\Xi(E,V,\mu)$ and $Z(\beta,V,z)$ is analogous to that between $\Omega(E,V,N)$ and $Q(\beta,V,N)$.

47 The grand canonical partition function in (2.9.22) is useful.

48 Assume that the probability $p(E) \propto \Omega(E)\,\mathrm{e}^{-\beta E}$ has a very narrow peak centered at $\bigl\langle E\bigr\rangle$. Can you justify this assumption?

49 This is a variant of the example in (2.5.22)–(2.5.34).

50 Remember Exercise 44.

51 No hint needed.

52 Apply (2.8.2) and (2.9.1) to $\log Z \propto \beta^{-\kappa} V z$, then draw conclusions.

53 Extract $Q(\beta, V, N)$ from $Z(\beta, V, z)$, then infer $F(T, V, N)$ and $U(S, V, N)$. What you get should look familiar, and then you know the heat capacities right away.

54 You should find $p_k = \frac{1}{\Xi}\Omega(E, N_k, V)$.

55, 56 Remember (3.1.7).

57 No hint needed.

58 The canonical partition function is most compactly expressed in terms of a modified Bessel* function from the $\mathrm{K}_n(\ )$ family,

$$\mathrm{K}_n(x) = \int_0^\infty \mathrm{d}\vartheta\, \cosh(n\vartheta)\, \mathrm{e}^{-x\cosh\vartheta}\,.$$

Useful approximations are

$$K_n(x) \cong \frac{1}{2}\left(\frac{2}{x}\right)^n \quad \text{for} \quad 0 < x \ll 1\,,$$

$$K_n(x) \cong \sqrt{\frac{\pi}{2x}}\,\mathrm{e}^{-x}\left(1 + \frac{4n^2 - 1}{8x}\right) \quad \text{for} \quad x \gg 1\,.$$

59 Note that the low-temperature limit is also the nonrelativistic limit, and that the high-temperature limit is also the limit of vanishing mass.

---

*Friedrich Wilhelm BESSEL (1784–1846)

60 Euler's factorial integral in (3.6.16) is useful. For the average energy per particle, you should find

$$\frac{\langle E\rangle}{N} = \frac{3}{2}k_{\mathrm{B}}T + \frac{3}{\nu}k_{\mathrm{B}}T\,.$$

61 Remember that a three-dimensional harmonic oscillator is composed of three independent one-dimensional harmonic oscillators.

62 Note that the canonical partition function in (3.1.4) is a product of three factors, one for rotations, one for vibrations, and and for electronic excitations. What does this factorization imply for the free energy?

63 Proceed from

$$\langle n_j n_{j'}\rangle = Z^{-1}\frac{\partial}{\partial(\beta\varepsilon_j)}\frac{\partial}{\partial(\beta\varepsilon_{j'})}Z = Z^{-1}\frac{\partial}{\partial(\beta\varepsilon_j)}\Big(Z\frac{\partial}{\partial(\beta\varepsilon_{j'})}\log Z\Big)$$

for the appropriate Z.

64 Remember that the factor $\hbar\omega$ in (3.4.9) is the energy per photon for the modes with circular frequency ω. You should get

$$\langle N\rangle = \frac{V}{\pi^2(\beta\hbar c)^3}\int\limits_0^\infty \frac{\mathrm{d}x\,x^2}{\mathrm{e}^x-1} = \frac{2\,\zeta(3)}{\pi^2}\frac{V}{(\beta\hbar c)^3}\,.$$

65 Note that $(\mathrm{e}^x-1)^{-1} = \sum\limits_{k=1}^{\infty}\mathrm{e}^{-kx}$ and use (3.6.16).

66 No hint needed.

67 When ignoring, as usual, the complications of Exercise 72, you simply have $z=1$ for subcritical temperatures.

68 The coexistence pressure is the pressure of either phase, so you can just take the condensate pressure for that. For the entropy, see Exercise 67.

69 Remember that $T_c \propto v^{-\frac{2}{3}}$. Your plot should look like this:

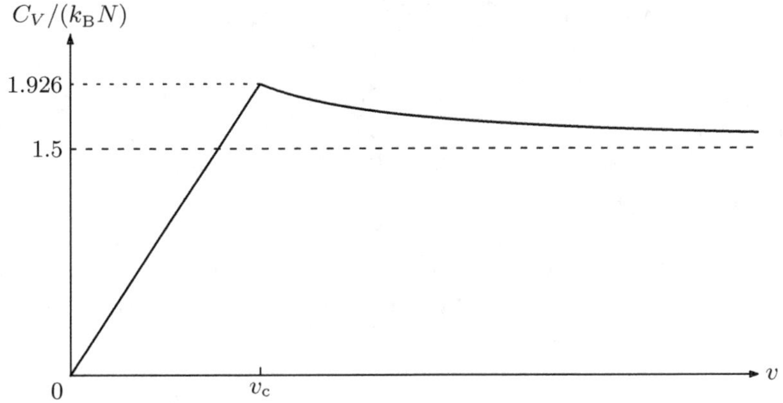

70 Note that C_V is continuous at $v = v_c$. As another exercise, confirm the statement between (1.17.18) and (1.17.19).

71 You could make use of what you found for C_P in Exercise 16, here for $\text{H} = \langle E\rangle + PV = -\beta\left(\dfrac{\partial(\beta^{-1}\log Z)}{\partial\beta}\right)_{V,z}$. Alternatively, you could find the isentropic compressibility $K_S = \dfrac{3}{5P} = \dfrac{3\lambda^3}{5g_{\frac{5}{2}}(z)}$, see Exercise 78, supplement it with the corresponding expression for the isothermal compressibility, then use (1.10.23).

72 The statement combines (3.6.19), (3.6.20), and (3.6.22). The few-term approximation $g_{\frac{3}{2}}(z) \cong \zeta(\frac{3}{2}) - 2\sqrt{-\pi\log z}$ is sufficiently accurate for $0 < 1 - z \ll 1$.

73 Rather than integrating over the (3 + 3)-dimensional phase space in (3.6.6), you integrate over a (1 + 1)-dimensional or a (2 + 2)-dimensional phase space. Eventually, this should take you to

$$\langle N\rangle - \langle N_0\rangle = \begin{cases} \dfrac{L}{\lambda}g_{\frac{1}{2}}(z) & \text{(one-dimensional gas)}, \\ \dfrac{A}{\lambda^2}g_1(z) & \text{(two-dimensional gas)}, \end{cases}$$

and both right-hind sides are arbitrarily large for $z < 1$ close enough to $z = 1$.

74 The analog of (3.6.20) is

$$\langle N\rangle - \langle N_0\rangle = \sum_{j=1}^{\infty} \binom{j+D-1}{D-1} \frac{z}{\mathrm{e}^{j\beta\hbar\omega} - z},$$

where the binomial factor is the count of oscillator states with energy $j\hbar\omega$ in D dimensions: 1 for $D=1$, $j+1$ for $D=2$, $\frac{1}{2}(j+1)(j+2)$ for $D=3$. Note that the sum converges for all positive temperatures and all $z \leq 1$, including in particular $z=1$.

75 Note that for the count of microstates in the condensate, we have $\Omega_0(E_0, N_0) = \delta_{E_0,0}$ since there is only one condensate state with N_0 bosons and that has energy $E_0 = 0$.

76 Confirm that the probability of finding exactly N_{ex} particles in the thermal cloud is $\dfrac{\Omega_{\mathrm{ex}}(E, N_{\mathrm{ex}})}{\Omega(E,N)}$, then proceed.

77 Remember that the replacement $f_\alpha(z) \to g_\alpha(z)$ converts most fermion expressions into the corresponding boson expressions. The identity (3.7.4) is useful.

78 Combine $U = \langle E\rangle = -\left(\dfrac{\partial \log Z}{\partial\beta}\right)_{V,z} = \dfrac{3}{2\beta}\log Z = \dfrac{3}{2}PV$ with $(\mathrm{d}U)_{S,N} = -P(\mathrm{d}V)_{S,N}$. You should find that $P^3V^5 = $ constant for isentropic changes; the particular form of $h(z)$ does not matter here.

79 Exploit $\cosh(\vartheta + \mathrm{i}\pi) = -\cosh\vartheta$.

80 Proceed from the expression for $\log\bigl(Z(\beta, V, z)\bigr)$ in (3.7.2).

81 No hint needed.

82 Establish first that $(\rho+\Delta\rho)^{5/3} = \rho^{5/3} + \frac{5}{3}\rho^{2/3}\Delta\rho + \frac{5}{9}(\rho+\epsilon\Delta\rho)^{-1/3}(\Delta\rho)^2$ with $0 \leq \epsilon \leq 1$ if both ρ and $\rho+\Delta\rho$ are permissible densities.

83 Exploit the minimum property of $E_{\mathrm{TF}}[\rho]$.

84 Proceed from $(F')^2 = (FF')' - FF''$ and $(F')^2 = \bigl[x(F')^2\bigr]' - 2xF'F''$.

85 You should find that $F(x_0) = 0$ and $-x_0F'(x_0) = 1 - \dfrac{N}{Z}$.

86 Write $\mu = -\dfrac{Ze^2}{r_0}$ in the version of (4.1.9) that applies here.

87 Euler's beta function integral,

$$\int_0^1 \mathrm{d}x\, x^a (1-x)^b = \frac{a!\, b!}{(a+b+1)!} \qquad \text{for} \quad a, b > -1\,,$$

is useful; remember that $(-\frac{1}{2})! = \sqrt{\pi}$. When the repulsion between the electrons is ignored, the energy is lower.

88 No hint needed.

89 Remember Exercise 83.

90 First consider the special cases of $E_0 = 0,\ J > 0$ and $E_0 > 0,\ J = 0$, before you deal with $E_0 > 0,\ J > 0$.

91 Consider the response of the partition function to infinitesimal changes of βE_0 or βJ.

92 A variant of the matrix method in Section 4.2 will work.

93 Compare the 2×2 matrix for the pair $(j, j+1)$ here with the matrix for the standard Ising model.

94, 95 Try a 3×3 analog of the matrix method in Section 4.2.

96, 97 Recognize a chain of triangles and its degenerate forms when $J_1 = 0$ or $J_2 = 0$.

98 No hint needed.

99 Follow the procedure in Section 4.3.

100 You should find that x_0 is a root of a 6th-order polynomial. As a check, confirm that $x_0 = 1$ is one root; a graph may help in establishing that the root of interest is near $x_0 = \frac{1}{2}$.

101 Exploit the relations in (4.4.26) together with the small-K and large-K approximations in (4.4.27).

102 For the expected value of the energy E in (4.8.1), assume that $\langle s_j s_{j'}\rangle = \langle s_j\rangle\langle s_{j'}\rangle$. Justify this assumption.

103 Proceed from $F = \langle E\rangle - TS$, and arrive at (2.7.4). On the way, you encounter $S(\beta) = -k_{\mathrm{B}}\mathrm{tr}\{\rho_\beta \log \rho_\beta\}$, the quantum version of Gibbs entropy formula.

104 No hint needed.

105 You should, of course, find Gibbs's answer (2.10.13).

106 You'll have a use for (5.2.25).

107 No hint needed.

108 When $J = 0$, you have two separate Ising chains.

109, 110 Remember that $s_{j-1}s_{j+1} = (s_{j-1}s_j)\,(s_j s_{j+1})$ and recall the parameterization of (4.2.13).

111 Follow the example treated in Section 5.5.

112 Generalize (5.5.18)–(5.5.20) suitably.

113 You should find that

$$\mathrm{e}^{\epsilon(A+B)} = \mathrm{e}^{\frac{1}{2}\epsilon A}\,\mathrm{e}^{\epsilon B}\,\mathrm{e}^{\frac{1}{2}\epsilon A}\left(1 - \frac{\epsilon^3}{24}[A,[A,B]] + \frac{\epsilon^3}{12}[[A,B],B] + \cdots\right),$$

where the ellipsis represents terms of order ϵ^4 and higher powers of ϵ.

114 Write the 2×2 matrix in (5.6.10) as

$$\begin{aligned}
&\cosh(\epsilon W)\left(\mathbf{1}_2 + s_x\sigma_x + s_z\sigma_z\right)\\
&= \cosh(\epsilon W)\left[\left(1+\sqrt{s_x^2+s_z^2}\right)\frac{1}{2}\left(\mathbf{1}_2 + \frac{s_x\sigma_x + s_z\sigma_z}{\sqrt{s_x^2+s_z^2}}\right)\right.\\
&\qquad\left. + \left(1-\sqrt{s_x^2+s_z^2}\right)\frac{1}{2}\left(\mathbf{1}_2 - \frac{s_x\sigma_x + s_z\sigma_z}{\sqrt{s_x^2+s_z^2}}\right)\right]
\end{aligned}$$

before raising it to the Nth power and recognize that, for example,

$$\left(1+\sqrt{s_x^2+s_z^2}\right)^N \xrightarrow[N\to\infty]{} \mathrm{e}^{\beta\sqrt{E^2+W^2}}.$$

115 No hint needed.

116 Confirm that only irreducible 4-clusters contribute to $a_4(\beta)$.

117 In each case expand $\beta P v$ in powers of v^{-1} and read off the coefficients.

118 Here we have $f(r) = -\eta(r_0 - r)$. In the double integral for $a_3(\beta)$,

$$a_3(\beta) = \frac{1}{3\lambda^6} \int (\mathrm{d}\boldsymbol{r})\,(\mathrm{d}\boldsymbol{r})\,\eta(r_0 - r)\eta(r_0 - r')\eta(r_0 - |\boldsymbol{r} - \boldsymbol{r}'|)\,,$$

you can use spherical coordinates for the $\boldsymbol{r}'$ integration with the polar axis in the direction specified by $\boldsymbol{r}$. You should find

$$a_2(\beta) = \frac{2\pi}{3}\left(\frac{r_0}{\lambda}\right)^3 \quad \text{and} \quad a_3(\beta) = \frac{5\pi}{18}\left(\frac{r_0}{\lambda}\right)^6 .$$

119 Recall (1.16.7) and (1.16.8) and account for the difference in the meaning of a and b in Section 1.16 and Section 6.5.

120 Proceed from $\log Z = \dfrac{V}{\lambda^3}\bigl(z + z^2 b_2(\beta)\bigr)$, remember that $Z(\beta, V, N)$ is a generating function for the $Q(\beta, V, N)$s, and treat the $b_2(\beta)$ contributions consistently as small corrections.

121 Rearrange the terms with the aid of $\log\dfrac{v}{\lambda^3} - \dfrac{b}{v} = \log\dfrac{v-b}{\lambda^3}$ (justify this!) and arrive at

$$f(T, v) = -k_{\mathrm{B}}T\left(1 + \log\frac{v-b}{\lambda^3}\right) - \frac{a}{v}\,.$$

You should find that $\left(P + \dfrac{a}{v^2}\right)^3 (v-b)^5$ is constant when S and N are.

122, 123 No hints needed.

124 Show that the probability of having m_1 clockwise jumps and m_2 anti-clockwise jumps in the lapse of t is

$$p(m_1, m_2, t) = \frac{(qrt)^{m_1}}{m_1!}\,\frac{\bigl((1-q)rt\bigr)^{m_2}}{m_2!}\,\mathrm{e}^{-rt}\,,$$

then proceed.

125 Try an integration by parts.

126 Recall Hamilton's equations of motion, $\dfrac{\mathrm{d}x_j}{\mathrm{d}t} = \{x_j, H(\boldsymbol{x})\}$ with the Poisson bracket of (7.7.10). You should conclude that

$$\left\langle \sum_{k=1}^{N} \frac{\boldsymbol{p}_k^2}{2m_k} \right\rangle = \frac{3}{2} N k_{\mathrm{B}} T\,, \quad \left\langle \sum_{k=1}^{N} \boldsymbol{r}_k \cdot \boldsymbol{\nabla}_k V \right\rangle = 3 N k_{\mathrm{B}} T\,,$$

with each summand contributing one-Nth to the sum, which is the statement of the equipartition theorem.

127 Note that the probability of finding the speed in the range $v \cdots v + \mathrm{d}v$ is proportional to $\mathrm{d}v\, v^2 \exp\Bigl(-\dfrac{mv^2}{2k_{\mathrm{B}}T}\Bigr)$.

128 Consider $\left\langle \left((\lambda X)^{\frac{1}{2}} - (\lambda X)^{-\frac{1}{2}} \right)^2 \right\rangle$ and adjust the value of $\lambda > 0$.

129 First find $\mathrm{H}_k(0)$ from (7.8.21), then use this to determine the k-dependent proportionality factor in $\langle p^k \rangle_\infty = \langle p^k \rangle_t \big|_{t \gg \tau} \propto Q^k$. From (7.8.5), you get $\langle p^k \rangle_\infty = (k-1)Q^2 \langle p^{k-2} \rangle_\infty$, from which you can infer $\langle p^k \rangle_\infty$ starting with the known values of $\langle p^0 \rangle_\infty = 1$ and $\langle p^1 \rangle_\infty = 0$.

130 You should find $\tilde{f}_t(\tilde{x}, \tilde{p}) = \mathrm{e}^{t/\tau} f_0\Bigl(\tilde{x} - (\tilde{p}\tau/M)\bigl(\mathrm{e}^{t/\tau} - 1\bigr), \tilde{p}\,\mathrm{e}^{t/\tau}\Bigr)$.

131 Rely on the generating function for the Hermite polynomials in (7.8.21).

132 No hints needed.

133 Treat the factorization as an ansatz and show that it works. Find first the late-time stationary expectation values of powers of p, then those of powers of x.

134 Note that the factorization of $G_t(a, b)$ at late times is implied by the findings of Exercise 133.

135 You could try an ansatz for $G_t(a, b)$ of the form in (7.8.37) and find $x(t)$, $p(t)$, and $\mathrm{Q}(t)^2$ such that $G_t(a, b)$ obeys its differential equation. Then use the procedure of Section 7.8.3 to obtain the Green's function.

Index

www.ingramcontent.com/pod-product-compliance
Lightning Source LLC
LaVergne TN
LVHW010332220926
848484LV00004BB/887